# Sobolev Spaces

# Pure and Applied Mathematics

## A Series of Monographs and Textbooks

Editors **Samuel Eilenberg and Hyman Bass**

Columbia University, New York

RECENT TITLES

GLEN E. BREDON. Introduction to Compact Transformation Groups

WERNER GREUB, STEPHEN HALPERIN, AND RAY VANSTONE. Connections, Curvature, and Cohomology: Volume I, De Rham Cohomology of Manifolds and Vector Bundles. Volume II, Lie Groups, Principal Bundles, and Characteristic Classes. *In preparation:* Volume III, Cohomology of Principal Bundles and Homogeneous Spaces

XIA DAO-XING. Measure and Integration Theory of Infinite-Dimensional Spaces: Abstract Harmonic Analysis

RONALD G. DOUGLAS. Banach Algebra Techniques in Operator Theory

WILLARD MILLER, JR. Symmetry Groups and Their Applications

ARTHUR A. SAGLE AND RALPH E. WALDE. Introduction to Lie Groups and Lie Algebras

T. BENNY RUSHING. Topological Embeddings

JAMES W. VICK. Homology Theory: An Introduction to Algebraic Topology

E. R. KOLCHIN. Differential Algebra and Algebraic Groups

GERALD J. JANUSZ. Algebraic Number Fields

A. S. B. HOLLAND. Introduction to the Theory of Entire Functions

WAYNE ROBERTS AND DALE VARBERG. Convex Functions

A. M. OSTROWSKI. Solution of Equations in Euclidean and Banach Spaces, Third Edition of Solution of Equations and Systems of Equations

H. M. EDWARDS. Riemann's Zeta Function

SAMUEL EILENBERG. Automata, Languages, and Machines: Volume A. *In preparation:* Volume B

MORRIS HIRSCH AND STEPHEN SMALE. Differential Equations, Dynamical Systems, and Linear Algebra

WILHELM MAGNUS. Noneuclidean Tesselations and Their Groups

J. DIEUDONNÉ. Treatise on Analysis, Volume IV

FRANÇOIS TREVES. Basic Linear Partial Differential Equations

WILLIAM M. BOOTHBY. An Introduction to Differentiable Manifolds and Riemannian Geometry

BRAYTON GRAY. Homotopy Theory: An Introduction to Algebraic Topology

ROBERT A. ADAMS. Sobolev Spaces

*In preparation*

D. V. WIDDER. The Heat Equation

IRVING E. SEGAL. Mathematical Cosmology and Extragalactic Astronomy

J. DIEUDONNÉ. Treatise on Analysis, Volume II, enlarged and corrected printing

# SOBOLEV SPACES

ROBERT A. ADAMS

*Department of Mathematics*
*The University of British Columbia*
*Vancouver, British Columbia, Canada*

ACADEMIC PRESS New York San Francisco London 1975

A Subsidiary of Harcourt Brace Jovanovich, Publishers

ACADEMIC PRESS, INC.
111 Fifth Avenue, New York, New York 10003

*United Kingdom Edition published by*
ACADEMIC PRESS, INC. (LONDON) LTD.
24/28 Oval Road, London NW1

**Library of Congress Cataloging in Publication Data**

Adams, Robert A
Sobolev spaces.

(Pure and applied mathematics series; v. 65)
Bibliography: p.
Includes index.
1. Sobolev spaces. I. Title.
II. Series: Pure and applied mathematics; a series of monographs and textbooks; v. 65
QA3.P8 [QA323] 510'.8s [515'.7] 74-17978
ISBN 0-12-044150-0

AMS (MOS) 1970 Subject Classifications: 46E30, 46E35

PRINTED IN THE UNITED STATES OF AMERICA

*To Anne*
*Marion, Valerie, Alex, and Andrew*

# Contents

# Preface

This monograph is devoted to a study of properties of certain Banach spaces of weakly differentiable functions of several real variables which arise in connection with numerous problems in the theory of partial differential equations and related areas of mathematical analysis, and which have become an essential tool in those disciplines. These spaces are now most often associated with the name of the Soviet mathematician S. L. Sobolev, though their origins predate his major contributions to their development in the late 1930s.

Sobolev spaces are very interesting mathematical structures in their own right, but their principal significance lies in the central role they, and their numerous generalizations, now play in partial differential equations. Accordingly, most of this book concentrates on those aspects of the theory of Sobolev spaces that have proven most useful in applications. Although no specific applications to problems in partial differential equations are discussed (these are to be found in almost any modern textbook on partial differential equations), this monograph is nevertheless intended mainly to serve as a textbook and reference on Sobolev spaces for graduate students and researchers in differential equations. Some of the material in Chapters III–VI has grown out of lecture notes [18] for a graduate course and seminar given by Professor Colin Clark at the University of British Columbia in 1967–1968.

The material is organized into eight chapters. Chapter I is a potpourri of standard topics from real and functional analysis, included, mainly without proofs, because they form a necessary background for what follows. Chapter II is also largely "background" but concentrates on a specific topic, the Lebesgue spaces $L^p(\Omega)$, of which Sobolev spaces are special subspaces. For completeness, proofs are included here. Most of the material in these first two chapters will be quite familiar to the reader and may be omitted, or simply given a superficial reading to settle questions of notation and such. (Possible exceptions are Sections 1.25–1.27, 1.31, and 2.21–2.22 which may be less familiar.) The inclusion of these elementary chapters makes the book fairly self-contained. Only a solid undergraduate background in mathematical analysis is assumed of the reader.

Chapters III–VI may be described as the heart of the book. These develop all the basic properties of Sobolev spaces of positive integral order and culminate in the very important Sobolev imbedding theorem (Theorem 5.4) and the corresponding compact imbedding theorem (Theorem 6.2). Sections 5.33–5.54 and 6.12–6.50 consist of refinements and generalizations of these basic imbedding theorems, and could be omitted from a first reading.

Chapter VII is concerned with generalization of ordinary Sobolev spaces to allow fractional orders of differentiation. Such spaces are often involved in research into nonlinear partial differential equations, for instance the Navier–Stokes equations of fluid mechanics. Several approaches to defining fractional-order spaces can be taken. We concentrate in Chapter VII on the trace-interpolation approach of J. L. Lions and E. Magenes, and discuss other approaches more briefly at the end of the chapter (Sections 7.59–7.74). It is necessary to develop a reasonable body of abstract functional analysis (the trace-interpolation theory) before introducing the fractional-order spaces. Most readers will find that a reading of this material (in Sections 7.2–7.34, possibly omitting proofs) is essential for an understanding of the discussion of fractional-order spaces that begins in Section 7.35.

Chapter VIII concerns Orlicz–Sobolev spaces and, for the sake of completeness, necessarily begins with a self-contained introduction to the theory of Orlicz spaces. These spaces are finding increasingly important applications in applied analysis. The main results of Chapter VIII are the theorem of N. S. Trudinger (Theorem 8.25) establishing a limiting case of the Sobolev imbedding theorem, and the imbedding theorems of Trudinger and T. K. Donaldson for Orlicz–Sobolev spaces given in Sections 8.29–8.40.

The existing mathematical literature on Sobolev spaces and their generalizations is vast, and it would be neither easy nor particularly desirable to include everything that was known about such spaces between the covers of one book. An attempt has been made in this monograph to present all the core material in sufficient generality to cover most applications, to give the reader an overview of the subject that is difficult to obtain by reading research papers, and finally, as mentioned above, to provide a ready reference for someone requiring a result about Sobolev spaces for use in some application. Complete proofs are given for most theorems, but some assertions are left for the interested reader to verify as exercises. Literature references are given in square brackets, equation numbers in parentheses, and sections are numbered in the form *m.n* with *m* denoting the chapter.

# Acknowledgments

We acknowledge with deep gratitude the considerable assistance we have received from Professor John Fournier in the preparation of this monograph. Also much appreciated are the helpful comments received from Professor Bui An Ton and the encouragement of Professor Colin Clark who originally suggested that this book be written. Thanks are also due to Mrs. Yit-Sin Choo for a superb job of typing a difficult manuscript. Finally, of course, we accept all responsibility for error or obscurity and welcome comments, or corrections, from readers.

# List of Spaces and Norms

The numbers at the right indicate the sections in which the symbols are introduced. In some cases the notations are not those used in other areas of analysis.

| | | |
|---|---|---|
| $B$ | $\lVert\cdot\rVert_B$ | 7.2 |
| $B_1 + B_2$ | $\lVert\cdot\,; B_1 + B_2\rVert$ | 7.11 |
| $B^{s,p}(\mathbb{R}^n)$ | $\lVert\cdot\,; B^{s,p}(\mathbb{R}^n)\rVert$ | 7.67 |
| $B^{s,p}(\Omega)$ | $\lVert\cdot\,; B^{s,p}(\Omega)\rVert$ | 7.72 |
| $\mathbb{C}$ | $\lvert\cdot\rvert$ | 1.1 |
| $C(\Omega)$ | | 1.25 |
| $C^m(\Omega)$ | | 1.25 |
| $C^\infty(\Omega)$ | | 1.25 |
| $C_0(\Omega)$ | | 1.25 |
| $C_0^\infty(\Omega)$ | | 1.25 |
| $C^m(\overline{\Omega})$ | $\lVert\cdot\,; C^m(\overline{\Omega})\rVert$ | 1.26 |
| $C^{m,\lambda}(\overline{\Omega})$ | $\lVert\cdot\,; C^{m,\lambda}(\overline{\Omega})\rVert$ | 1.27 |
| $C_B^j(\Omega)$ | $\lVert\cdot\,; C_B^j(\Omega)\rVert$ | 5.2 |
| $C_\mu(\overline{\Omega})$ | $\lVert\cdot\,; C_\mu(\overline{\Omega})\rVert$ | 8.37 |
| $\mathscr{D}(\Omega)$ | | 1.51 |
| $\mathscr{D}'(\Omega)$ | | 1.52 |
| $D(\Lambda)$ | $\lVert\cdot\,; D(\Lambda)\rVert$ | 7.7, 7.9 |

| | | |
|---|---|---|
| $D(\Lambda^k)$ | $\lVert\cdot; D(\Lambda^k)\rVert$ | 7.29 |
| $E_A(\Omega)$ | | 8.14 |
| $H^{m,p}(\Omega)$ | $\lVert\cdot\rVert_{m,p} = \lVert\cdot\rVert_{m,p,\Omega}$ | 3.1 |
| $H^{-m,p'}(\Omega)$ | $\lVert\cdot\rVert_{-m,p'}$ | 3.12 |
| $H^{s,p}(\Omega)$ | $\lVert\cdot; H^{s,p}(\Omega)\rVert$ | 7.33 |
| $K_A(\Omega)$ | | 8.7 |
| $L^1(A)$ | | 1.40, 1.46 |
| $L^1_{loc}(\Omega)$ | | 1.53 |
| $L^p(\Omega)$ | $\lVert\cdot\rVert_p = \lVert\cdot\rVert_{p,\Omega}$ | 2.1 |
| | $\lVert\cdot\rVert_{0,p} = \lVert\cdot\rVert_{0,p,\Omega}$ | 3.1 |
| $L^\infty(\Omega)$ | $\lVert\cdot\rVert_\infty = \lVert\cdot\rVert_{\infty,\Omega}$ | 2.5 |
| $L^p(\text{bdry }\Omega)$ | | 5.22 |
| $L^p(a,b;B)$ | $\lVert\cdot; L^p(a,b;B)\rVert$ | 7.3 |
| $L^p_{loc}(a,b;B)$ | | 7.3 |
| $L(B)$ | $\lVert\cdot\rVert_{L(B)}$ | 7.5 |
| $L^{s,p}(\mathbb{R}^n)$ | $\lVert\cdot; L^{s,p}(\mathbb{R}^n)\rVert$ | 7.62 |
| $L^{s,p}(\Omega)$ | | 7.66 |
| $L_A(\Omega)$ | $\lVert\cdot\rVert_A = \lVert\cdot\rVert_{A,\Omega}$ | 8.9 |
| $\mathbb{R}^n$ | $\lvert\cdot\rvert$ | 1.1 |
| $T = T(p,\nu;B_1,B_2)$ | $\lVert\cdot\rVert_T = \lVert\cdot; T(p,\nu;B_1,B_2)\rVert$ | 7.14 |
| $T^0$ | $\lVert\cdot\rVert_{T^0}$ | 7.24, 7.26, 7.32 |
| $T^k = T^k(p,\nu;\Lambda;B)$ | $\lVert\cdot\rVert_{T^k}$ | 7.32 |
| $T^{\theta,p}(\Omega)$ | $\lVert\cdot; T^{\theta,p}(\Omega)\rVert$ | 7.35 |
| $\tilde{T}^{\theta,p}(\Omega)$ | $\lVert\cdot; \tilde{T}^{\theta,p}(\Omega)\rVert$ | 7.43 |
| $W^{m,p}(\Omega)$ | $\lVert\cdot\rVert_{m,p} = \lVert\cdot\rVert_{m,p,\Omega}$ | 3.1 |
| $W_0^{m,p}(\Omega)$ | | 3.1 |
| $W^{-m,p'}(\Omega)$ | $\lVert\cdot; W^{-m,p'}(\Omega)\rVert$ | 3.11 |
| $W_0^{m,2;\mu}(\Omega)$ | $\lVert\cdot\rVert_{m,2;\mu}$ | 6.54 |
| $W = W(p,\nu;B_1,B_2)$ | $\lVert\cdot\rVert_W = \lVert\cdot; W(p,\nu;B_1,B_2)\rVert$ | 7.11 |
| $W^m = W^m(p,\nu;\Lambda;B)$ | $\lVert\cdot\rVert_{W^m}$ | 7.30 |
| $W^{s,p}(\Omega)$ | $\lVert\cdot\rVert_{s,p} = \lVert\cdot\rVert_{s,p,\Omega}$ | 7.36, 7.39 |
| $W_0^{s,p}(\Omega)$ | | 7.39 |
| $\tilde{W}^{s,p}(\Omega)$ | $\lVert\cdot\rVert^{\sim}_{s,p,\Omega}$ | 7.48 |
| $W^{s,\infty}(\Omega)$ | $\lVert\cdot\rVert_{s,\infty,\Omega}$ | 7.49 |
| $W^m L_A(\Omega)$ | $\lVert\cdot\rVert_{m,A} = \lVert\cdot\rVert_{m,A,\Omega}$ | 8.27 |
| $W^m E_A(\Omega)$ | | 8.27 |
| $W_0{}^m L_A(\Omega)$ | | 8.27 |
| $X$ | $\lVert\cdot; X\rVert$, $\lvert\lvert\lvert\cdot\rvert\rvert\rvert$ | 1.4, 1.6, 6.51 |
| $X'$ | $\lVert\cdot; X'\rVert$ | 1.5, 1.10 |
| $\prod_{j=1}^n X_j$ | $\lVert\cdot\rVert_{(p)}$ | 1.22 |

# I

# Introductory Topics

## Notation

**1.1** Throughout this monograph the term *domain* and the symbol $\Omega$ shall be reserved for an open set in $n$-dimensional, real Euclidean space $\mathbb{R}^n$. We shall be concerned with differentiability and integrability of functions defined on $\Omega$—these functions are allowed to be complex valued unless the contrary is stated explicitly. The complex field is denoted by $\mathbb{C}$. For $c \in \mathbb{C}$ and two functions $u$ and $v$ the scalar multiple $cu$, the sum $u+v$, and the product $uv$ are always taken to be defined pointwise as

$$\begin{aligned}(cu)(x) &= cu(x),\\ (u+v)(x) &= u(x)+v(x),\\ (uv)(x) &= u(x)\,v(x),\end{aligned}$$

at all points $x$ where the right sides make sense.

A typical point in $\mathbb{R}^n$ is denoted by $x = (x_1, \ldots, x_n)$; its norm $|x| = (\sum_{j=1}^n x_j^2)^{1/2}$. The inner product of $x$ and $y$ is $x \cdot y = \sum_{j=1}^n x_j y_j$.

If $\alpha = (\alpha_1, \ldots, \alpha_n)$ is an $n$-tuple of nonnegative integers $\alpha_j$, we call $\alpha$ a *multi-index* and denote by $x^\alpha$ the monomial $x_1^{\alpha_1} \cdots x_n^{\alpha_n}$, which has degree $|\alpha| = \sum_{j=1}^n \alpha_j$. Similarly, if $D_j = \partial/\partial x_j$ for $1 \le j \le n$, then

$$D^\alpha = D_1^{\alpha_1} \cdots D_n^{\alpha_n}$$

denotes a differential operator of order $|\alpha|$. $D^{(0,\ldots,0)}u = u$.

If $\alpha$ and $\beta$ are two multi-indices, we say $\beta \le \alpha$ provided $\beta_j \le \alpha_j$ for $1 \le j \le n$. In this case $\alpha - \beta$ is also a multi-index and $|\alpha-\beta|+|\beta|=|\alpha|$. We also denote

$$\alpha! = \alpha_1! \cdots \alpha_n!$$

and if $\beta \le \alpha$,

$$\binom{\alpha}{\beta} = \frac{\alpha!}{\beta!(\alpha-\beta)!} = \binom{\alpha_1}{\beta_1}\cdots\binom{\alpha_n}{\beta_n}.$$

The reader may wish to verify the Leibniz formula

$$D^\alpha(uv)(x) = \sum_{\beta\le\alpha}\binom{\alpha}{\beta} D^\beta u(x)\, D^{\alpha-\beta}v(x)$$

valid for functions $u$ and $v$ that are $|\alpha|$ times continuously differentiable near $x$.

**1.2** If $G \subset \mathbb{R}^n$, we denote by $\bar{G}$ the closure of $G$ in $\mathbb{R}^n$. We shall write $G \subset\subset \Omega$ provided $\bar{G} \subset \Omega$ and $\bar{G}$ is a compact (i.e., closed and bounded) subset of $\mathbb{R}^n$. If $u$ is a function defined on $G$, we define the *support* of $u$ as

$$\operatorname{supp} u = \overline{\{x \in G : u(x) \ne 0\}}.$$

We say that $u$ has compact support in $\Omega$ if $\operatorname{supp} u \subset\subset \Omega$. We shall denote by "bdry $G$" the boundary of $G$ in $\mathbb{R}^n$, that is, the set $\bar{G} \cap \overline{G^c}$ where $G^c = \mathbb{R}^n \sim G = \{x \in \mathbb{R}^n : x \notin G\}$ is the complement of $G$.

If $x \in \mathbb{R}^n$ and $G \subset \mathbb{R}^n$, we denote by "dist$(x, G)$" the distance from $x$ to $G$, that is, the number $\inf_{y\in G}|x-y|$. Similarly, if $F, G \subset \mathbb{R}^n$,

$$\operatorname{dist}(F,G) = \inf_{y\in F} \operatorname{dist}(y,G) = \inf_{\substack{x\in G\\ y\in F}} |x-y|.$$

## Topological Vector Spaces

**1.3** We assume that the reader is familiar with the concept of a vector space over the real or complex scalar field, and with the related notions of dimension, subspace, linear transformation, and convex set. We also assume familiarity with the basic concepts of general topology, Hausdorff topological spaces, weaker and stronger topologies, continuous functions, convergent sequences, topological product spaces, subspaces, and relative topology.

Let it be assumed throughout this monograph that all vector spaces referred to are taken over the complex field unless the contrary is explicitly stated.

**1.4** A *topological vector space*, hereafter abbreviated TVS, is a Hausdorff topological space that is also a vector space for which the vector space operations of addition and scalar multiplication are continuous. That is, if $X$ is a TVS, then the mappings

$$(x, y) \to x + y \qquad \text{and} \qquad (c, x) \to cx$$

from the topological product spaces $X \times X$ and $\mathbb{C} \times X$, respectively, into $X$ are continuous. $X$ is a *locally convex* TVS if each neighborhood of the origin in $X$ contains a convex neighborhood.

We shall outline below, mainly omitting proofs and details, those aspects of the theory of topological and normed vector spaces that play a significant role in the study of Sobolev spaces. For a more thorough discussion of these topics the reader is referred to standard textbooks on functional analysis, for example, those by Yosida [69] or Rudin [59].

**1.5** By a *functional* on a vector space $X$ we mean a scalar-valued function $f$ defined on $X$. The functional $f$ is linear provided

$$f(ax+by) = af(x) + bf(y), \qquad x, y \in X, \quad a, b \in \mathbb{C}.$$

If $X$ is a TVS, a functional on $X$ is continuous if it is continuous from $X$ into $\mathbb{C}$ where $\mathbb{C}$ has its usual topology, induced by the Euclidean metric.

The set of all continuous, linear functionals on $X$ is called the *dual of X* and is denoted by $X'$. Under pointwise addition and scalar multiplication $X'$ is a vector space:

$$(f+g)(x) = f(x) + g(x), \qquad (cf)(x) = cf(x), \qquad f, g \in X', \quad x \in X, \quad c \in \mathbb{C}.$$

$X'$ will be a TVS provided a suitable topology is specified for it. One such topology is the *weak-star topology*, the weakest topology with respect to which the functional $F_x$ defined on $X'$ by $F_x(f) = f(x)$ for each $f \in X'$ is continuous for each $x \in X$. This topology is used, for instance, in the space of Schwartz distributions introduced in Section 1.52. The dual of a normed space can be given a stronger topology with respect to which it is a normed space itself (Section 1.10).

## Normed Spaces

**1.6** A *norm* on a vector space $X$ is a real-valued functional $f$ on $X$ which satisfies

(i) $f(x) \geq 0$ for all $x \in X$ with equality if and only if $x = 0$,
(ii) $f(cx) = |c| f(x)$ for every $x \in X$ and $c \in \mathbb{C}$,
(iii) $f(x+y) \leq f(x) + f(y)$ for every $x, y \in X$.

A *normed space* is a vector space $X$ which is provided with a norm. The norm will be denoted by $\|\cdot\,; X\|$ except where simpler notations are introduced. If $r > 0$, the set

$$B_r(x) = \{y \in X : \|y-x; X\| < r\}$$

is called the *open ball* of radius $r$ with center $x \in X$. Any subset $A$ of $X$ is called *open* if for every $x \in A$ there exists $r > 0$ such that $B_r(x) \subset A$. The open sets thus defined constitute a topology for $X$ with respect to which $X$ is a TVS. This topology is called the *norm topology* on $X$. The closure of $B_r(x)$ in this topology is

$$\overline{B_r(x)} = \{y \in X : \|y-x; X\| \le r\}.$$

A TVS $X$ is *normable* if its topology coincides with the topology induced by some norm on $X$. Two different norms on a vector space are *equivalent* if they induce the same topology on $X$, that is, if for some constant $c > 0$

$$c\,\|x\|_1 \le \|x\|_2 \le (1/c)\,\|x\|_1$$

for all $x \in X$, $\|\cdot\|_1$ and $\|\cdot\|_2$ being the two norms.

If $X$ and $Y$ are two normed spaces and if there exists a one-to-one linear operator $L$ mapping $X$ onto $Y$ and having the property $\|L(x); Y\| = \|x; X\|$ for every $x \in X$, then $L$ is called an *isometric isomorphism* between $X$ and $Y$, and $X$ and $Y$ are said to be *isometrically isomorphic*; we write $X \cong Y$. Such spaces are often identified since they have identical structures and differ only in the nature of their elements.

**1.7** A sequence $\{x_n\}$ in a normed space $X$ is convergent to the limit $x_0$ if and only if $\lim_{n\to\infty} \|x_n - x_0; X\| = 0$ in $\mathbb{R}$. The norm topology of $X$ is completely determined by the sequences it renders convergent.

A subset $S$ of a normed space $X$ is said to be *dense* in $X$ if each $x \in X$ is the limit of a sequence of elements of $S$. The normed space $X$ is called *separable* if it has a countable dense subset.

**1.8** A sequence $\{x_n\}$ in a normed space $X$ is called a *Cauchy sequence* if and only if $\lim_{m,n\to\infty} \|x_m - x_n; X\| = 0$. If every Cauchy sequence in $X$ converges to a limit in $X$ ,then $X$ is *complete* and a *Banach space*. Every normed space $X$ is either a Banach space or a dense subset of a Banach space $Y$ whose norm satisfies

$$\|x; Y\| = \|x; X\| \qquad \text{for every} \quad x \in X.$$

In this latter case $Y$ is called the *completion* of $X$.

**1.9** If $X$ is a vector space, a functional $(\cdot\,,\cdot)_X$ defined on $X \times X$ is called an *inner product* on $X$ provided that for every $x, y, z \in X$ and $a, b \in \mathbb{C}$

(i) $(x, y)_X = \overline{(y, x)_X}$,
(ii) $(ax+by, z)_X = a(x, z)_X + b(y, z)_X$,
(iii) $(x, x)_X = 0$ if and only if $x = 0$,

where $\bar{c}$ denotes the complex conjugate of $c \in \mathbb{C}$. Given such an inner product, a norm on $X$ can be defined by

$$\|x; X\| = (x, x)_X^{1/2}. \tag{1}$$

If $X$ is a Banach space under this norm, it is called a *Hilbert space*. The parallelogram law

$$\|x+y; X\|^2 + \|x-y; X\|^2 = 2\|x; X\|^2 + 2\|y; X\|^2 \tag{2}$$

holds in any normed space whose norm is obtained from an inner product via (1).

## The Normed Dual

**1.10** A norm on the dual $X'$ of a normed space $X$ can be defined by setting, for $x' \in X'$,

$$\|x'; X'\| = \sup_{\substack{x \in X \\ x \neq 0}} \frac{|x'(x)|}{\|x; X\|}.$$

Since $\mathbb{C}$ is complete, $X'$, with the topology induced by this norm, is a Banach space (whether or not $X$ is) and is called the *normed dual* of $X$. If $X$ is infinite dimensional, the norm topology of $X'$ is stronger (i.e., has more open sets) than the weak-star topology defined in Section 1.5.

If $X$ is a Hilbert space, it can be identified with its normed dual $X'$ in a natural way.

**1.11 THEOREM** (*Riesz representation theorem*) Let $X$ be a Hilbert space. A linear functional $x'$ on $X$ belongs to $X'$ if and only if there exists $x \in X$ such that for every $y \in X$ we have

$$x'(y) = (y, x)_X,$$

and in this case $\|x'; X'\| = \|x; X\|$. Moreover, $x$ is uniquely determined by $x' \in X'$.

A vector subspace $M$ of a normed space $X$ is itself a normed space under the norm of $X$, and so normed is called a *subspace* of $X$. A closed subspace of a Banach space is a Banach space.

**1.12 THEOREM** (*Hahn–Banach extension theorem*) Let $M$ be a subspace of the normed space $X$. If $m' \in M'$, there exists $x' \in X'$ such that $\|x'; X'\| = \|m'; M'\|$ and $x'(m) = m'(m)$ for every $m \in M$.

**1.13** A natural linear injection of a normed space $X$ into its second dual space $X'' = (X')'$ is provided by the mapping $J_X$ whose value at $x \in X$ is given by

$$J_X x(x') = x'(x), \qquad x' \in X'.$$

Since $|J_X x(x')| \le \|x'; X'\| \, \|x; X\|$ we have

$$\|J_X x; X''\| \le \|x; X\|.$$

On the other hand, the Hahn–Banach theorem assures us that for any $x \in X$ we can find an $x' \in X'$ such that $\|x'; X'\| = 1$ and $x'(x) = \|x; X\|$. Hence

$$\|J_X x; X''\| = \|x; X\|,$$

and $J_X$ is an isometric isomorphism from $X$ into $X''$.

If the range of the isomorphism is the entire space $X''$, we say that the normed space $X$ is *reflexive*. A reflexive space must be complete and hence a Banach space.

**1.14 THEOREM** Let $X$ be a normed space. $X$ is reflexive if and only if $X'$ is reflexive. $X$ is separable if $X'$ is separable. Hence if $X$ is separable and reflexive, so is $X'$.

## Weak Topology and Weak Convergence

**1.15** The *weak topology* of a normed space $X$ is the weakest topology on $X$ that still renders continuous each $x' \in X'$. Unless $X$ is finite dimensional the weak topology is weaker than the norm topology on $X$. It is a consequence of the Hahn–Banach theorem that a closed, convex set in a normed space is also closed in the weak topology of that space. A sequence convergent with respect to the weak topology on $X$ is said to *converge weakly*. Thus $x_n$ converges weakly to $x$ in $X$ provided $x'(x_n) \to x'(x)$ in $\mathbb{C}$ for every $x' \in X'$. We denote norm convergence of a sequence $\{x_n\}$ to $x$ in $X$ by $x_n \to x$; weak convergence by $x_n \rightharpoonup x$. Since $|x'(x_n - x)| \le \|x'; X'\| \, \|x_n - x; X\|$ we see that $x_n \to x$ implies $x_n \rightharpoonup x$. The converse is generally not true.

## Compact Sets

**1.16** A subset $A$ of a normed space $X$ is called *compact* if every sequence of points in $A$ has a subsequence converging in $X$ to an element of $A$. Compact sets are closed and bounded, but closed and bounded sets need not be compact unless $X$ is finite dimensional. $A$ is called *precompact* if its closure $\bar{A}$ (in the norm topology) is compact. $A$ is called *weakly sequentially compact* if every sequence in $A$ has a subsequence converging weakly in $X$ to a point in $A$. The reflexivity of a Banach space can be characterized in terms of this property.

**1.17 THEOREM** A Banach space $X$ is reflexive if and only if its closed unit ball $\overline{B_1(0)} = \{x \in X : \|x; X\| \le 1\}$ is weakly sequentially compact.

**1.18 THEOREM** A set $A$ is precompact in a Banach space $X$ if and only if for every positive number $\varepsilon$ there is a finite subset $N_\varepsilon$ of points of $X$ with the property

$$A \subset \bigcup_{y \in N_\varepsilon} B_\varepsilon(y).$$

A set $N_\varepsilon$ with this property is called a *finite $\varepsilon$-net* for $A$.

## Uniform Convexity

**1.19** Any normed space is locally convex with respect to its norm topology. The norm on $X$ is called *uniformly convex* if for every number $\varepsilon$ satisfying $0 < \varepsilon \le 2$ there exists a number $\delta(\varepsilon) > 0$ such that if $x, y \in X$ satisfy $\|x; X\| = \|y; X\| = 1$ and $\|x-y; X\| \ge \varepsilon$, then $\|(x+y)/2; X\| \le 1 - \delta(\varepsilon)$. The normed space $X$ is itself called "uniformly convex" in this case. It should, however, be noted that uniform convexity is a property of the norm—$X$ may possess an equivalent norm that is not uniformly convex. Any normable space is called "uniformly convex" if it possesses a uniformly convex norm. The parallelogram law (2) shows that a Hilbert space is uniformly convex.

**1.20 THEOREM** A uniformly convex Banach space is reflexive.

The following two theorems will be used to establish the separability, reflexivity, and uniform convexity of the Sobolev spaces introduced in Chapter III.

**1.21 THEOREM** Let $X$ be a Banach space and $M$ a subspace of $X$ closed with respect to the norm topology on $X$. Then $M$ is itself a Banach space under the norm inherited from $X$. Furthermore,

(i) $M$ is separable if $X$ is separable,
(ii) $M$ is reflexive if $X$ is reflexive,
(iii) $M$ is uniformly convex if $X$ is uniformly convex.

The completeness, separability, and uniform convexity of $M$ follow easily from the corresponding properties for $X$. The reflexivity of $M$ is a consequence of Theorem 1.17 and the fact that $M$, being closed and convex, is closed in the weak topology of $X$.

**1.22 THEOREM** For $j = 1, 2, \ldots, n$ let $X_j$ be a Banach space with norm $\|\cdot\|_j$. The Cartesian product $X = \prod_{j=1}^n X_j$, consisting of points $x = (x_1, \ldots, x_n)$ with $x_j \in X_j$, is a vector space under the definitions

$$x + y = (x_1 + y_1, \ldots, x_n + y_n), \qquad cx = (cx_1, \ldots, cx_n),$$

and is a Banach space with respect to any of the equivalent norms

$$\|x\|_{(p)} = \left( \sum_{j=1}^{n} \|x_j\|_j^p \right)^{1/p}, \qquad 1 \le p < \infty,$$

$$\|x\|_{(\infty)} = \max_{1 \le j \le n} \|x_j\|_j.$$

Furthermore,

(i) if $X_j$ is separable for $1 \le j \le n$, then $X$ is separable;
(ii) if $X_j$ is reflexive for $1 \le j \le n$, then $X$ is reflexive;
(iii) if $X_j$ is uniformly convex for $1 \le j \le n$, then $X$ is uniformly convex. More precisely, $\|\cdot\|_{(p)}$ is a uniformly convex norm on $X$ provided $1 < p < \infty$.

The reader may verify that the functionals $\|\cdot\|_{(p)}$, $1 \le p \le \infty$, are in fact norms on $X$ and that $X$ is complete with respect to each of them. Equivalance of the norms follows from the inequalities

$$\|x\|_{(\infty)} \le \|x\|_{(p)} \le \|x\|_{(1)} \le n\,\|x\|_{(\infty)}.$$

The separability and uniform convexity of $X$ are readily deduced from the corresponding properties of the spaces $X_j$. The reflexivity of $X$ follows from that of $X_j$, $1 \le j \le n$, via Theorem 1.17 or via a natural isomorphism between $X'$ and $\prod_{j=1}^n X_j'$ (see, for example, Lemma 3.7).

## Operators and Imbeddings

**1.23** Since the topology of a normed space $X$ is determined by its convergent sequences, an operator $f$ defined on $X$ into a topological space $Y$ is continuous if and only if $f(x_n) \to f(x)$ in $Y$ whenever $x_n \to x$ in $X$. Such is also the case for any topological space $X$ whose topology is determined by the sequences it renders convergent (first countable spaces).

Let $X$, $Y$ be normed spaces and $f$ an operator from $X$ into $Y$. The operator $f$ is called *compact* if $f(A)$ is precompact in $Y$ whenever $A$ is bounded in $X$. [A bounded set in a normed space is one which is contained in the ball $B_R(0)$ for some $R$.] $f$ is *completely continuous* if it is continuous and compact. $f$ is *bounded* if $f(A)$ is bounded in $Y$ whenever $A$ is bounded in $X$.

Every compact operator is bounded. Every bounded linear operator is continuous. Hence every compact linear operator is completely continuous.

**1.24** We say that the normed space $X$ is *imbedded* in the normed space $Y$, and write $X \to Y$ to designate this imbedding, provided

(i) $X$ is a vector subspace of $Y$, and
(ii) the identity operator $I$ defined on $X$ into $Y$ by $Ix = x$ for all $x \in X$ is continuous.

Since $I$ is linear, (ii) is equivalent to the existence of a constant $M$ such that

$$\|Ix; Y\| \leq M \|x; X\|, \qquad x \in X.$$

In some circumstances the requirement that $X$ be a subspace of $Y$ and $I$ be the identity map is weakened to allow as imbeddings certain canonical linear transformations of $X$ into $Y$. (Examples are trace imbeddings of Sobolev spaces as well as imbeddings of these spaces into spaces of continuous functions. See Chapter V.)

We say that $X$ is *compactly imbedded* in $Y$ if the imbedding operator $I$ is compact.

## Spaces of Continuous Functions

**1.25** Let $\Omega$ be a domain in $\mathbb{R}^n$. For any nonnegative integer $m$ let $C^m(\Omega)$ be the vector space consisting of all functions $\phi$ which, together with all their partial derivatives $D^\alpha \phi$ of orders $|\alpha| \leq m$, are continuous on $\Omega$. We abbreviate $C^0(\Omega) \equiv C(\Omega)$. Let $C^\infty(\Omega) = \bigcap_{m=0}^{\infty} C^m(\Omega)$. The subspaces $C_0(\Omega)$ and $C_0^\infty(\Omega)$ consist of all those functions in $C(\Omega)$ and $C^\infty(\Omega)$, respectively, which have compact support in $\Omega$.

**1.26** Since $\Omega$ is open, functions in $C^m(\Omega)$ need not be bounded on $\Omega$. If $\phi \in C(\Omega)$ is bounded and uniformly continuous on $\Omega$, then it possesses a unique, bounded, continuous extension to the closure $\overline{\Omega}$ of $\Omega$. Accordingly, we define the vector space $C^m(\overline{\Omega})$ to consist of all those functions $\phi \in C^m(\Omega)$ for which $D^\alpha \phi$ is bounded and uniformly continuous on $\Omega$ for $0 \leq |\alpha| \leq m$. [Note that $C^m(\overline{\mathbb{R}^n}) \neq C^m(\mathbb{R}^n)$.] $C^m(\overline{\Omega})$ is a Banach space with norm given by

$$\|\phi; C^m(\overline{\Omega})\| = \max_{0 \leq |\alpha| \leq m} \sup_{x \in \Omega} |D^\alpha \phi(x)|.$$

**1.27** If $0 < \lambda \le 1$, we define $C^{m,\lambda}(\overline{\Omega})$ to be the subspace of $C^m(\overline{\Omega})$ consisting of those functions $\phi$ for which, for $0 \le |\alpha| \le m$, $D^\alpha \phi$ satisfies in $\Omega$ a Hölder condition of exponent $\lambda$, that is, there exists a constant $K$ such that

$$|D^\alpha \phi(x) - D^\alpha \phi(y)| \le K|x-y|^\lambda, \qquad x, y \in \Omega.$$

$C^{m,\lambda}(\overline{\Omega})$ is a Banach space with norm given by

$$\|\phi; C^{m,\lambda}(\overline{\Omega})\| = \|\phi; C^m(\overline{\Omega})\| + \max_{0 \le |\alpha| \le m} \sup_{\substack{x, y \in \Omega \\ x \ne y}} \frac{|D^\alpha \phi(x) - D^\alpha \phi(y)|}{|x-y|^\lambda}.$$

It should be noted that for $0 < \nu < \lambda \le 1$,

$$C^{m,\lambda}(\overline{\Omega}) \subsetneqq C^{m,\nu}(\overline{\Omega}) \subsetneqq C^m(\overline{\Omega}).$$

It is also clear that $C^{m,1}(\overline{\Omega}) \not\subset C^{m+1}(\overline{\Omega})$. In general $C^{m+1}(\overline{\Omega}) \not\subset C^{m,1}(\overline{\Omega})$ either, but the inclusion is possible for some domains $\Omega$, for instance convex ones as can be seen by appealing to the mean value theorem (see Theorem 1.31).

If $\Omega$ is bounded, the following two well-known theorems provide useful criteria for denseness and compactness of subsets of $C(\overline{\Omega})$. If $\phi \in C(\overline{\Omega})$, we may regard $\phi$ as defined on $\overline{\Omega}$, that is, we identify $\phi$ with its unique continuous extension to the closure of $\Omega$.

**1.28 THEOREM** (*Stone–Weierstrass theorem*) Let $\Omega$ be a bounded domain in $\mathbb{R}^n$. A subset $\mathscr{A}$ of $C(\overline{\Omega})$ is dense in $C(\overline{\Omega})$ if it has the following four properties:

(i) If $\phi, \psi \in \mathscr{A}$ and $c \in \mathbb{C}$, then $\phi + \psi$, $\phi\psi$, and $c\phi$ all belong to $\mathscr{A}$.
(ii) If $\phi \in \mathscr{A}$, then $\bar{\phi} \in \mathscr{A}$, where $\bar{\phi}$ is the complex conjugate of $\phi$.
(iii) If $x, y \in \overline{\Omega}$, $x \ne y$, there exists $\phi \in \mathscr{A}$ such that $\phi(x) \ne \phi(y)$.
(iv) If $x \in \overline{\Omega}$, there exists $\phi \in \mathscr{A}$ such that $\phi(x) \ne 0$.

**1.29 COROLLARY** If $\Omega$ is bounded in $\mathbb{R}^n$, then the set $P$ of all polynomials in $x = (x_1, \ldots, x_n)$ having rational-complex coefficients is dense in $C(\overline{\Omega})$. ($c \in \mathbb{C}$ is rational complex if $c = c_1 + ic_2$, where $c_1$ and $c_2$ are rational numbers.) Hence $C(\overline{\Omega})$ is separable.

PROOF The set of all polynomials in $x$ is dense in $C(\overline{\Omega})$ by the Stone–Weierstrass theorem. Any polynomial can be uniformly approximated on the compact set $\overline{\Omega}$ by elements of the countable set $P$, which is therefore also dense in $C(\overline{\Omega})$. ∎

**1.30 THEOREM** (*Ascoli–Arzela theorem*) Let $\Omega$ be a bounded domain in $\mathbb{R}^n$. A subset $K$ of $C(\overline{\Omega})$ is precompact in $C(\overline{\Omega})$ providing the following two conditions hold:

(i) There exists a constant $M$ such that for every $\phi \in K$ and $x \in \Omega$, $|\phi(x)| \leq M$.

(ii) For every $\varepsilon > 0$ there exists $\delta > 0$ such that if $\phi \in K$, $x, y \in \Omega$, and $|x-y| < \delta$, then $|\phi(x) - \phi(y)| < \varepsilon$.

The following is a straightforward imbedding theorem for the spaces introduced above.

**1.31 THEOREM** Let $m$ be a nonnegative integer and let $0 < \nu < \lambda \leq 1$. Then the following imbeddings exist:

$$C^{m+1}(\overline{\Omega}) \to C^m(\overline{\Omega}), \tag{3}$$

$$C^{m,\lambda}(\overline{\Omega}) \to C^m(\overline{\Omega}), \tag{4}$$

$$C^{m,\lambda}(\overline{\Omega}) \to C^{m,\nu}(\overline{\Omega}). \tag{5}$$

If $\Omega$ is bounded, then imbeddings (4) and (5) are compact. If $\Omega$ is convex, we have the further imbeddings

$$C^{m+1}(\overline{\Omega}) \to C^{m,1}(\overline{\Omega}), \tag{6}$$

$$C^{m+1}(\overline{\Omega}) \to C^{m,\nu}(\overline{\Omega}). \tag{7}$$

If $\Omega$ is convex and bounded, then imbeddings (3) and (7) are compact.

PROOF The existence of imbeddings (3) and (4) follows from the obvious inequalities

$$\|\phi; C^m(\overline{\Omega})\| \leq \|\phi; C^{m+1}(\overline{\Omega})\|,$$

$$\|\phi; C^m(\overline{\Omega})\| \leq \|\phi; C^{m,\lambda}(\overline{\Omega})\|.$$

To establish (5) we note that for $|\alpha| \leq m$,

$$\sup_{\substack{x,y\in\Omega \\ 0<|x-y|<1}} \frac{|D^\alpha\phi(x) - D^\alpha\phi(y)|}{|x-y|^\nu} \leq \sup_{x,y\in\Omega} \frac{|D^\alpha\phi(x) - D^\alpha\phi(y)|}{|x-y|^\lambda}$$

and

$$\sup_{\substack{x,y\in\Omega \\ |x-y|\geq 1}} \frac{|D^\alpha\phi(x) - D^\alpha\phi(y)|}{|x-y|^\nu} \leq 2\sup_{x\in\Omega} |D^\alpha\phi(x)|,$$

from which we conclude that

$$\|\phi; C^{m,\nu}(\overline{\Omega})\| \leq 3\,\|\phi; C^{m,\lambda}(\overline{\Omega})\|.$$

If $\Omega$ is convex and $x, y \in \Omega$, then by the mean value theorem there is a point $z \in \Omega$ on the line segment joining $x$ and $y$ such that $D^\alpha\phi(x) - D^\alpha\phi(y) =$

$(x-y)\cdot\nabla D^\alpha\phi(z)$, where $\nabla u = (D_1 u, D_2 u, \ldots, D_n u)$. Thus

$$|D^\alpha\phi(x) - D^\alpha\phi(y)| \le n|x-y|\,\|\phi; C^{m+1}(\overline{\Omega})\|, \tag{8}$$

and so

$$\|\phi; C^{m,1}(\overline{\Omega})\| \le n\,\|\phi; C^{m+1}(\overline{\Omega})\|.$$

Thus (6) is proved and (7) follows from (5) and (6).

Now suppose that $\Omega$ is bounded. If $A$ is a bounded set in $C^{0,\lambda}(\overline{\Omega})$, then there exists $M$ such that $\|\phi; C^{0,\lambda}(\overline{\Omega})\| \le M$ for all $\phi \in A$. But then $|\phi(x) - \phi(y)| \le M|x-y|^\lambda$ for all $\phi \in A$ and all $x, y \in \Omega$, whence $A$ is precompact in $C(\overline{\Omega})$ by Theorem 1.30. This proves the compactness of (4) for $m = 0$. If $m \ge 1$ and $A$ is bounded in $C^{m,\lambda}(\overline{\Omega})$, then $A$ is bounded in $C^{0,\lambda}(\overline{\Omega})$ and there is a sequence $\{\phi_j\} \subset A$ such that $\phi_j \to \phi$ in $C(\overline{\Omega})$. But $\{D_1\phi_j\}$ is also bounded in $C^{0,\lambda}(\overline{\Omega})$ so there exists a subsequence of $\{\phi_j\}$ which we again denote by $\{\phi_j\}$ such that $D_1\phi_j \to \psi_1$ in $C(\overline{\Omega})$. Convergence in $C(\overline{\Omega})$ being uniform convergence on $\Omega$, we have $\psi_1 = D_1\phi$. We may continue to extract subsequences in this manner until we obtain one for which $D^\alpha\phi_j \to D^\alpha\phi$ in $C(\overline{\Omega})$ for each $\alpha$, $0 \le |\alpha| \le m$. This proves the compactness of (4). For (5) we argue as follows:

$$\begin{aligned}\frac{|D^\alpha\phi(x) - D^\alpha\phi(y)|}{|x-y|^\nu} &= \left(\frac{|D^\alpha\phi(x) - D^\alpha\phi(y)|}{|x-y|^\lambda}\right)^{\nu/\lambda} |D^\alpha\phi(x) - D^\alpha\phi(y)|^{1-\nu/\nu} \\ &\le \text{const}\,|D^\alpha\phi(x) - D^\alpha\phi(y)|^{1-\nu/\lambda}\end{aligned} \tag{9}$$

for all $\phi$ in a bounded subset of $C^{m,\lambda}(\overline{\Omega})$. Since (9) shows that any sequence bounded in $C^{m,\lambda}(\overline{\Omega})$ and converging in $C^m(\overline{\Omega})$ also converges in $C^{m,\nu}(\overline{\Omega})$, the compactness of (5) follows from that of (4).

Finally, if $\Omega$ is convex and bounded, the compactness of (3) and (7) follows from composing the continuous imbedding (6) with the compact imbeddings (4) and (5) for the case $\lambda = 1$. ▮

The existence of imbeddings (6) and (7), as well as the compactness of (3) and (7), can be obtained under less restrictive hypotheses than the convexity of $\Omega$. For instance, if every pair of points $x, y \in \Omega$ can be joined by a rectifiable arc in $\Omega$ having length not exceeding some fixed multiple of $|x-y|$, then we can obtain an inequality similar to (8) and carry out the proof. We leave it to the reader to show that (6) is not compact.

## The Lebesgue Measure in $\mathbb{R}^n$

**1.32** Many of the spaces of functions considered in this monograph consist of functions integrable in the Lebesgue sense over domains in $\mathbb{R}^n$. While we assume that most readers are familiar with Lebesgue measure and integration, we nevertheless include here a brief discussion of that theory, especially those aspects of it relevant to the study of the $L^p$-spaces and Sobolev spaces considered hereafter. All proofs are omitted. For a more complete and systematic discussion of the Lebesgue theory, as well as more general measures and integrals, the reader is referred to any of the standard works on integration theory, for example, the book by Munroe [48].

**1.33** A collection $\Sigma$ of subsets of $\mathbb{R}^n$ is called a *σ-algebra* if the following conditions hold:

(i) $\mathbb{R}^n \in \Sigma$.
(ii) If $A \in \Sigma$, then $A^c = \{x \in \mathbb{R}^n : x \notin A\} \in \Sigma$.
(iii) If $A_j \in \Sigma$, $j = 1, 2, \ldots$, then $\bigcup_{j=1}^{\infty} A_j \in \Sigma$.

It follows from (i)–(iii) that:

(iv) $\emptyset \in \Sigma$.
(v) If $A_j \in \Sigma$, $j = 1, 2, \ldots$, then $\bigcap_{j=1}^{\infty} A_j \in \Sigma$.
(vi) If $A, B \in \Sigma$, then $A - B = A \cap B^c \in \Sigma$.

**1.34** By a *measure* $\mu$ on $\Sigma$ we mean a function on $\Sigma$ taking values in either $\mathbb{R} \cup \{+\infty\}$ (a *positive measure*) or $\mathbb{C}$ (a *complex measure*) which is *countably additive* in the sense that

$$\mu\left(\bigcup_{j=1}^{\infty} A_j\right) = \sum_{j=1}^{\infty} \mu(A_j)$$

whenever $A_j \in \Sigma$, $j = 1, 2, \ldots$, and $A_j \cap A_k = \emptyset$ for $j \neq k$. (For a complex measure the series on the right, being convergent for any such sequence $\{A_j\}$, is absolutely convergent.) If $\mu$ is a positive measure and if $A, B \in \Sigma$ and $A \subset B$, then $\mu(A) \leq \mu(B)$. Also, if $A_j \in \Sigma$, $j = 1, 2, \ldots$ and $A_1 \subset A_2 \subset \cdots$, then $\mu(\bigcup_{j=1}^{\infty} A_j) = \lim_{j\to\infty} \mu(A_j)$.

**1.35 THEOREM** There exists a σ-algebra $\Sigma$ of subsets of $\mathbb{R}^n$ and a positive measure $\mu$ on $\Sigma$ having the following properties:

(i) Every open set in $\mathbb{R}^n$ belongs to $\Sigma$.
(ii) If $A \subset B$, $B \in \Sigma$, and $\mu(B) = 0$, then $A \in \Sigma$ and $\mu(A) = 0$.
(iii) If $A = \{x \in \mathbb{R}^n : a_j \leq x_j \leq b_j, j = 1, 2, \ldots, n\}$, then $A \in \Sigma$ and $\mu(A) = \prod_{j=1}^{n} (b_j - a_j)$.
(iv) $\mu$ is translation invariant, that is, if $x \in \mathbb{R}^n$ and $A \in \Sigma$, then $x + A = \{x + y : y \in A\} \in \Sigma$ and $\mu(x + A) = \mu(A)$.

The elements of $\Sigma$ are called (*Lebesgue*) *measurable subsets* of $\mathbb{R}^n$; $\mu$ is called the (*Lebesgue*) *measure* in $\mathbb{R}^n$. (We shall normally suppress the word "Lebesgue" in these terms as it is the only measure on $\mathbb{R}^n$ which we shall require for our purposes.) For $A \in \Sigma$ we call $\mu(A)$ the *measure of* $A$ or the *volume of* $A$, since Lebesgue measure is a natural generalization of the concept of volume in $\mathbb{R}^3$. While we make no formal distinction between "measure" and "volume" we shall often prefer the latter term for sets that are easily visualized geometrically (balls, cubes, domains) and shall write $\operatorname{vol} A$ in place of $\mu(A)$ in these cases. In $\mathbb{R}^1$ and $\mathbb{R}^2$ the terms *length* and *area* are more appropriate than *volume*.

**1.36** If $B \subset A \subset \mathbb{R}^n$ and $\mu(B) = 0$, then any condition that holds at every point of the set $A - B$ is said to hold *almost everywhere* (a.e.) in $A$. It is easily seen that every countable set in $\mathbb{R}^n$ has measure zero. The converse is, however, not true.

A function $f$ defined on a measurable set and having values in $\mathbb{R} \cup \{+\infty, -\infty\}$ is itself called *measurable* if the set

$$\{x : f(x) > a\}$$

is measurable for every real $a$. Some of the more important aspects of this definition are listed in the following theorem.

**1.37 THEOREM** (a) If $f$ is measurable, so is $|f|$.

(b) If $f$ and $g$ are measurable and real valued, then so are $f+g$ and $fg$.

(c) If $\{f_n\}$ is a sequence of measurable functions, then $\sup_n f_n$, $\inf_n f_n$, $\limsup_{n\to\infty} f_n$, and $\liminf_{n\to\infty} f_n$ are measurable.

(d) If $f$ is continuous and defined on a measurable set, then $f$ is measurable.

(e) If $f$ is continuous on $\mathbb{R}$ into $\mathbb{R}$ and $g$ is measurable and real valued, then the composite function $f \circ g$ defined by $f \circ g(x) = f(g(x))$ is measurable.

(f) (*Lusin's theorem*) If $f$ is measurable and $f(x) = 0$ for $x \in A^c$ where $\mu(A) < \infty$, and if $\varepsilon > 0$, then there exists a function $g \in C_0(A)$ such that $\sup_{x \in \mathbb{R}^n} |g(x)| \le \sup_{x \in \mathbb{R}^n} |f(x)|$ and $\mu\{x \in \mathbb{R}^n : f(x) \ne g(x)\} < \varepsilon$.

**1.38** Let $A \subset \mathbb{R}^n$. The function $\chi_A$ defined by

$$\chi_A(x) = \begin{cases} 1 & \text{if } x \in A \\ 0 & \text{if } x \notin A \end{cases}$$

is called the *characteristic function* of $A$. A real-valued function $s$ on $\mathbb{R}^n$ is called a *simple function* if its range is a finite set of real numbers. If for every $x$, $s(x) \in \{a_1, \ldots, a_m\}$, then clearly $s = \sum_{j=1}^m a_j \chi_{A_j}$, where $A_j = \{x \in \mathbb{R}^n : s(x) = a_j\}$, and $s$ is measurable if and only if $A_1, A_2, \ldots, A_m$ are measurable. Because of the following approximation theorem simple functions are a very useful tool in integration theory.

**1.39 THEOREM** Given a real-valued function $f$ with domain $A \subset \mathbb{R}^n$ there is a sequence $\{s_n\}$ of simple functions converging pointwise to $f$ on $A$. If $f$ is bounded, $\{s_n\}$ may be chosen so that the convergence is uniform. If $f$ is measurable, each $s_n$ may be chosen measurable. If $f$ is nonnegative valued, the sequence $\{s_n\}$ may be chosen to be monotonically increasing at each point.

## The Lebesgue Integral

**1.40** We are now in a position to define the (*Lebesgue*) *integral* of a measurable, real-valued function defined on a measurable set $A \subset \mathbb{R}^n$. For a simple function $s = \sum_{j=1}^{m} a_j \chi_{A_j}$, where $A_j \subset A$, $A_j$ is measurable, we define

$$\int_A s(x)\,dx = \sum_{j=1}^{m} a_j \mu(A_j). \tag{10}$$

If $f$ is measurable and nonnegative valued, we define

$$\int_A f(x)\,dx = \sup \int_A s(x)\,dx, \tag{11}$$

the supremum being taken over measurable simple functions $s$ vanishing outside $A$ and satisfying $0 \le s(x) \le f(x)$ in $A$. If $f$ is a nonnegative simple function, then the two definitions of $\int_A f(x)\,dx$ given by (10) and (11) coincide. Note that the integral of a nonnegative function may be $+\infty$.

If $f$ is measurable and real valued, we set $f = f^+ - f^-$, where $f^+ = \max(f, 0)$ and $f^- = -\min(f, 0)$ are both measurable and nonnegative. We define

$$\int_A f(x)\,dx = \int_A f^+(x)\,dx - \int_A f^-(x)\,dx$$

provided at least one of the integrals on the right is finite. If *both integrals are finite*, we say that $f$ is (Lebesgue) *integrable on* $A$. The class of integrable functions on $A$ is denoted $L^1(A)$.

**1.41 THEOREM** Assume all of the functions and sets appearing below are measurable.

(a) If $f$ is bounded on $A$ and $\mu(A) < \infty$, then $f \in L^1(A)$.
(b) If $a \le f(x) \le b$ for all $x \in A$ and if $\mu(A) < \infty$, then

$$a\mu(A) \le \int_A f(x)\,dx \le b\mu(A).$$

(c) If $f(x) \le g(x)$ for all $x \in A$, and if both integrals exist, then

$$\int_A f(x)\,dx \le \int_A g(x)\,dx.$$

(d) If $f, g \in L^1(A)$, then $f+g \in L^1(A)$ and

$$\int_A (f+g)(x)\,dx = \int_A f(x)\,dx + \int_A g(x)\,dx.$$

(e) If $f \in L^1(A)$ and $c \in \mathbb{R}$, then $cf \in L^1(A)$ and

$$\int_A (cf)(x)\,dx = c \int_A f(x)\,dx.$$

(f) If $f \in L^1(A)$, then $|f| \in L^1(A)$ and

$$\left| \int_A f(x)\,dx \right| \le \int_A |f(x)|\,dx.$$

(g) If $f \in L^1(A)$ and $B \subset A$, then $f \in L^1(B)$; if in addition $f(x) \ge 0$ for all $x \in A$, then

$$\int_B f(x)\,dx \le \int_A f(x)\,dx.$$

(h) If $\mu(A) = 0$, then $\int_A f(x)\,dx = 0$.

(i) If $f \in L^1(A)$ and if $\int_B f(x)\,dx = 0$ for every $B \subset A$, then $f(x) = 0$ a.e. on $A$.

**1.42 THEOREM** If $f$ is either an element of $L^1(\mathbb{R}^n)$ or measurable and nonnegative on $\mathbb{R}^n$, then the set function $\lambda$ defined by

$$\lambda(A) = \int_A f(x)\,dx$$

is countably additive, and hence a measure, on the $\sigma$-algebra of Lebesgue measurable subsets of $\mathbb{R}^n$.

One consequence of this additivity of the integral is that sets of measure zero may be ignored for purposes of integration, that is, if $f$ and $g$ are measurable on $A$ and if $f(x) = g(x)$ a.e. on $A$, then $\int_A f(x)\,dx = \int_A g(x)\,dx$. Accordingly, two elements of $L^1(A)$ are considered identical if they are equal almost everywhere.

The following three theorems are concerned with the interchange of integration and limit processes.

**1.43 THEOREM** (*Monotone convergence theorem*) Let $A \subset \mathbb{R}^n$ be measurable, and $\{f_n\}$ a sequence of measurable functions satisfying $0 \le f_1(x) \le f_2(x) \le \cdots$ for every $x \in A$. Then

$$\lim_{n\to\infty} \int_A f_n(x)\,dx = \int_A \left( \lim_{n\to\infty} f_n(x) \right) dx.$$

**1.44 THEOREM** (*Fatou's lemma*) Let $A \subset \mathbb{R}^n$ be measurable and let $\{f_n\}$ be a sequence of nonnegative, measurable functions. Then

$$\int_A \left( \liminf_{n\to\infty} f_n(x) \right) dx \le \liminf_{n\to\infty} \int_A f_n(x)\, dx.$$

**1.45 THEOREM** (*Dominated convergence theorem*) Let $A \subset \mathbb{R}^n$ be measurable and let $\{f_n\}$ be a sequence of measurable functions converging to a limit pointwise on $A$. If there is a function $g \in L^1(A)$ such that $|f_n(x)| \le g(x)$ for every $n$ and all $x \in A$, then

$$\lim_{n\to\infty} \int_A f_n(x)\, dx = \int_A \left( \lim_{n\to\infty} f_n(x) \right) dx.$$

**1.46** The integral of a complex-valued function over a measurable set $A \subset \mathbb{R}^n$ is defined as follows. Set $f = u + iv$, where $u$ and $v$ are real valued and call $f$ measurable if and only if $u$ and $v$ are measurable. We shall say that $f$ is integrable over $A$, and write $f \in L^1(A)$, provided $|f| = (u^2 + v^2)^{1/2}$ belongs to $L^1(A)$ in the sense described in Section 1.40. For $f \in L^1(A)$, and only for such $f$, the integral is defined by

$$\int_A f(x)\, dx = \int_A u(x)\, dx + i \int_A v(x)\, dx.$$

It is easily checked that $f \in L^1(A)$ if and only if $u, v \in L^1(A)$. Theorem 1.37(a, b, d–f), Theorem 1.41(a, d–i), Theorem 1.42 [assuming $f \in L^1(\mathbb{R}^n)$], and Theorem 1.45 all extend to cover the case of complex $f$.

The following theorem enables us to express certain complex measures in terms of Lebesgue measure $\mu$. It is the converse of Theorem 1.42.

**1.47 THEOREM** (*The Radon–Nikodym theorem*) Let $\lambda$ be a complex measure defined on the $\sigma$-algebra $\Sigma$ of Lebesgue measurable subsets of $\mathbb{R}^n$. Suppose that $\lambda(A) = 0$ for every $A \in \Sigma$ for which $\mu(A) = 0$. Then there exists $f \in L^1(\mathbb{R}^n)$ such that for every $A \in \Sigma$

$$\lambda(A) = \int_A f(x)\, dx.$$

The function $f$ is uniquely determined by $\lambda$ up to sets of measure zero.

**1.48** If $f$ is a function defined on a subset $A$ of $\mathbb{R}^{n+m}$, we may regard $f$ as depending on the pair of variables $(x, y)$ with $x \in \mathbb{R}^n$ and $y \in \mathbb{R}^m$. The integral of $f$ over $A$ is then commonly denoted by

$$\int_A f(x,y)\,dx\,dy$$

or, if it is desired to have the integral extend over all of $\mathbb{R}^{n+m}$,

$$\int_{\mathbb{R}^{n+m}} f(x,y)\,\chi_A(x,y)\,dx\,dy,$$

where $\chi_A$ is the characteristic function of $A$. In particular, if $A \subset \mathbb{R}^n$, we may write

$$\int_A f(x)\,dx = \int_A f(x_1, \dots, x_n)\,dx_1 \cdots dx_n.$$

**1.49 THEOREM** (*Fubini's theorem*) Let $f$ be a measurable function on $\mathbb{R}^{n+m}$ and suppose that at least one of the integrals

$$\begin{aligned} I_1 &= \int_{\mathbb{R}^{n+m}} |f(x,y)|\,dx\,dy, \\ I_2 &= \int_{\mathbb{R}^m} \left( \int_{\mathbb{R}^n} |f(x,y)|\,dx \right) dy, \qquad (12) \\ I_3 &= \int_{\mathbb{R}^n} \left( \int_{\mathbb{R}^m} |f(x,y)|\,dy \right) dx \end{aligned}$$

exists and is finite. Then

(a) $f(\cdot, y) \in L^1(\mathbb{R}^n)$ for almost all $y \in \mathbb{R}^m$,
(b) $f(x, \cdot) \in L^1(\mathbb{R}^m)$ for almost all $x \in \mathbb{R}^n$,
(c) $\int_{\mathbb{R}^n} f(x, \cdot)\,dx \in L^1(\mathbb{R}^m)$,
(d) $\int_{\mathbb{R}^m} f(\cdot, y)\,dy \in L^1(\mathbb{R}^n)$, and
(e) $I_1 = I_2 = I_3$.

## Distributions and Weak Derivatives

**1.50** We shall require in subsequent chapters some of the basic concepts and techniques of the Schwartz theory of distributions [60], and we present here a brief description of those aspects of the theory that are relevant for our purposes. Of special importance is the notion of weak or distributional derivative of an integrable function. One of the standard definitions of Sobolev spaces is phrased in terms of such derivatives (Section 3.1). In addition to Ref. [60] the reader is referred to Rudin [59] and Yosida [69] for more complete treatments of the spaces $\mathscr{D}(\Omega)$ and $\mathscr{D}'(\Omega)$ introduced below, as well as useful generalizations of these spaces.

**1.51** Let $\Omega$ be a domain in $\mathbb{R}^n$. A sequence $\{\phi_n\}$ of functions belonging to $C_0^\infty(\Omega)$ is said to *converge in the sense of the space* $\mathscr{D}(\Omega)$ to the function $\phi \in C_0^\infty(\Omega)$ provided the following conditions are satisfied:

(i) there exists $K \subset\subset \Omega$ such that $\operatorname{supp}(\phi_n - \phi) \subset K$ for every $n$, and
(ii) $\lim_{n\to\infty} D^\alpha \phi_n(x) = D^\alpha \phi(x)$ uniformly on $K$ for each multi-index $\alpha$.

There exists a locally convex topology on the vector space $C_0^\infty(\Omega)$ with respect to which a linear functional $T$ is continuous if and only if $T(\phi_n) \to T(\phi)$ in $\mathbb{C}$ whenever $\phi_n \to \phi$ in the sense of the space $\mathscr{D}(\Omega)$. This TVS is called $\mathscr{D}(\Omega)$ and its elements *testing functions*. $\mathscr{D}(\Omega)$ is not a normable space. (We ignore the question of uniqueness of the topology asserted above. It uniquely determines the dual of $\mathscr{D}(\Omega)$ which is sufficient for our purposes.)

**1.52** The dual space $\mathscr{D}'(\Omega)$ of $\mathscr{D}(\Omega)$ is called the *space of* (Schwartz) *distributions*. $\mathscr{D}'(\Omega)$ is given the weak-star topology as dual of $\mathscr{D}(\Omega)$, and is a locally convex TVS with that topology. We summarize the vector space and convergence operations in $\mathscr{D}'(\Omega)$ as follows: if $S$, $T$, $T_n \in \mathscr{D}'(\Omega)$ and $c \in \mathbb{C}$, then

$$(S+T)(\phi) = S(\phi) + T(\phi), \qquad \phi \in \mathscr{D}(\Omega),$$

$$(cT)(\phi) = cT(\phi), \qquad \phi \in \mathscr{D}(\Omega),$$

$T_n \to T$ in $\mathscr{D}'(\Omega)$ if and only if $T_n(\phi) \to T(\phi)$ in $\mathbb{C}$ for every $\phi \in \mathscr{D}(\Omega)$.

**1.53** A function $u$ defined almost everywhere on $\Omega$ is said to be *locally integrable* on $\Omega$ provided $u \in L^1(A)$ for every measurable $A \subset\subset \Omega$. In this case we write $u \in L^1_{\text{loc}}(\Omega)$. Corresponding to every $u \in L^1_{\text{loc}}(\Omega)$ there is a distribution $T_u \in \mathscr{D}'(\Omega)$ defined by

$$T_u(\phi) = \int_\Omega u(x)\,\phi(x)\,dx, \qquad \phi \in \mathscr{D}(\Omega). \tag{13}$$

It is clear that $T_u$, thus defined, is a linear functional on $\mathscr{D}(\Omega)$. To see that it is continuous suppose that $\phi_n \to \phi$ in $\mathscr{D}(\Omega)$. Then there exists $K \subset\subset \Omega$ such that $\operatorname{supp}(\phi_n - \phi) \subset K$ for $n = 1, 2, 3, \ldots$. Thus

$$|T_u(\phi_n) - T_u(\phi)| \le \sup_{x\in K} |\phi_n(x) - \phi(x)| \int_K |u(x)|\,dx.$$

The right side of the inequality above tends to zero as $n \to \infty$ since $\phi_n \to \phi$ uniformly on $K$.

**1.54** Not every distribution $T \in \mathscr{D}'(\Omega)$ is of the form $T = T_u$ [defined by (13)] for some $u \in L^1_{\text{loc}}(\Omega)$. Indeed, assuming $0 \in \Omega$, the reader may wish to convince himself that there can be no locally integrable function $\delta$ on $\Omega$ such that for every $\phi \in \mathscr{D}(\Omega)$

$$\int_\Omega \delta(x)\,\phi(x)\,dx = \phi(0).$$

The linear functional $\delta$ defined on $\mathscr{D}(\Omega)$ by

$$\delta(\phi) = \phi(0) \tag{14}$$

is, however, easily seen to be continuous, and hence a distribution on $\Omega$.

**1.55** Let $u \in C^1(\Omega)$ and $\phi \in \mathscr{D}(\Omega)$. Since $\phi$ vanishes identically outside some compact subset of $\Omega$, we obtain by integration by parts in the variable $x_j$

$$\int_\Omega \left(\frac{\partial}{\partial x_j} u(x)\right) \phi(x)\,dx = -\int_\Omega u(x)\left(\frac{\partial}{\partial x_j}\phi(x)\right) dx.$$

Similarly, integration by parts $|\alpha|$ times leads to

$$\int_\Omega (D^\alpha u(x))\,\phi(x)\,dx = (-1)^{|\alpha|}\int_\Omega u(x)\,D^\alpha\phi(x)\,dx$$

if $u \in C^{|\alpha|}(\Omega)$. This motivates the following definition of the derivative $D^\alpha T$ of a distribution $T \in \mathscr{D}'(\Omega)$:

$$(D^\alpha T)(\phi) = (-1)^{|\alpha|}T(D^\alpha\phi). \tag{15}$$

Since $D^\alpha\phi \in \mathscr{D}(\Omega)$ whenever $\phi \in \mathscr{D}(\Omega)$, $D^\alpha T$ is a functional on $\mathscr{D}(\Omega)$. Clearly $D^\alpha T$ is linear on $\mathscr{D}(\Omega)$. We show that $D^\alpha T$ is continuous, and hence a distribution on $\Omega$. To this end suppose $\phi_n \to \phi$ in $\mathscr{D}(\Omega)$. Then

$$\operatorname{supp}(D^\alpha(\phi_n - \phi)) \subset \operatorname{supp}(\phi_n - \phi) \subset K$$

for some $K \subset\subset \Omega$. Moreover,

$$D^\beta[D^\alpha(\phi_n - \phi)] = D^{\beta+\alpha}(\phi_n - \phi)$$

converges to zero uniformly on $K$ as $n \to \infty$ for each multi-index $\beta$. Hence $D^\alpha\phi_n \to D^\alpha\phi$ in $\mathscr{D}(\Omega)$. Since $T \in \mathscr{D}'(\Omega)$ it follows that

$$D^\alpha T(\phi_n) = (-1)^{|\alpha|}T(D^\alpha\phi_n) \to (-1)^{|\alpha|}T(D^\alpha\phi) = D^\alpha T(\phi)$$

in $\mathbb{C}$. Thus $D^\alpha T \in \mathscr{D}'(\Omega)$.

We have shown that every distribution in $\mathscr{D}'(\Omega)$ possesses derivatives of arbitrary orders in $\mathscr{D}'(\Omega)$ in the sense of the definition (15). Furthermore, the mapping $D^\alpha$ from $\mathscr{D}'(\Omega)$ into $\mathscr{D}'(\Omega)$ is continuous. If $T_n \to T$ in $\mathscr{D}'(\Omega)$ and if $\phi \in \mathscr{D}(\Omega)$, then

$$D^\alpha T_n(\phi) = (-1)^{|\alpha|}T_n(D^\alpha\phi) \to (-1)^{|\alpha|}T(D^\alpha\phi) = D^\alpha T(\phi).$$

**1.56 EXAMPLES** (1) If $0 \in \Omega$ and $\delta \in \mathscr{D}'(\Omega)$ is defined by (14), then $D^\alpha \delta$ is given by

$$D^\alpha \delta(\phi) = (-1)^{|\alpha|} D^\alpha \phi(0).$$

(2) If $\Omega = \mathbb{R}$ (i.e., $n = 1$) and $H \in L^1_{\mathrm{loc}}(\mathbb{R})$ is defined by

$$H(x) = \begin{cases} 1 & \text{if } x \geq 0 \\ 0 & \text{if } x < 0, \end{cases}$$

then the derivative $(T_H)'$ is $\delta$, for if $\phi \in \mathscr{D}(\mathbb{R})$ has support in the interval $[-a, a]$, then

$$(T_H)'(\phi) = -T_H(\phi') = -\int_0^a \phi'(x)\,dx = \phi(0) = \delta(\phi).$$

**1.57** We now define the concept of weak derivative of a locally integrable function. Let $u \in L^1_{\mathrm{loc}}(\Omega)$. There may or may not exist a function $v_\alpha \in L^1_{\mathrm{loc}}(\Omega)$ such that $T_{v_\alpha} = D^\alpha(T_u)$ in $\mathscr{D}'(\Omega)$. If such a $v_\alpha$ exists, it is unique up to sets of measure zero and it is called the *weak* or *distributional partial derivative* of $u$ and is denoted by $D^\alpha u$. Thus $D^\alpha u = v_\alpha$ in the weak (distributional) sense provided $v_\alpha \in L^1_{\mathrm{loc}}(\Omega)$ satisfies

$$\int_\Omega u(x)\, D^\alpha \phi(x)\,dx = (-1)^{|\alpha|} \int_\Omega v_\alpha(x)\,\phi(x)\,dx$$

for every $\phi \in \mathscr{D}(\Omega)$.

If $u$ is sufficiently smooth to have a continuous partial derivative $D^\alpha u$ in the usual (classical) sense, then $D^\alpha u$ is also a distributional partial derivative of $u$. Of course $D^\alpha u$ may exist in the distributional sense without existing in the classical sense. For example a function $u$, continuous on $\mathbb{R}$, which has a bounded derivative $u'$ except at finitely many points, has a derivative in the distributional sense. We shall show in Theorem 3.16 that functions having weak derivatives can be suitably approximated by smooth functions.

**1.58** Let us note in conclusion that distributions on $\Omega$ can be multiplied by smooth functions. If $T \in \mathscr{D}'(\Omega)$ and $\omega \in C^\infty(\Omega)$, the product $\omega T \in \mathscr{D}'(\Omega)$ is defined by

$$(\omega T)(\phi) = T(\omega\phi), \qquad \phi \in \mathscr{D}(\Omega).$$

If $T = T_u$ for some $u \in L^1_{\mathrm{loc}}(\Omega)$, then $\omega T = T_{\omega u}$. The Leibniz rule (see Section 1.1) is easily checked to hold for $D^\alpha(\omega T)$.

II

# The Spaces $L^p(\Omega)$

## Definition and Basic Properties

**2.1** Let $\Omega$ be a domain in $\mathbb{R}^n$ and let $p$ be a positive real number. We denote by $L^p(\Omega)$ the class of all measurable functions $u$, defined on $\Omega$, for which

$$\int_\Omega |u(x)|^p \, dx < \infty. \tag{1}$$

We identify in $L^p(\Omega)$ functions that are equal almost everywhere on $\Omega$. The elements of $L^p(\Omega)$ are thus actually equivalence classes of measurable functions satisfying (1), two functions being equivalent if they are equal a.e. in $\Omega$. For convenience, however, we ignore this distinction and write $u \in L^p(\Omega)$ if $u$ satisfies (1), and $u = 0$ in $L^p(\Omega)$ if $u(x) = 0$ a.e. in $\Omega$. It is clear that if $u \in L^p(\Omega)$ and $c \in \mathbb{C}$, then $cu \in L^p(\Omega)$. Moreover, if $u, v \in L^p(\Omega)$, then since

$$|u(x) + v(x)|^p \le (|u(x)| + |v(x)|)^p \le 2^p(|u(x)|^p + |v(x)|^p),$$

$u + v \in L^p(\Omega)$, so $L^p(\Omega)$ is a vector space.

**2.2** We shall verify presently that the functional $\|\cdot\|_p$ defined by

$$\|u\|_p = \left\{\int_\Omega |u(x)|^p \, dx\right\}^{1/p}$$

is a norm on $L^p(\Omega)$ provided $1 \le p < \infty$. (It is not a norm if $0 < p < 1$.) In arguments where confusion of domains might occur we shall use $\|\cdot\|_{p,\Omega}$ in place of $\|\cdot\|_p$. It is clear that $\|u\|_p \geq 0$ and equality occurs if and only if $u = 0$ in $L^p(\Omega)$. Moreover,

$$\|cu\|_p = |c|\,\|u\|_p, \qquad c \in \mathbb{C}.$$

It remains to be shown, then, that if $1 \leq p < \infty$,

$$\|u+v\|_p \leq \|u\|_p + \|v\|_p, \tag{2}$$

which is known as *Minkowski's inequality*. Condition (2) certainly holds for $p = 1$ since

$$\int_\Omega |u(x)+v(x)|\,dx \leq \int_\Omega |u(x)|\,dx + \int_\Omega |v(x)|\,dx.$$

If $1 < p < \infty$, we denote by $p'$ the number $p/(p-1)$ so that $1 < p' < \infty$ and

$$(1/p) + (1/p') = 1.$$

$p'$ is called the *exponent conjugate to p*.

**2.3 THEOREM** (*Hölder's inequality*) If $1 < p < \infty$ and $u \in L^p(\Omega)$, $v \in L^{p'}(\Omega)$, then $uv \in L^1(\Omega)$ and

$$\int_\Omega |u(x)\,v(x)|\,dx \leq \|u\|_p\,\|v\|_{p'}. \tag{3}$$

PROOF The function $f(t) = (t^p/p) + (1/p') - t$ has, for $t \geq 0$, the minimum value zero, and this minimum is attained only at $t = 1$. Setting $t = ab^{-p'/p}$, we conclude, for nonnegative numbers $a$ and $b$, that

$$ab \leq (a^p/p) + (b^{p'}/p') \tag{4}$$

with equality occurring if and only if $a^p = b^{p'}$. If either $\|u\|_p = 0$ or $\|v\|_{p'} = 0$, then $u(x)\,v(x) = 0$ a.e. in $\Omega$ so (3) is satisfied. Otherwise we obtain (3) by setting $a = |u(x)|/\|u\|_p$ and $b = |v(x)|/\|v\|_{p'}$ in (4) and integrating over $\Omega$. Equality occurs in (3) if and only if $|u(x)|^p$ and $|v(x)|^{p'}$ are proportional a.e. in $\Omega$. ∎

We remark that a form of Hölder's inequality for finite or infinite sums,

$$\sum |a_k b_k| \leq \{\sum |a_k|^p\}^{1/p}\{\sum |b_k|^{p'}\}^{1/p'},$$

can be proved in the same manner.

**2.4 THEOREM** (*Minkowski's inequality*) If $1 \leq p < \infty$, then

$$\|u+v\|_p \leq \|u\|_p + \|v\|_p. \tag{5}$$

PROOF We have already done the case in which $p = 1$ so we assume $1 < p < \infty$. We may also assume that $u, v \in L^p(\Omega)$, for otherwise the right side of (5) is infinite. Now

$$\int_\Omega |u(x)+v(x)|^p\,dx \le \int_\Omega |u(x)+v(x)|^{p-1}(|u(x)|+|v(x)|)\,dx$$

$$\le \left\{\int_\Omega |u(x)+v(x)|^p\,dx\right\}^{1/p'} (\|u\|_p + \|v\|_p)$$

by separate applications of Hölder's inequality. Inequality (5) follows by cancellation, which is valid since $\|u+v\|_p < \infty$. ∎

**2.5** A function $u$, measurable on $\Omega$, is said to be *essentially bounded* on $\Omega$ provided there exists a constant $K$ for which $|u(x)| \le K$ a.e. on $\Omega$. The greatest lower bound of such constants $K$ is called the *essential supremum* of $|u|$ on $\Omega$ and is denoted by ess $\sup_{x\in\Omega}|u(x)|$. We denote by $L^\infty(\Omega)$ the vector space consisting of all functions $u$ that are essentially bounded on $\Omega$, functions being once again identified if they are equal a.e. on $\Omega$. It is easily verified that the functional $\|\cdot\|_\infty$ defined by

$$\|u\|_\infty = \operatorname*{ess\,sup}_{x\in\Omega} |u(x)|$$

is a norm on $L^\infty(\Omega)$. Moreover, Hölder's inequality (3) clearly extends to cover the two cases $p = 1$, $p' = \infty$, and $p = \infty$, $p' = 1$.

The following pair of theorems establishes reverse forms of Hölder's and Minkowski's inequalities for the case $0 < p < 1$. The latter inequality will be used later in establishing the uniform convexity of certain $L^p$-spaces.

**2.6 THEOREM** Let $0 < p < 1$ so that $p' = p/(p-1) < 0$. Suppose $f \in L^p(\Omega)$ and

$$0 < \int_\Omega |g(x)|^{p'}\,dx < \infty.$$

Then

$$\int_\Omega |f(x)g(x)|\,dx \ge \left\{\int_\Omega |f(x)|^p\,dx\right\}^{1/p} \left\{\int_\Omega |g(x)|^{p'}\,dx\right\}^{1/p'}. \tag{6}$$

PROOF We may assume $fg \in L^1(\Omega)$; otherwise the left side of (6) is infinite. Set $\phi = |g|^{-p}$ and $\psi = |fg|^p$ so that $\phi\psi = |f|^p$. Then $\psi \in L^q(\Omega)$ where $q = 1/p > 1$, and since $p' = -pq'$ where $q' = q/(q-1)$ we have $\phi \in L^{q'}(\Omega)$. By the direct version of Hölder's inequality (3) we have

$$\int_\Omega |f(x)|^p\,dx = \int_\Omega \phi(x)\psi(x)\,dx \le \|\psi\|_q\|\phi\|_{q'}$$

$$= \left\{\int_\Omega |f(x)g(x)|\,dx\right\}^p \left\{\int_\Omega |g(x)|^{p'}\,dx\right\}^{1-p}.$$

Taking $p$th roots and dividing by the last factor on the right side we obtain (6). ∎

**2.7 THEOREM** Let $0 < p < 1$. If $u, v \in L^p(\Omega)$, then

$$\|\,|u| + |v|\,\|_p \ge \|u\|_p + \|v\|_p. \tag{7}$$

PROOF If $u = v = 0$ in $L^p(\Omega)$, then (7) is trivial. Otherwise the left-hand side is greater than zero. Applying the reverse Hölder's inequality (6), we obtain

$$\|\,|u| + |v|\,\|_p^p = \int_\Omega (|u(x)| + |v(x)|)^{p-1}(|u(x)| + |v(x)|)\,dx$$

$$\ge \left\{\int_\Omega (|u(x)| + |v(x)|)^{(p-1)p'}\,dx\right\}^{1/p'} (\|u\|_p + \|v\|_p)$$

$$= \|\,|u| + |v|\,\|_p^{p/p'}(\|u\|_p + \|v\|_p)$$

and (7) follows by cancellation. ∎

The following theorem gives a useful imbedding result for $L^p$-spaces over domains with finite volume, and some consequences of this imbedding.

**2.8 THEOREM** Suppose $\operatorname{vol}\Omega = \int_\Omega 1\,dx < \infty$ and $1 \le p \le q \le \infty$. If $u \in L^q(\Omega)$, then $u \in L^p(\Omega)$ and

$$\|u\|_p \le (\operatorname{vol}\Omega)^{(1/p)-(1/q)}\,\|u\|_q. \tag{8}$$

Hence

$$L^q(\Omega) \to L^p(\Omega). \tag{9}$$

If $u \in L^\infty(\Omega)$, then

$$\lim_{p\to\infty} \|u\|_p = \|u\|_\infty. \tag{10}$$

Finally, if $u \in L^p(\Omega)$ for $1 \le p < \infty$ and if there is a constant $K$ such that for all such $p$

$$\|u\|_p \le K, \tag{11}$$

then $u \in L^\infty(\Omega)$ and

$$\|u\|_\infty \le K. \tag{12}$$

PROOF If $p = q$, (8) and (9) are trivial. If $1 \le p < q \le \infty$ and $u \in L^q(\Omega)$, Hölder's inequality gives

$$\int_\Omega |u(x)|^p\,dx \le \left\{\int_\Omega |u(x)|^q\,dx\right\}^{p/q} \left\{\int_\Omega 1\,dx\right\}^{1-(p/q)}$$

from which (8) and (9) follow immediately. If $u \in L^\infty(\Omega)$, we obtain from (8)

$$\limsup_{p\to\infty} \|u\|_p \le \|u\|_\infty. \tag{13}$$

On the other hand, for any $\varepsilon > 0$ there exists a set $A \subset \Omega$ having positive measure $\mu(A)$ such that

$$|u(x)| \ge \|u\|_\infty - \varepsilon \qquad \text{if} \quad x \in A.$$

Hence

$$\int_\Omega |u(x)|^p\,dx \ge \int_A |u(x)|^p\,dx \ge \mu(A)\,(\|u\|_\infty - \varepsilon)^p.$$

It follows that $\|u\|_p \ge (\mu(A))^{1/p}(\|u\|_\infty - \varepsilon)$, whence

$$\liminf_{p\to\infty} \|u\|_p \ge \|u\|_\infty. \tag{14}$$

Equation (10) now follows from (13) and (14).

Now suppose (11) holds for $1 \le p < \infty$. If $u \notin L^\infty(\Omega)$ or else if (12) does not hold, then we can find a constant $K_1 > K$ and a set $A \subset \Omega$ with $\mu(A) > 0$ such that for $x \in A$, $|u(x)| \ge K_1$. The same argument used to obtain (14) now shows that

$$\liminf_{p\to\infty} \|u\|_p \ge K_1,$$

which contradicts (11). ∎

**2.9 COROLLARY** $L^p(\Omega) \subset L^1_{\mathrm{loc}}(\Omega)$ for $1 \le p \le \infty$ and any domain $\Omega$.

## Completeness of $L^p(\Omega)$

**2.10 THEOREM** $L^p(\Omega)$ is a Banach space if $1 \le p \le \infty$.

PROOF First assume $1 \le p < \infty$ and let $\{u_n\}$ be a Cauchy sequence in $L^p(\Omega)$.

There is a subsequence $\{u_{n_j}\}$ of $\{u_n\}$ such that

$$\|u_{n_{j+1}} - u_{n_j}\|_p \le 1/2^j, \qquad j = 1, 2, \ldots.$$

Let $v_m(x) = \sum_{j=1}^{m} |u_{n_{j+1}}(x) - u_{n_j}(x)|$. Then

$$\|v_m\|_p \le \sum_{j=1}^{m} (1/2^j) < 1, \qquad m = 1, 2, \ldots.$$

Putting $v(x) = \lim_{m\to\infty} v_m(x)$, which may be infinite for some $x$, we obtain by Fatou's lemma 1.44

$$\int_\Omega |v(x)|^p\,dx \le \liminf_{m\to\infty} \int_\Omega |v_m(x)|^p\,dx \le 1.$$

Hence $v(x) < \infty$ a.e. in $\Omega$ and the series

$$u_{n_1}(x) + \sum_{j=1}^{\infty} (u_{n_{j+1}}(x) - u_{n_j}(x)) \tag{15}$$

converges to a limit $u(x)$ a.e. in $\Omega$. Let $u(x) = 0$ whenever it is undefined as the limit of (15). Since (15) telescopes we have

$$\lim_{m\to\infty} u_{n_m}(x) = u(x) \qquad \text{a.e. in } \Omega.$$

For any $\varepsilon > 0$ there exists $N$ such that if $m, n \ge N$, then $\|u_m - u_n\|_p < \varepsilon$. Hence, by Fatou's lemma again

$$\begin{aligned}\int_\Omega |u(x) - u_n(x)|^p\,dx &= \int_\Omega \lim_{j\to\infty} |u_{n_j}(x) - u_n(x)|^p\,dx \\ &\le \liminf_{j\to\infty} \int_\Omega |u_{n_j}(x) - u_n(x)|^p\,dx \le \varepsilon^p\end{aligned}$$

if $n \ge M$. Thus $u = (u - u_n) + u_n \in L^p(\Omega)$ and $\|u - u_n\|_p \to 0$ as $n \to \infty$. Therefore $L^p(\Omega)$ is complete.

Finally, if $\{u_n\}$ is a Cauchy sequence in $L^\infty(\Omega)$, then there exists a set $A \subset \Omega$ having measure zero such that if $x \notin A$, then for every $n, m = 1, 2, \ldots$

$$|u_n(x)| \le \|u_n\|_\infty, \qquad |u_n(x) - u_m(x)| \le \|u_n - u_m\|_\infty.$$

Since $\{\|u_n\|_\infty\}$ is bounded in $\mathbb{R}$, $u_n$ converges uniformly on $\Omega \sim A$ to a bounded function $u$. Setting $u(x) = 0$ for $x \in A$, we have $u \in L^\infty(\Omega)$ and $\|u_n - u\|_\infty \to 0$ as $n \to \infty$. Thus $L^\infty(\Omega)$ is complete. ∎

**2.11 COROLLARY** If $1 \le p \le \infty$, a Cauchy sequence in $L^p(\Omega)$ has a subsequence converging pointwise almost everywhere on $\Omega$.

**2.12 COROLLARY** $L^2(\Omega)$ is a Hilbert space with respect to the inner product

$$(u, v) = \int_\Omega u(x)\overline{v(x)}\,dx.$$

Hölder's inequality for $L^2(\Omega)$ is actually just the well-known Schwarz inequality

$$|(u, v)| \le \|u\|_2 \|v\|_2.$$

### Approximation by Continuous Functions, Separability

**2.13 THEOREM** $C_0(\Omega)$ is dense in $L^p(\Omega)$ if $1 \le p < \infty$.

PROOF Let $u \in L^p(\Omega)$ and let $\varepsilon > 0$. We show that there exists a function $\phi \in C_0(\Omega)$ such that $\|u - \phi\|_p < \varepsilon$. Setting $u = u_1 - u_2 + i(u_3 - u_4)$ where each $u_j$, $1 \le j \le 4$, is real valued and nonnegative, we find $\phi_j \in C_0(\Omega)$ such that $\|\phi_j - u_j\|_p < \varepsilon/4$, $1 \le j \le 4$. Then $\|u - \phi_1 + \phi_2 - i(\phi_3 - \phi_4)\|_p < \varepsilon$. We assume without loss of generality, therefore, that $u$ is real valued and nonnegative. By Theorem 1.39 there exists a monotonically increasing sequence $\{s_n\}$ of nonnegative simple functions converging pointwise to $u$ on $\Omega$. Since $0 \le s_n(x) \le u(x)$ we have $s_n \in L^p(\Omega)$. Since $(u(x) - s_n(x))^p \le (u(x))^p$ we have $s_n \to u$ in $L^p(\Omega)$ by the dominated convergence theorem 1.45. We may thus pick $s \in \{s_n\}$ such that $\|u - s\|_p < \varepsilon/2$. Since $s$ is simple and $p < \infty$ the support of $s$ must have finite volume. We may also assume that $s(x) = 0$ for all $x \in \Omega^c$. Applying Lusin's theorem 1.37(f) we obtain a function $\phi \in C_0(\Omega)$ such that

$$|\phi(x)| \le \|s\|_\infty \qquad \text{for all} \quad x \in \Omega,$$

and

$$\operatorname{vol}\{x \in \Omega : s(x) \ne \phi(x)\} < (\varepsilon/4\|s\|_\infty)^p.$$

Hence by Theorem 2.8 we have

$$\begin{aligned}\|s - \phi\|_p &\le \|s - \phi\|_\infty (\operatorname{vol}\{x \in \Omega : s(x) \ne \phi(x)\})^{1/p} \\ &< 2\|s\|_\infty (\varepsilon/4\|s\|_\infty) = \varepsilon/2.\end{aligned}$$

It follows that $\|u - \phi\|_p < \varepsilon$. ∎

**2.14** The above proof shows that in fact the set of simple functions in $L^p(\Omega)$ is dense in $L^p(\Omega)$ for $1 \le p < \infty$. That this is also true for $L^\infty(\Omega)$ is a direct consequence of Theorem 1.39.

**2.15 THEOREM** $L^p(\Omega)$ is separable if $1 \le p < \infty$.

PROOF For $m = 1, 2, \ldots$ let

$$\overline{\Omega}_m = \{x \in \Omega : \operatorname{dist}(x, \operatorname{bdry}\Omega) \ge 1/m \text{ and } |x| \le m\}.$$

Thus $\overline{\Omega}_m$ is a compact subset of $\Omega$. Let $P$ be the set of all polynomials on $\mathbb{R}^n$ having rational-complex coefficients. Let $P_m = \{\chi_{\overline{\Omega}_m} f : f \in P\}$ where $\chi_{\overline{\Omega}_m}$ is the characteristic function of $\overline{\Omega}_m$. By Corollary 1.29, $P_m$ is dense in $C(\overline{\Omega}_m)$. Moreover, $\bigcup_{m=1}^{\infty} P_m$ is countable.

If $u \in L^p(\Omega)$ and $\varepsilon > 0$, there exists $\phi \in C_0(\Omega)$ such that $\|u-\phi\|_p < \varepsilon/2$. If $1/m < \operatorname{dist}(\operatorname{supp}\phi, \operatorname{bdry}\Omega)$, there exists $f \in P_m$ such that $\|\phi - f\|_\infty < (\varepsilon/2)(\operatorname{vol}\overline{\Omega}_m)^{-1/p}$. It follows that

$$\|\phi-f\|_p \leq \|\phi-f\|_\infty (\operatorname{vol}\overline{\Omega}_m)^{1/p} < \varepsilon/2$$

and so $\|u-f\|_p < \varepsilon$. Thus the countable set $\bigcup_{m=1}^{\infty} P_m$ is dense in $L^p(\Omega)$ and $L^p(\Omega)$ is separable. ▮

**2.16** $C(\Omega)$, being a proper closed subspace of $L^\infty(\Omega)$, is not dense in that space. Thus neither is $C_0(\Omega)$ nor $C_0^\infty(\Omega)$, and $L^\infty(\Omega)$ is not separable.

## Mollifiers, Approximation by Smooth Functions

**2.17** Let $J$ be a nonnegative, real-valued function belonging to $C_0^\infty(\mathbb{R}^n)$ and having the properties

(i) $J(x) = 0$ if $|x| \geq 1$, and
(ii) $\int_{\mathbb{R}^n} J(x)\,dx = 1$.

For example, we may take

$$J(x) = \begin{cases} k \exp[-1/(1-|x|^2)] & \text{if } |x| < 1 \\ 0 & \text{if } |x| \geq 1, \end{cases}$$

where $k > 0$ is so chosen that condition (ii) is satisfied. If $\varepsilon > 0$, the function $J_\varepsilon(x) = \varepsilon^{-n} J(x/\varepsilon)$ is nonnegative, belongs to $C_0^\infty(\mathbb{R}^n)$, and satisfies

(i) $J_\varepsilon(x) = 0$ if $|x| \geq \varepsilon$, and
(ii) $\int_{\mathbb{R}^n} J_\varepsilon(x)\,dx = 1$.

$J_\varepsilon$ is called a *mollifier*, and the convolution

$$J_\varepsilon * u(x) = \int_{\mathbb{R}^n} J_\varepsilon(x-y)\,u(y)\,dy, \tag{16}$$

defined for functions $u$ for which the right side of (16) makes sense, is called a *mollification* or *regularization* of $u$. We summarize some properties of mollification in the following lemma.

**2.18 LEMMA** Let $u$ be a function which is defined on $\mathbb{R}^n$ and vanishes identically outside the domain $\Omega$.

(a) If $u \in L^1_{\text{loc}}(\bar{\Omega})$, then $J_\varepsilon * u \in C^\infty(\mathbb{R}^n)$.
(b) If also $\operatorname{supp} u \subset\subset \Omega$, then $J_\varepsilon * u \in C_0^\infty(\Omega)$ provided

$$\varepsilon < \operatorname{dist}(\operatorname{supp} u, \operatorname{bdry} \Omega).$$

(c) If $u \in L^p(\Omega)$ where $1 \le p < \infty$, then $J_\varepsilon * u \in L^p(\Omega)$. Moreover,

$$\|J_\varepsilon * u\|_p \le \|u\|_p \qquad \text{and} \qquad \lim_{\varepsilon \to 0+} \|J_\varepsilon * u - u\|_p = 0.$$

(d) If $u \in C(\Omega)$ and $G \subset\subset \Omega$, then $\lim_{\varepsilon \to 0+} J_\varepsilon u(x) = u(x)$ uniformly on $G$.
(e) If $u \in C(\bar{\Omega})$, then $\lim_{\varepsilon \to 0+} J_\varepsilon u(x) = u(x)$ uniformly on $\Omega$.

PROOF Since $J_\varepsilon(x-y)$ is an infinitely differentiable function of $x$ and vanishes if $|y-x| \ge \varepsilon$, and since for every multi-index $\alpha$ and every function $u$ that is integrable on compact sets in $\mathbb{R}^n$ we have

$$D^\alpha (J_\varepsilon * u)(x) = \int_{\mathbb{R}^n} D_x{}^\alpha J_\varepsilon(x-y) u(y)\, dy,$$

it follows that conclusions (a) and (b) are valid.

Suppose $u \in L^p(\Omega)$. If $1 < p < \infty$, we let $p' = p/(p-1)$ and obtain by Hölder's inequality

$$\begin{aligned} |J_\varepsilon * u(x)| &= \left| \int_{\mathbb{R}^n} J_\varepsilon(x-y) u(y)\, dy \right| \\ &\le \left\{ \int_{\mathbb{R}^n} J_\varepsilon(x-y)\, dy \right\}^{1/p'} \left\{ \int_{\mathbb{R}^n} J_\varepsilon(x-y) |u(y)|^p\, dy \right\}^{1/p} \\ &= \left\{ \int_{\mathbb{R}^n} J_\varepsilon(x-y) |u(y)|^p\, dy \right\}^{1/p}. \end{aligned} \tag{17}$$

Hence by Fubini's theorem

$$\begin{aligned} \int_\Omega |J_\varepsilon * u(x)|^p\, dx &\le \int_{\mathbb{R}^n} \int_{\mathbb{R}^n} J_\varepsilon(x-y) |u(y)|^p\, dy\, dx \\ &= \int_{\mathbb{R}^n} |u(y)|^p\, dy \int_{\mathbb{R}^n} J_\varepsilon(x-y)\, dx = \|u\|_p^p. \end{aligned} \tag{18}$$

Let $\eta > 0$. By Theorem 2.13 there exists $\phi \in C_0(\Omega)$ such that $\|u - \phi\|_p < \eta/3$. Thus by (18), $\|J_\varepsilon * u - J_\varepsilon * \phi\|_p < \eta/3$. Now

$$\begin{aligned} |J_\varepsilon * \phi(x) - \phi(x)| &= \left| \int_{\mathbb{R}^n} J_\varepsilon(x-y)(\phi(y) - \phi(x))\, dy \right| \\ &\le \sup_{|y-x| < \varepsilon} |\phi(y) - \phi(x)|. \end{aligned} \tag{19}$$

Since $\phi$ is uniformly continuous on $\Omega$ the right side of (19) tends to zero as

$\varepsilon \to 0+$. Since $\operatorname{supp} \phi$ is compact we may therefore arrange to have $\|J_\varepsilon * \phi - \phi\|_p < \eta/3$ by choosing $\varepsilon$ sufficiently small. For such $\varepsilon$ we therefore have $\|J_\varepsilon * u - u\|_p < \eta$ and (c) follows. If $p = 1$, (18) follows directly from (16) without use of Hölder's inequality, and the rest of the proof of (c) is the same as above. The proofs of (d) and (e) may be obtained by replacing $\phi$ by $u$ in (19). ∎

**2.19 THEOREM** $C_0^\infty(\Omega)$ is dense in $L^p(\Omega)$ if $1 \le p < \infty$.

The proof is an immediate consequence of Theorem 2.13 and Lemma 2.18(b,e).

## Precompact Sets in $L^p(\Omega)$

**2.20** The following theorem plays a role in the study of $L^p$-spaces similar to that played by the Ascoli–Arzela theorem 1.30 in the study of spaces of continuous functions. If $u$ is a function defined (a.e.) on a domain $\Omega \subset \mathbb{R}^n$, we denote by $\tilde{u}$ the zero extension of $u$ outside $\Omega$, that is,

$$\tilde{u}(x) = \begin{cases} u(x) & \text{if } x \in \Omega \\ 0 & \text{if } x \in \mathbb{R}^n \sim \Omega. \end{cases}$$

**2.21 THEOREM** Let $1 \le p < \infty$. A bounded subset $K \subset L^p(\Omega)$ is precompact in $L^p(\Omega)$ if and only if for every number $\varepsilon > 0$ there exists a number $\delta > 0$ and a subset $G \subset\subset \Omega$ such that for every $u \in K$ and every $h \in \mathbb{R}^n$ with $|h| < \delta$

$$\int_\Omega |\tilde{u}(x+h) - \tilde{u}(x)|^p \, dx < \varepsilon^p \tag{20}$$

and

$$\int_{\Omega \sim \bar{G}} |u(x)|^p \, dx < \varepsilon^p. \tag{21}$$

PROOF It is sufficient to prove the theorem for the special case $\Omega = \mathbb{R}^n$, as the theorem follows for general $\Omega$ from its application in this special case to the set $\tilde{K} = \{\tilde{u} : u \in K\}$.

Let us assume first that $K$ is precompact in $L^p(\mathbb{R}^n)$. Let $\varepsilon > 0$ be given. Since $K$ has a finite $(\varepsilon/6)$-net (Theorem 1.18), and since $C_0(\mathbb{R}^n)$ is dense in $L^p(\mathbb{R}^n)$ (Theorem 2.13), there exists a finite set $S$ of continuous functions having compact support, such that for each $u \in K$ there exists $\phi \in S$ satisfying $\|u - \phi\|_p < \varepsilon/3$. Since $S$ is finite there exists $r > 0$ such that $\operatorname{supp} \phi \subset \bar{B}_r$ for every $\phi \in S$, where $B_r$ is the ball $\{x \in \mathbb{R}^n : |x| < r\}$. Setting $G = B_r$, we obtain (21). Also, $\phi(x+h) - \phi(x)$ is uniformly continuous for all $x$ and vanishes identically outside $B_{r+1}$ provided $|h| < 1$. Hence

$$\lim_{|h| \to 0} \int_{\mathbb{R}^n} |\phi(x+h) - \phi(x)|^p \, dx = 0. \tag{22}$$

Since $S$ is finite, (22) is uniform for $\phi \in S$. For $u \in K$ let $T_h u$ be the translate of $u$ by $h$:

$$T_h u(x) = u(x+h). \tag{23}$$

If $\phi \in S$ satisfies $\|u - \phi\|_p < \varepsilon/3$, then also $\|T_h u - T_h \phi\|_p < \varepsilon/3$. Hence by (22) we have for $|h|$ sufficiently small (independent of $u \in K$),

$$\begin{aligned} \|T_h u - u\|_p &\le \|T_h u - T_h \phi\|_p + \|T_h \phi - \phi\|_p + \|\phi - u\|_p \\ &< (2\varepsilon/3) + \|T_h \phi - \phi\|_p < \varepsilon, \end{aligned}$$

and (20) follows. [This argument shows translation is continuous in $L^p(\Omega)$.]

To prove the converse let $\varepsilon > 0$ be given and choose $G \subset\subset \mathbb{R}^n$ such that for all $u \in K$

$$\int_{\mathbb{R}^n \sim \bar{G}} |u(x)|^p \, dx < \varepsilon/3. \tag{24}$$

For any $\eta > 0$ the function $J_\eta * u$ defined as in (16) belongs to $C^\infty(\mathbb{R}^n)$ and in particular to $C(\bar{G})$. If $\phi \in C_0(\mathbb{R}^n)$, then by Hölder's inequality

$$\begin{aligned} |J_\eta * \phi(x) - \phi(x)|^p &= \left| \int_{\mathbb{R}^n} J_\eta(y)(\phi(x-y) - \phi(x)) \, dy \right|^p \\ &\le \int_{B_\eta} J_\eta(y) |T_{-y} \phi(x) - \phi(x)|^p \, dy \end{aligned}$$

where $T_h \phi$ is given as in (23). Hence

$$\|J_\eta * \phi - \phi\|_p \le \sup_{h \in B_\eta} \|T_h \phi - \phi\|_p. \tag{25}$$

If $u \in L^p(\mathbb{R}^n)$, let $\{\phi_n\}$ be a sequence in $C_0(\mathbb{R}^n)$ converging to $u$ in $L^p$-norm. By Lemma 2.18(c), $\{J_\eta * \phi_n\}$ is a Cauchy sequence converging to $J_\eta * u$ in $L^p(\mathbb{R}^n)$. Since also $T_h \phi_n \to T_h u$ in $L^p(\mathbb{R}^n)$, (25) extends to all $u \in L^p(\mathbb{R}^n)$:

$$\|J_\eta * u - u\|_p \le \sup_{h \in B_\eta} \|T_h u - u\|_p.$$

Now (20) implies that $\lim_{|h| \to 0} \|T_h u - u\|_p = 0$ uniformly for $u \in K$. Hence

$\lim_{\eta \to 0} \|J_\eta * u - u\|_p = 0$ uniformly for $u \in K$. We now fix $\eta > 0$ so that

$$\int_{\bar{G}} |J_\eta * u(x) - u(x)|^p \, dx < \frac{\varepsilon}{3 \cdot 2^p} \tag{26}$$

for all $u \in K$.

We show that $\{J_\eta * u : u \in K\}$ satisfies the conditions of the Ascoli–Arzela theorem 1.30 on $\bar{G}$ and hence is precompact in $C(\bar{G})$. By (19) we have

$$|J_\eta * u(x)| \le \left( \sup_{x \in \mathbb{R}^n} J_\eta(x) \right)^{1/p} \|u\|_p$$

which is bounded uniformly for $x \in \mathbb{R}^n$ and $u \in K$ since $K$ is a bounded set in $L^p(\Omega)$ and $\eta$ is fixed. Similarly

$$|J_\eta * u(x+h) - J_\eta * u(x)| \le \left( \sup_{x \in \mathbb{R}^n} J_\eta(x) \right)^{1/p} \|T_h u - u\|_p$$

and so $\lim_{|h| \to 0} J_\eta * u(x+h) = J_\eta * u(x)$ uniformly for $x \in \mathbb{R}^n$ and $u \in K$.

Thus $\{J_\eta * u : u \in K\}$ is precompact in $C(\bar{G})$ and by Theorem 1.18 there exists a finite set $\{\psi_1, \ldots, \psi_m\}$ of functions in $C(\bar{G})$ such that if $u \in K$, then for some $j$, $1 \le j \le m$, and all $x \in \bar{G}$ we have

$$|\psi_j(x) - J_\eta * u(x)|^p < \frac{\varepsilon}{3 \cdot 2^p \cdot \operatorname{vol} \bar{G}}. \tag{27}$$

Denoting by $\tilde{\psi}_j$ the zero extension of $\psi_j$ outside $\bar{G}$, we obtain from (24), (26), (27), and the inequality $(|a| + |b|)^p \le 2^p (|a|^p + |b|^p)$

$$\begin{aligned}
\int_{\mathbb{R}^n} |u(x) - \tilde{\psi}_j(x)|^p \, dx &= \int_{\mathbb{R}^n \sim \bar{G}} |u(x)|^p \, dx + \int_{\bar{G}} |u(x) - \psi_j(x)|^p \, dx \\
&< \frac{\varepsilon}{3} + 2^p \int_{\bar{G}} (|u(x) - J_\eta * u(x)|^p + |J_\eta * u(x) - \psi_j(x)|^p) \, dx \\
&< \frac{\varepsilon}{3} + 2^p \left( \frac{\varepsilon}{3 \cdot 2^p} + \frac{\varepsilon}{3 \cdot 2^p \cdot \operatorname{vol} \bar{G}} \operatorname{vol} \bar{G} \right) = \varepsilon.
\end{aligned}$$

Hence $K$ has a finite $\varepsilon$-net in $L^p(\mathbb{R}^n)$, namely $\{\tilde{\psi}_j : 1 \le j \le m\}$, and so is precompact by Theorem 1.18. ∎

**2.22 THEOREM** Let $1 \le p < \infty$ and let $K \subset L^p(\Omega)$. Suppose there exists a sequence $\{\Omega_j\}$ of subdomains of $\Omega$ having the following properties:

(a) for each $j$, $\Omega_j \subset \Omega_{j+1}$;

(b) for each $j$ the set of restrictions to $\Omega_j$ of the functions in $K$ is precompact in $L^p(\Omega_j)$;

(c) for every $\varepsilon > 0$, there exists $j$ such that

$$\int_{\Omega \sim \Omega_j} |u(x)|^p \, dx < \varepsilon \qquad \text{for every} \quad u \in K.$$

Then $K$ is precompact in $L^p(\Omega)$.

PROOF Let $\{u_n\}$ be a sequence in $K$. Then by (b) there exists a subsequence $\{u_n^{(1)}\}$ such that the restrictions $\{u_n^{(1)}|_{\Omega_1}\}$ converge in $L^p(\Omega_1)$. Having selected $\{u_n^{(1)}\}, \ldots, \{u_n^{(k)}\}$, we may select a subsequence $\{u_n^{(k+1)}\}$ of $\{u_n^{(k)}\}$ such that $\{u_n^{(k+1)}|_{\Omega_{k+1}}\}$ converges in $L^p(\Omega_{k+1})$. Hence also $\{u_n^{(k+1)}|_{\Omega_j}\}$ converges in $L^p(\Omega_j)$ for $1 \le j \le k+1$ by (a).

Let $v_n = u_n^{(n)}$ for $n = 1, 2, \ldots$. Clearly $\{v_n\}$ is a subsequence of $\{u_n\}$. Given $\varepsilon > 0$, there exists $j$ [by (c)] such that

$$\int_{\Omega \sim \Omega_j} |v_n(x) - v_m(x)|^p \, dx < \varepsilon/2 \tag{28}$$

for all $n, m = 1, 2, \ldots$. Except for the first $j-1$ terms, $\{v_n\}$ is a subsequence of $\{u_n^{(j)}\}$ and so $\{v_n|_{\Omega_j}\}$ is a Cauchy sequence in $L^p(\Omega_j)$. Thus for $n, m$ sufficiently large we have

$$\int_{\Omega_j} |v_n(x) - v_m(x)|^p \, dx < \varepsilon/2. \tag{29}$$

Combining (28) and (29) we see that $\{v_n\}$ is a Cauchy sequence in $L^p(\Omega)$ and so converges there. Hence $K$ is precompact in $L^p(\Omega)$. ∎

We remark that Theorem 2.22 is just a setting, suitable for our purposes, of a well-known theorem stating that the operator-norm limit of a sequence of compact operators is compact.

### The Uniform Convexity of $L^p(\Omega)$

**2.23** For $1 < p < \infty$ the space $L^p(\Omega)$ is uniformly convex, its norm $\|\cdot\|_p$ satisfying the condition prescribed in Section 1.19. This result, due to Clarkson [19], is obtained via a set of inequalities for $L^p(\Omega)$ that generalizes the parallelogram law in $L^2(\Omega)$. These inequalities are given in Theorem 2.28, for the proof of which we prepare the following lemmas.

**2.24 LEMMA** If $1 \le p < \infty$ and $a \ge 0$, $b \ge 0$, then

$$(a+b)^p \le 2^{p-1}(a^p + b^p). \tag{30}$$

PROOF If $a = 0$, (30) clearly holds. If $a > 0$, (30) may be rewritten in the form

$$(1+x)^p \le 2^{p-1}(1+x^p) \tag{31}$$

where $0 \le x = b/a$. The function $f(x) = (1+x)^p/(1+x^p)$ satisfies $f(0) = 1 = \lim_{x\to\infty} f(x)$ and $f(x) > 1$ if $0 < x < \infty$. Hence $f$ has its maximum for $x \ge 0$ at its only critical point, $x = 1$. Since $f(1) = 2^{p-1}$, (31) follows. ∎

**2.25 LEMMA** If $0 < s < 1$, the function $f(x) = (1-s^x)/x$ is a decreasing function of $x > 0$.

PROOF $f'(x) = (1/x^2)(g(s^x)-1)$ where $g(t) = t - t\ln t$. Since $0 < s^x < 1$ and since $g'(t) = -\ln t \ge 0$ for $0 < t \le 1$, it follows that $g(s^x) < g(1) = 1$ whence $f'(x) < 0$. ∎

**2.26 LEMMA** If $1 < p \le 2$ and $0 \le t \le 1$, then

$$\left|\frac{1+t}{2}\right|^{p'} + \left|\frac{1-t}{2}\right|^{p'} \le \left(\frac{1}{2} + \frac{1}{2}t^p\right)^{1/(p-1)}, \tag{32}$$

where $p' = p/(p-1)$ is the exponent conjugate to $p$.

PROOF Since equality clearly holds in (32) if either $p = 2$ or $t = 0$ or $t = 1$, we may assume that $1 < p < 2$ and $0 < t < 1$. Under the transformation $t = (1-s)/(1+s)$, which maps the interval $0 < t < 1$ onto the interval $1 > s > 0$, (32) reduces to the equivalent form

$$\tfrac{1}{2}[(1+s)^p + (1-s)^p] - (1+s^{p'})^{p-1} \ge 0. \tag{33}$$

If we denote

$$\binom{p}{0} = 1 \qquad \text{and} \qquad \binom{p}{k} = \frac{p(p-1)(p-2)\cdots(p-k+1)}{k!}, \qquad k \ge 1,$$

the power series expansion of the left side of (33) takes the form

$$\frac{1}{2}\sum_{k=0}^{\infty}\binom{p}{k}s^k + \frac{1}{2}\sum_{k=0}^{\infty}\binom{p}{k}(-s)^k - \sum_{k=0}^{\infty}\binom{p-1}{k}s^{p'k}$$

$$= \sum_{k=0}^{\infty}\binom{p}{2k}s^{2k} - \sum_{k=0}^{\infty}\binom{p-1}{k}s^{p'k}$$

$$= \sum_{k=1}^{\infty}\left\{\binom{p}{2k}s^{2k} - \binom{p-1}{2k-1}s^{p'(2k-1)} - \binom{p-1}{2k}s^{2p'k}\right\}.$$

The latter series is convergent for $0 \le s \le 1$. We prove (33) by showing that each term in the series is positive for $0 < s < 1$. The $k$th term can be written in the form

$$\frac{p(p-1)(2-p)(3-p)\cdots(2k-1-p)}{(2k)!}s^{2k}$$

$$-\frac{(p-1)(2-p)(3-p)\cdots(2k-1-p)}{(2k-1)!}s^{p'(2k-1)}+\frac{(p-1)(2-p)\cdots(2k-p)}{(2k)!}s^{2kp'}$$

$$=\frac{(2-p)(3-p)\cdots(2k-p)}{(2k-1)!}s^{2k}\left[\frac{p(p-1)}{2k(2k-p)}-\frac{p-1}{2k-p}s^{p'(2k-1)-2k}+\frac{p-1}{2k}s^{2kp'-2k}\right]$$

$$=\frac{(2-p)(3-p)\cdots(2k-p)}{(2k-1)!}s^{2k}\left[\frac{1-s^{(2k-p)/(p-1)}}{(2k-p)/(p-1)}-\frac{1-s^{2k/(p-1)}}{2k/(p-1)}\right].$$

The first factor above is positive since $p<2$; the factor in brackets is positive by Lemma 2.25 since $0<(2k-p)/(p-1)<2k/(p-1)$. Thus (33) and hence (32) is established. ∎

**2.27 LEMMA** Let $z, w \in \mathbb{C}$. If $1<p\le 2$, then

$$\left|\frac{z+w}{2}\right|^{p'}+\left|\frac{z-w}{2}\right|^{p'}\le\left(\frac{1}{2}|z|^p+\frac{1}{2}|w|^p\right)^{1/(p-1)},\tag{34}$$

where $p'=p/(p-1)$. If $2\le p<\infty$, then

$$\left|\frac{z+w}{2}\right|^{p}+\left|\frac{z-w}{2}\right|^{p}\le\frac{1}{2}|z|^p+\frac{1}{2}|w|^p.\tag{35}$$

PROOF Since (34) obviously holds if $z=0$ or $w=0$ and is symmetric in $z$ and $w$, we may assume that $|z|\ge|w|>0$. In this case (34) can be rewritten in the form

$$\left|\frac{1+re^{i\theta}}{2}\right|^{p'}+\left|\frac{1-re^{i\theta}}{2}\right|^{p'}\le\left(\frac{1}{2}+\frac{1}{2}r^p\right)^{1/(p-1)},\tag{36}$$

where $w/z=re^{i\theta}$, $r\ge 0$, $0\le\theta<2\pi$. If $\theta=0$, (36) is already proved in Lemma 2.26. We complete the proof of (36) by showing that for fixed $r$ the function

$$f(\theta)=|1+re^{i\theta}|^{p'}+|1-re^{i\theta}|^{p'}$$

has a maximum value for $0\le\theta<2\pi$ at $\theta=0$. Since

$$f(\theta)=(1+r^2+2r\cos\theta)^{p'/2}+(1+r^2-2r\cos\theta)^{p'/2},$$

it is clear that $f(2\pi-\theta)=f(\pi-\theta)=f(\theta)$, so that we need consider $f$ only on the interval $0\le\theta\le\pi/2$. Since $p'\ge 2$, we have on that interval

$$f'(\theta)=-p'r\sin\theta[(1+r^2+2r\cos\theta)^{(p'/2)-1}-(1+r^2-2r\cos\theta)^{(p'/2)-1}]\le 0.$$

Thus the maximum value of $f$ occurs at $\theta=0$ and (36) is proved.

If $2 \le p < \infty$, then $1 < p' \le 2$ and we have, interchanging $p$ and $p'$ in (34) and using Lemma 2.24,

$$\left|\frac{z+w}{2}\right|^p + \left|\frac{z-w}{2}\right|^p \le \left(\frac{1}{2}|z|^{p'} + \frac{1}{2}|w|^{p'}\right)^{1/(p'-1)}$$
$$= \left(\frac{1}{2}|z|^{p'} + \frac{1}{2}|w|^{p'}\right)^{p/p'}$$
$$\le 2^{(p/p')-1}\left(\left(\frac{1}{2}\right)^{p/p'}|z|^p + \left(\frac{1}{2}\right)^{p/p'}|w|^p\right)$$
$$= \frac{1}{2}|z|^p + \frac{1}{2}|w|^p,$$

so that (35) is also proved. ∎

**2.28 THEOREM** (*Clarkson's inequalities*) Let $u, v \in L^p(\Omega)$. For $1 < p < \infty$ let $p' = p/(p-1)$. If $2 \le p < \infty$, then

$$\left\|\frac{u+v}{2}\right\|_p^p + \left\|\frac{u-v}{2}\right\|_p^p \le \frac{1}{2}\|u\|_p^p + \frac{1}{2}\|v\|_p^p, \tag{37}$$

$$\left\|\frac{u+v}{2}\right\|_p^{p'} + \left\|\frac{u-v}{2}\right\|_p^{p'} \ge \left(\frac{1}{2}\|u\|_p^p + \frac{1}{2}\|v\|_p^p\right)^{p'-1}. \tag{38}$$

If $1 < p \le 2$, then

$$\left\|\frac{u+v}{2}\right\|_p^{p'} + \left\|\frac{u-v}{2}\right\|_p^{p'} \le \left(\frac{1}{2}\|u\|_p^p + \frac{1}{2}\|v\|_p^p\right)^{p'-1}, \tag{39}$$

$$\left\|\frac{u+v}{2}\right\|_p^p + \left\|\frac{u-v}{2}\right\|_p^p \ge \frac{1}{2}\|u\|_p^p + \frac{1}{2}\|v\|_p^p. \tag{40}$$

PROOF For $2 \le p < \infty$, (37) is obtained by taking $z = u(x)$ and $w = v(x)$ in (35) and integrating over $\Omega$. To prove (39) for $1 < p \le 2$ we first note that $\|\,|u|^{p'}\|_{p-1} = \|u\|_p^{p'}$ for any $u \in L^p(\Omega)$. Using the reverse Minkowski inequality (7) corresponding to the exponent $p-1 < 1$, and (34) with $z = u(x)$, $w = v(x)$, we obtain

$$\left\|\frac{u+v}{2}\right\|_p^{p'} + \left\|\frac{u-v}{2}\right\|_p^{p'} = \left\|\left|\frac{u+v}{2}\right|^{p'}\right\|_{p-1} + \left\|\left|\frac{u-v}{2}\right|^{p'}\right\|_{p-1}$$
$$\le \left(\int_\Omega \left(\left|\frac{u(x)+v(x)}{2}\right|^{p'} + \left|\frac{u(x)-v(x)}{2}\right|^{p'}\right)^{p-1} dx\right)^{1/(p-1)}$$

*equation continues*

$$\leq \left( \int_\Omega \left( \frac{1}{2} |u(x)|^p + \frac{1}{2} |v(x)|^p \right) dx \right)^{p'-1}$$

$$= \left( \frac{1}{2} \|u\|_p^p + \frac{1}{2} \|v\|_p^p \right)^{p'-1}$$

which is (39).

Inequality (38) can be proved for $2 \leq p < \infty$ by the same method used to prove (39), except that the direct Minkowski inequality (5), corresponding to $p-1 \geq 1$, is used in place of the reverse inequality, and in place of (34) is used the inequality

$$\left( \left| \frac{\xi+\eta}{2} \right|^{p'} + \left| \frac{\xi-\eta}{2} \right|^{p'} \right)^{p-1} \geq \frac{1}{2} |\xi|^p + \frac{1}{2} |\eta|^p,$$

which is obtained from (34) by replacing $p$ by $p'$, $z$ by $\xi+\eta$, and $w$ by $\xi-\eta$. Finally, (40) can be obtained from a similar revision of (35). We remark that all four of Clarkson's inequalities reduce to the parallelogram law

$$\|u+v\|_2^2 + \|u-v\|_2^2 = 2\|u\|_2^2 + 2\|v\|_2^2$$

in the case $p = 2$. ∎

**2.29 COROLLARY** If $1 < p < \infty$, $L^p(\Omega)$ is uniformly convex.

PROOF Let $u, v \in L^p(\Omega)$ satisfy $\|u\|_p = \|v\|_p = 1$ and $\|u-v\|_p \geq \varepsilon > 0$. If $2 \leq p < \infty$, we have from (37)

$$\left\| \frac{u+v}{2} \right\|_p^p \leq 1 - \frac{\varepsilon^p}{2^p}.$$

If $1 < p \leq 2$, we have from (39)

$$\left\| \frac{u+v}{2} \right\|_p^{p'} \leq 1 - \frac{\varepsilon^{p'}}{2^{p'}}.$$

In either case there exists $\delta = \delta(\varepsilon) > 0$ such that $\|(u+v)/2\|_p \leq 1-\delta$. ∎

Being uniformly convex, $L^p(\Omega)$ is reflexive for $1 < p < \infty$ by Theorem 1.20. We shall give a direct proof of this reflexivity after computing the dual of $L^p(\Omega)$.

## The Normed Dual of $L^p(\Omega)$

**2.30** Let $1 \leq p \leq \infty$ and let $p'$ denote the exponent conjugate to $p$. For each element $v \in L^{p'}(\Omega)$ we can define a linear functional $L_v$ on $L^p(\Omega)$ via

$$L_v(u) = \int_\Omega u(x)\, v(x)\, dx, \qquad u \in L^p(\Omega).$$

By Hölder's inequality $|L_v(u)| \le \|u\|_p \|v\|_{p'}$ so that $L_v \in [L^p(\Omega)]'$ and

$$\|L_v; [L^p(\Omega)]'\| \le \|v\|_{p'}. \tag{41}$$

We show that equality must hold in (41). If $1 < p \le \infty$, let $u(x) = |v(x)|^{p'-2}\overline{v(x)}$ if $v(x) \ne 0$ and $u(x) = 0$ otherwise. Then $u \in L^p(\Omega)$ and $L_v(u) = \|u\|_p \|v\|_{p'}$. Now suppose $p = 1$ so $p' = \infty$. If $\|v\|_{p'} = 0$, let $u(x) = 0$. Otherwise let $0 < \varepsilon < \|v\|_\infty$ and let $A$ be a measurable subset of $\Omega$ such that $0 < \mu(A) < \infty$ and $|v(x)| \ge \|v\|_\infty - \varepsilon$ on $A$. Let $u(x) = |v(x)|^{-1}\overline{v(x)}$ for $x \in A$; $u(x) = 0$ otherwise. Then $u \in L^1(\Omega)$ and $L_v(u) \ge \|u\|_1(\|v\|_\infty - \varepsilon)$. Thus we have shown that

$$\|L_v; [L^p(\Omega)]'\| = \|v\|_{p'} \tag{42}$$

so that the operator $L$ mapping $v$ to $L_v$ is an isometric isomorphism of $L^{p'}(\Omega)$ onto a subspace of $[L^p(\Omega)]'$.

**2.31** It is natural to ask whether the range of the isomorphism $L$ is all of $[L^p(\Omega)]'$, that is, whether every continuous linear functional on $L^p(\Omega)$ is of the form $L_v$ for some $v \in L^{p'}(\Omega)$. We shall show that such is the case provided $1 \le p < \infty$. For $p = 2$ this is an immediate consequence of the Riesz representation theorem for Hilbert spaces. For general $p$ a direct proof can be given using the Radon–Nikodym theorem (see Rudin [58] or Theorem 8.18). We shall give a more elementary proof based on the uniform convexity of $L^p(\Omega)$ and a variational argument. This method of proof is also used by Hewitt and Stromberg [32]. Finally we shall use a limiting argument to obtain the case $p = 1$ from the case $p > 1$.

**2.32 LEMMA** Let $1 < p < \infty$. If $L \in [L^p(\Omega)]'$ and $\|L; [L^p(\Omega)]'\| = 1$, then there exists unique $w \in L^p(\Omega)$ such that $\|w\|_p = L(w) = 1$. Dually, if $w \in L^p(\Omega)$ is given and $\|w\|_p = 1$, then there exists unique $L \in [L^p(\Omega)]'$ such that $\|L; [L^p(\Omega)]'\| = L(w) = 1$.

PROOF First assume that $L \in [L^p(\Omega)]'$ is given and $\|L\| = 1$. There exists a sequence $\{w_n\} \in L^p(\Omega)$ such that $\|w_n\| = 1$ and $\lim_{n\to\infty}|L(w_n)| = 1$. We may assume that $|L(w_n)| > \frac{1}{2}$ for each $n$, and, replacing $w_n$ by a suitable multiple of $w_n$ by a complex number of unit modulus, that $L(w_n) > 0$. Suppose the sequence $\{w_n\}$ is not a Cauchy sequence in $L^p(\Omega)$. Then there exists $\varepsilon > 0$ such that $\|w_n - w_m\|_p \ge \varepsilon$ for some arbitrarily large values of $m$ and $n$, so that by uniform convexity we have $\|\frac{1}{2}(w_n + w_m)\|_p \le 1 - \delta$, where $\delta$ is a fixed positive number. Thus

$$\begin{aligned} 1 \ge L\left(\frac{w_n + w_m}{\|w_n + w_m\|_p}\right) &= \left\|\frac{w_n + w_m}{2}\right\|_p^{-1} L\left(\frac{w_n + w_m}{2}\right) \\ &\ge \frac{1}{1-\delta} \cdot \frac{1}{2}[L(w_n) + L(w_m)]. \end{aligned} \tag{43}$$

Since the last expression approaches $1/(1-\delta)$ as $n, m \to \infty$, we have a contradiction. Thus $\{w_n\}$ is a Cauchy sequence in $L^p(\Omega)$ and so converges to an element $w$ of that space. Clearly $\|w\|_p = 1$ and $L(w) = \lim_{n\to\infty} L(w_n) = 1$. The uniqueness of $w$ follows from (43) applied to two distinct candidates.

Now suppose $w \in L^p(\Omega)$ is given and $\|w\|_p = 1$. As noted in Section 2.30 the functional $L_v$ defined by

$$L_v(u) = \int_\Omega u(x)\,v(x)\,dx, \qquad u \in L^p(\Omega), \tag{44}$$

where

$$v(x) = \begin{cases} |w(x)|^{p-2}\overline{w(x)} & \text{if } w(x) \neq 0 \\ 0 & \text{otherwise} \end{cases} \tag{45}$$

satisfies $L_v(w) = \|w\|_p^p = 1$ and $\|L_v; [L^p(\Omega)]'\| = \|v\|_{p'} = \|w\|_p^{p/p'} = 1$. It remains to be shown, therefore, that if $L_1, L_2 \in [L^p(\Omega)]'$ satisfy $\|L_1\| = \|L_2\| = L_1(w) = L_2(w) = 1$, then $L_1 = L_2$. Suppose not. Then there exists $u \in L^p(\Omega)$ such that $L_1(u) \neq L_2(u)$. Replacing $u$ by a suitable multiple of $u$, we may assume that $L_1(u) - L_2(u) = 2$. Then replacing $u$ by its sum with a suitable multiple of $w$, we can arrange that $L_1(u) = 1$ and $L_2(u) = -1$. If $t > 0$, then $L_1(w+tu) = 1+t$; since $\|L_1\| = 1$, therefore $\|w+tu\|_p \geq 1+t$. Similarly, $L_2(w-tu) = 1+t$ so $\|w-tu\|_p \geq 1+t$. If $1 < p \leq 2$, Clarkson's inequality (40) gives

$$\begin{aligned} 1 + t^p \|u\|_p^p &= \left\| \frac{(w+tu)+(w-tu)}{2} \right\|_p^p + \left\| \frac{(w+tu)-(w-tu)}{2} \right\|_p^p \\ &\geq \frac{1}{2}\|w+tu\|_p^p + \frac{1}{2}\|w-tu\|_p^p \geq (1+t)^p. \end{aligned} \tag{46}$$

If $2 \leq p < \infty$, Clarkson's inequality (38) gives

$$\begin{aligned} 1 + t^{p'} \|u\|_p^{p'} &= \left\| \frac{(w+tu)+(w-tu)}{2} \right\|_p^{p'} + \left\| \frac{(w+tu)-(w-tu)}{2} \right\|_p^{p'} \\ &\geq \left(\frac{1}{2}\|w+tu\|_p^p + \frac{1}{2}\|w-tu\|_p^p\right)^{p'-1} \geq (1+t)^{p'}. \end{aligned} \tag{47}$$

Equations (46) and (47) are not possible for all $t > 0$ unless $\|u\|_p = 0$ which is impossible. Thus $L_1 = L_2$. ▮

**2.33 THEOREM** (*The Riesz representation theorem for* $L^p(\Omega)$) Let $1 < p < \infty$ and let $L \in [L^p(\Omega)]'$. Then there exists $v \in L^{p'}(\Omega)$ such that for all

$u \in L^p(\Omega)$

$$L(u) = \int_\Omega u(x)\, v(x)\, dx.$$

Moreover, $\|v\|_{p'} = \|L; [L^p(\Omega)]'\|$. Thus $[L^p(\Omega)]' \cong L^{p'}(\Omega)$.

PROOF Lf $L = 0$, we may take $v = 0$. Accordingly, we assume $L \neq 0$ and, without loss of generality, that $\|L; [L^p(\Omega)]'\| = 1$. By Lemma 2.32 there exists $w \in L^p(\Omega)$ with $\|w\|_p = 1$ such that $L(w) = 1$. Let $v$ be given by (45). Then $L_v$, defined by (44), satisfies $\|L_v; [L^p(\Omega)]'\| = 1$ and $L_v(w) = 1$. By Lemma 2.32, again we have $L = L_v$. Since $\|v\|_{p'} = 1$, the proof is complete. ∎

**2.34 THEOREM** (*Riesz representation theorem for $L^1(\Omega)$*) Let $L \in [L^1(\Omega)]'$. Then there exists $v \in L^\infty(\Omega)$ such that for all $u \in L^1(\Omega)$

$$L(u) = \int_\Omega u(x)\, v(x)\, dx$$

and $\|v\|_\infty = \|L; [L^1(\Omega)]'\|$. Thus $[L^1(\Omega)]' \cong L^\infty(\Omega)$.

PROOF Once again we may assume that $L \neq 0$ and $\|L; [L^1(\Omega)]'\| = 1$. Let us suppose, for the moment, that $\Omega$ has finite volume. Then by Theorem 2.8 if $1 < p < \infty$, we have $L^p(\Omega) \subset L^1(\Omega)$ and

$$|L(u)| \leq \|u\|_1 \leq (\operatorname{vol} \Omega)^{1-(1/p)} \|u\|_p$$

for any $u \in L^p(\Omega)$. Hence $L \in [L^p(\Omega)]'$ and by Theorem 2.33 there exists $v_p \in L^{p'}(\Omega)$ such that

$$\|v_p\|_{p'} \leq (\operatorname{vol} \Omega)^{1-(1/p)} \tag{48}$$

and for every $u \in L^p(\Omega)$

$$L(u) = \int_\Omega u(x)\, v_p(x)\, dx. \tag{49}$$

Since $C_0^\infty(\Omega)$ is dense in $L^p(\Omega)$ for $1 < p < \infty$, and since for any $p, q$ satisfying $1 < p, q < \infty$ and any $\phi \in C_0^\infty(\Omega)$ we have

$$\int_\Omega \phi(x)\, v_p(x)\, dx = L(\phi) = \int_\Omega \phi(x)\, v_q(x)\, dx,$$

it follows that $v_p = v_q$ a.e. on $\Omega$. Hence we may replace $v_p$ in (49) bv a function $v$ belonging to $L^p(\Omega)$ for each $p$, $1 < p < \infty$, and satisfying, following (48),

$$\|v\|_{p'} \leq (\operatorname{vol} \Omega)^{1-(1/p)}.$$

It follows by Theorem 2.8 again that $v \in L^\infty(\Omega)$ and

$$\|v\|_\infty \leq \lim_{p' \to \infty} (\operatorname{vol} \Omega)^{1-(1/p)} = 1. \tag{50}$$

The argument at the beginning of Section 2.30 shows that equality must prevail in (50).

If $\Omega$ does not have finite volume, we may nevertheless write $\Omega = \bigcup_{j=1}^{\infty} G_j$, where $G_j = \{x \in \Omega : j-1 \le |x| < j\}$ has finite volume. The sets $G_j$ are mutually disjoint. Let $\chi_j(x)$ be the characteristic function of $G_j$. If $u_j \in L^1(G_j)$, let $\tilde{u}_j$ denote the zero extension of $u_j$ outside $G_j$, that is, $\tilde{u}_j(x) = u_j(x)$ if $x \in G_j$, $\tilde{u}_j(x) = 0$ otherwise. Let $L_j(u_j) = L(\tilde{u}_j)$. Then $L_j \in [L^1(G_j)]'$ and $\|L_j; [L^1(G_j)]'\| \le 1$. By the finite volume case considered above there exists $v_j \in L^\infty(G_j)$ such that $\|v_j\|_{\infty, G_j} \le 1$ and

$$L_j(u_j) = \int_{G_j} u_j(x)\, v_j(x)\, dx = \int_\Omega \tilde{u}_j(x)\, v(x)\, dx,$$

where $v(x) = v_j(x)$ for $x \in G_j$ $(j = 1, 2, \ldots)$, so that $\|v\|_\infty \le 1$. If $u \in L^1(\Omega)$, we put $u = \sum_{j=1}^{\infty} \chi_j u$; the series converging in norm in $L^1(\Omega)$ by dominated convergence. Since

$$L\left(\sum_{j=1}^{k} \chi_j u\right) = \sum_{j=1}^{k} L_j(\chi_j u) = \int_\Omega \sum_{j=1}^{k} \chi_j(x)\, u(x)\, v(x)\, dx,$$

we obtain, passing to the limit by dominated convergence,

$$L(u) = \int_\Omega u(x)\, v(x)\, dx.$$

It then follows, as in the finite volume case, that $\|v\|_\infty = 1$. ∎

**2.35 THEOREM** $L^p(\Omega)$ is reflexive if and only if $1 < p < \infty$.

PROOF Let $X = L^p(\Omega)$, where $1 < p < \infty$. Since $X' \cong L^{p'}(\Omega)$, to any $w \in X''$ there corresponds $\tilde{w} \in [L^{p'}(\Omega)]'$ such that $w(v) = \tilde{w}(\tilde{v})$, where

$$v(u) = \int_\Omega \tilde{v}(x)\, u(x)\, dx, \qquad u \in X.$$

Similarly, corresponding to $\tilde{w} \in [L^{p'}(\Omega)]'$ there exists $u \in X$ such that

$$\tilde{w}(\tilde{v}) = \int_\Omega \tilde{v}(x)\, u(x)\, dx, \qquad \tilde{v} \in L^{p'}(\Omega).$$

It follows that

$$w(v) = \tilde{w}(\tilde{v}) = \int_\Omega \tilde{v}(x)\, u(x)\, dx = v(u) = J_X u(v)$$

for all $v \in X'$, $J_X$ being the natural isometric isomorphism (see Section 1.13) of $X$ into $X''$. This shows that $J_X$ maps $X$ onto $X''$ so that $X = L^p(\Omega)$ is reflexive.

Since $L^1(\Omega)$ is separable while its dual, which is isometrically isomorphic to $L^\infty(\Omega)$, is not separable, neither $L^1(\Omega)$ nor $L^\infty(\Omega)$ can be reflexive. ∎

**2.36** The Riesz representation theorem cannot hold for the space $L^\infty(\Omega)$ in a form analogous to Theorem 2.33, for if so, then the argument of Theorem 2.35 would show that $L^1(\Omega)$ was reflexive. The dual of $L^\infty(\Omega)$ is larger than $L^1(\Omega)$. It may be identified with a space of absolutely continuous, finitely additive set functions of bounded total variation on $\Omega$. The reader is referred to Yosida [69, p. 118] for details.

# III

# The Spaces $W^{m,p}(\Omega)$

## Definitions and Basic Properties

In this chapter we introduce Sobolev spaces of integer order and establish some of their basic properties. These spaces are defined over an arbitrary domain $\Omega \subset \mathbb{R}^n$ and are vector subspaces of various spaces $L^p(\Omega)$.

**3.1** We define a functional $\|\cdot\|_{m,p}$, where $m$ is a nonnegative integer and $1 \le p \le \infty$, as follows:

$$\|u\|_{m,p} = \left\{ \sum_{0 \le |\alpha| \le m} \|D^\alpha u\|_p^p \right\}^{1/p} \qquad \text{if} \quad 1 \le p < \infty, \tag{1}$$

$$\|u\|_{m,\infty} = \max_{0 \le |\alpha| \le m} \|D^\alpha u\|_\infty \tag{2}$$

for any function $u$ for which the right side makes sense, $\|\cdot\|_p$ being, of course, the $L^p(\Omega)$-norm. (In situations where confusion of domains may occur we shall write $\|u\|_{m,p,\Omega}$ in place of $\|u\|_{m,p}$.) It is clear that (1) or (2) defines a norm on any vector space of functions on which the right side takes finite values provided functions are identified in the space if they are equal almost everywhere in $\Omega$. We consider three such spaces corresponding to any given values of $m$ and $p$:

$H^{m,p}(\Omega) \equiv$ the completion of $\{u \in C^m(\Omega) : \|u\|_{m,p} < \infty\}$ with respect to the norm $\|\cdot\|_{m,p}$,

$W^{m,p}(\Omega) \equiv \{u \in L^p(\Omega) : D^\alpha u \in L^p(\Omega)$ for $0 \le |\alpha| \le m$, where $D^\alpha u$ is the weak (or distributional) partial derivative of Section 1.57$\}$, and

$W_0^{m,p}(\Omega) \equiv$ the closure of $C_0^\infty(\Omega)$ in the space $W^{m,p}(\Omega)$.

Equipped with the appropriate norm (1) or (2), these are called *Sobolev spaces* over $\Omega$. Clearly $W^{0,p}(\Omega) = L^p(\Omega)$, and if $1 \le p < \infty$, $W_0^{0,p}(\Omega) = L^p(\Omega)$ by Theorem 2.19. For any $m$ the chain of imbeddings

$$W_0^{m,p}(\Omega) \to W^{m,p}(\Omega) \to L^p(\Omega)$$

is also clear. We shall show in Theorem 3.16 that $H^{m,p}(\Omega) = W^{m,p}(\Omega)$ for every domain $\Omega$. This result, published in 1964 by Meyers and Serrin [46] dispelled considerable confusion about the relationship of these spaces that had existed in the literature before that time. It is surprising that this elementary result remained undiscovered for so long.

The spaces $W^{m,p}(\Omega)$ were introduced by Sobolev [62, 63] with many related spaces being studied by other writers, in particular Morrey [47] and Deny and Lions [21]. Many different symbols ($W^{m,p}, H^{m,p}, P^{m,p}, L_m{}^p$, etc.) have been (and are being) used to denote these spaces and their variants, and before they became generally associated with the name of Sobolev they were sometimes referred to under other names, for example, as "Beppo Levi spaces."

Numerous generalizations and extensions of the basic spaces $W^{m,p}(\Omega)$ have been made in recent times, the great bulk of the literature originating in the Soviet Union. In particular we mention extensions that allow arbitrary real values of $m$ (see Chapter VII) and are interpreted as corresponding to fractional orders of differentiation, weighted spaces that introduce weight functions into the $L^p$-norms, spaces $W^{\mathbf{m},\mathbf{p}}$ that involve different orders of differentiation and different $L^p$-norms in the various coordinate directions (anisotropic spaces), and Orlicz–Sobolev spaces (Chapter VIII) modeled on the generalizations of $L^p$-spaces known as "Orlicz spaces."

It will not be possible to investigate the complete spectrum of possible generalizations in this monograph.

**3.2 THEOREM** $W^{m,p}(\Omega)$ is a Banach space.

PROOF Let $\{u_n\}$ be a Cauchy sequence in $W^{m,p}(\Omega)$. Then $\{D^\alpha u_n\}$ is a Cauchy sequence in $L^p(\Omega)$ for $0 \le |\alpha| \le m$. Since $L^p(\Omega)$ is complete there exist functions $u$ and $u_\alpha$, $0 < |\alpha| \le m$, in $L^p(\Omega)$ such that $u_n \to u$ and $D^\alpha u_n \to u_\alpha$ in $L^p(\Omega)$ as $n \to \infty$. Now $L^p(\Omega) \subset L^1_{\mathrm{loc}}(\Omega)$ and so $u_n$ determines a distribution $T_{u_n} \in \mathscr{D}'(\Omega)$ as in Section 1.53. For any $\phi \in \mathscr{D}(\Omega)$ we have

$$|T_{u_n}(\phi) - T_u(\phi)| \le \int_\Omega |u_n(x) - u(x)|\,|\phi(x)|\,dx \le \|\phi\|_{p'} \|u_n - u\|_p$$

by Hölder's inequality, where $p' = p/(p-1)$ (or $p' = \infty$ if $p = 1$, $p' = 1$ if $p = \infty$). Hence $T_{u_n}(\phi) \to T_u(\phi)$ for every $\phi \in \mathscr{D}(\Omega)$ as $n \to \infty$. Similarly, $T_{D^\alpha u_n}(\phi) \to T_{u_\alpha}(\phi)$ for every $\phi \in \mathscr{D}(\Omega)$. It follows that

$$T_{u_\alpha}(\phi) = \lim_{n\to\infty} T_{D^\alpha u_n}(\phi) = \lim_{n\to\infty} (-1)^{|\alpha|} T_{u_n}(D^\alpha \phi) = (-1)^{|\alpha|} T_u(D^\alpha \phi)$$

for every $\phi \in \mathscr{D}(\Omega)$. Thus $u_\alpha = D^\alpha u$ in the ditributional sense on $\Omega$ for $0 \le |\alpha| \le m$, whence $u \in W^{m,p}(\Omega)$. Since $\lim_{n\to\infty} \|u_n - u\|_{m,p} = 0$, $W^{m,p}(\Omega)$ is complete. ∎

Distributional and classical partial derivatives coincide when the latter exist and are continuous; thus it is clear that the set

$$S = \{\phi \in C^m(\Omega) : \|\phi\|_{m,p} < \infty\}$$

is contained in $W^{m,p}(\Omega)$. Since $W^{m,p}(\Omega)$ is complete, the identity operator in $S$ extends to an isometric isomorphism between $H^{m,p}(\Omega)$, the completion of $S$, and the closure of $S$ in $W^{m,p}(\Omega)$. It is thus natural to identify $H^{m,p}(\Omega)$ with this closure and so obtain the following corollary.

**3.3 COROLLARY** $H^{m,p}(\Omega) \subset W^{m,p}(\Omega)$.

**3.4** Several important properties of the spaces $W^{m,p}(\Omega)$ are most easily obtained by regarding $W^{m,p}(\Omega)$ as a closed subspace of a Cartesian product of spaces $L^p(\Omega)$. Let $N \equiv N(n,m) = \sum_{0 \le |\alpha| \le m} 1$ be the number of multi-indices $\alpha$ satisfying $0 \le |\alpha| \le m$. For $1 \le p \le \infty$ let $L_N{}^p = \prod_{j=1}^N L^p(\Omega)$, the norm of $u = (u_1, \ldots, u_N)$ in $L_N{}^p$ being given by

$$\|u; L_N{}^p\| = \begin{cases} \left(\sum_{j=1}^N \|u_j\|_p^p\right)^{1/p} & \text{if } 1 \le p < \infty \\ \max_{1 \le j \le N} \|u_j\|_\infty & \text{if } p = \infty. \end{cases}$$

By Theorems 1.22, 2.10, 2.17, 2.25, and 2.31, $L_N{}^p$ s a Banach space that is separable if $1 \le p < \infty$ and reflexive and uniformly convex if $1 < p < \infty$.

Let us suppose that the $N$ multi-indices $\alpha$ satisfying $0 \le |\alpha| \le m$ are linearly ordered in some convenient fashion so that to each $u \in W^{m,p}(\Omega)$ we may associate the well-defined vector $Pu$ in $L_N{}^p$ given by

$$Pu = (D^\alpha u)_{0 \le |\alpha| \le m}. \tag{3}$$

Since $\|Pu; L_N{}^p\| = \|u\|_{m,p}$, $P$ is an isometric isomorphism of $W^{m,p}(\Omega)$ onto a subspace $W \subset L_N{}^p$. Since $W^{m,p}(\Omega)$ is complete, $W$ is a closed subspace of $L_N{}^p$. By Theorem 1.21, $W$ is separable if $1 \le p < \infty$ and reflexive and uniformly convex if $1 < p < \infty$. The same conclusions must therefore hold for $W^{m,p}(\Omega) = P^{-1}(W)$.

**3.5 THEOREM** $W^{m,p}(\Omega)$ is separable if $1 \le p < \infty$, and is reflexive and uniformly convex if $1 < p < \infty$. In particular, therefore, $W^{m,2}(\Omega)$ is a separable Hilbert space with inner product

$$(u, v)_m = \sum_{0 \le |\alpha| \le m} (D^\alpha u, D^\alpha v),$$

where $(u, v) = \int_\Omega u(x)\overline{v(x)}\, dx$ is the inner product in $L^2(\Omega)$.

## Duality, the Spaces $W^{-m,p'}(\Omega)$

**3.6** In the following sections we shall take, for fixed $\Omega$, $m$, and $p$, the number $N$, the spaces $L_N{}^p$ and $W$, and the operator $P$ to be as specified in Section 3.4. We also define

$$\langle u, v\rangle = \int_\Omega u(x)v(x)\, dx$$

for any functions $u$, $v$ for which the right side makes sense. For given $p$, $p'$ shall always designate the conjugate exponent:

$$p' = \begin{cases} \infty & \text{if } \; p = 1 \\ p/(p-1) & \text{if } \; 1 < p < \infty \\ 1 & \text{if } \; p = \infty. \end{cases}$$

**3.7 LEMMA** Let $1 \le p < \infty$. To every $L \in (L_N{}^p)'$ there corresponds unique $v \in L_N^{p'}$ such that for every $u \in L_N{}^p$

$$L(u) = \sum_{j=1}^{N} \langle u_j, v_j\rangle.$$

Moreover,

$$\|L;(L_N{}^p)'\| = \|v; L_N^{p'}\|. \tag{4}$$

Thus $(L_N{}^p)' \cong L_N^{p'}$.

PROOF If $1 \le j \le N$ and $w \in L^p(\Omega)$, let $w_{(j)} = (0, \ldots, 0, w, 0, \ldots, 0)$ be that element of $L_N{}^p$ whose $j$th component is $w$, all other components being zero. Setting $L_j(w) = L(w_{(j)})$, we see that $L_j \in (L^p(\Omega))'$ and so by Theorems 2.33 and 2.34 there exists (unique) $v_j \in L^{p'}(\Omega)$ such that for every $w \in L^p(\Omega)$

$$L(w_{(j)}) = L_j(w) = \langle w, v_j\rangle.$$

If $u \in L_N{}^p$, then

$$L(u) = L\left(\sum_{j=1}^{N} u_{j(j)}\right) = \sum_{j=1}^{N} L(u_{j(j)}) = \sum_{j=1}^{N} \langle u_j, v_j\rangle.$$

By Hölder's inequality (for functions and for finite sums), we have

$$|L(u)| \le \sum_{j=1}^{N} \|u_j\|_p \|v_j\|_{p'} \le \|u; L_N{}^p\| \, \|v; L_N^{p'}\|$$

so that $\|L; (L_N{}^p)'\| \le \|v; L_N^{p'}\|$. We show that these norms are in fact equal as follows: if $1 < p < \infty$ and $1 \le j \le N$, let

$$u_j(x) = \begin{cases} |v_j(x)|^{p'-2}\overline{v_j(x)} & \text{if} \quad v_j(x) \ne 0 \\ 0 & \text{if} \quad v_j(x) = 0. \end{cases}$$

It is easily checked that $|L(u_1, \ldots, u_N)| = \|v; L_N^{p'}\|^{p'} = \|u; L_N{}^p\| \, \|v; L_N^{p'}\|$. If $p = 1$, we choose $k$ so that $\|v_k\|_\infty = \max_{1 \le j \le N} \|v_j\|_\infty$. For any $\varepsilon > 0$ there is a measurable subset $A \subset \Omega$ having finite, nonzero volume, such that $|v_k(x)| \ge \|v_k\|_\infty - \varepsilon$ for $x \in A$. Set

$$u(x) = \begin{cases} \overline{v_k(x)}/v_k(x) & \text{if} \quad x \in A \quad \text{and} \quad v_k(x) \ne 0 \\ 0 & \text{otherwise.} \end{cases}$$

Then

$$L(u_{(k)}) = \langle u, v_k \rangle = \int_A |v_k(x)| \, dx \ge (\|v_k\|_\infty - \varepsilon) \, \|u\|_1$$
$$= (\|v; L_N{}^\infty\| - \varepsilon) \, \|u_{(k)}; L_N{}^1\|.$$

Since $\varepsilon$ is arbitrary, (4) must follow in this case also. ▮

**3.8 THEOREM** Let $1 \le p < \infty$. For every $L \in (W^{m,p}(\Omega))'$ there exists an element $v \in L_N^{p'}$ such that, writing the vector $v$ in the form $(v_\alpha)_{0 \le |\alpha| \le m}$, we have for all $u \in W^{m,p}(\Omega)$

$$L(u) = \sum_{0 \le |\alpha| \le m} \langle D^\alpha u, v_\alpha \rangle. \tag{5}$$

Moreover,

$$\|L; (W^{m,p}(\Omega))'\| = \inf \|v; L_N^{p'}\| = \min \|v; L_N^{p'}\|, \tag{6}$$

the infimum being taken over, and *attained on* the set of all $v \in L_N^{p'}$ for which (5) holds for every $u \in W^{m,p}(\Omega)$.

PROOF A linear functional $L^*$ is defined as follows on the range $W$ of the operator $P$ defined by (3):

$$L^*(Pu) = L(u), \qquad u \in W^{m,p}(\Omega).$$

Since $P$ is an isometric isomorphism, $L^* \in W'$ and

$$\|L^*; W'\| = \|L; (W^{m,p}(\Omega))'\|.$$

By the Hahn–Banach theorem there exists a norm preserving extension $\tilde{L}$ of $L^*$ to all of $L_N{}^p$, and by Lemma 3.7 there exists $v \in L_N^{p'}$ such that if $u = (u_\alpha)_{0 \le |\alpha| \le m} \in L_N{}^p$, then

$$\tilde{L}(u) = \sum_{0 \le |\alpha| \le m} \langle u_\alpha, v_\alpha \rangle.$$

Thus for $u \in W^{m,p}(\Omega)$ we obtain

$$L(u) = L^*(Pu) = \tilde{L}(Pu) = \sum_{0 \le |\alpha| \le m} \langle D^\alpha u, v_\alpha \rangle.$$

Moreover,

$$\|L; (W^{m,p}(\Omega))'\| = \|L^*; W'\| = \|\tilde{L}; (L_N{}^p)'\| = \|v; L_N^{p'}\|. \tag{7}$$

Now any element $v \in L_N^{p'}$ for which (5) holds for every $u \in W^{m,p}(\Omega)$ corre-ponds to an extension $L$ of $L^*$ and so will have norm $\|v; L_N^{p'}\|$ not less than $\|L; (W^{m,p}(\Omega))'\|$. Combining this with (7), we obtain (6). ∎

We remark that, at least if $1 < p < \infty$, the element $v \in L_N^{p'}$ satisfying (5) and (6) is unique. Since $L_N{}^p$ and $L_N^{p'}$ are uniformly convex it follows by an argument similar to that of Lemma 2.32 that linear functionals defined on closed subspaces of $L_N{}^p$ have unique norm preserving extensions to $L_N{}^p$.

**3.9** For $1 \le p < \infty$ every element $L$ of the space $(W^{m,p}(\Omega))'$ is an extension to $W^{m,p}(\Omega)$ of a distribution $T \in \mathscr{D}'(\Omega)$. To see this suppose $L$ is given by (5) for some $v \in L_N^{p'}$ and define $T_{v_\alpha}$, $T \in \mathscr{D}'(\Omega)$, by

$$T_{v_\alpha}(\phi) = \langle \phi, v_\alpha \rangle, \qquad \phi \in \mathscr{D}(\Omega), \quad 0 \le |\alpha| \le m,$$

$$T = \sum_{0 \le |\alpha| \le m} (-1)^{|\alpha|} D^\alpha T_{v_\alpha}. \tag{8}$$

For every $\phi \in \mathscr{D}(\Omega) \subset W^{m,p}(\Omega)$ we have

$$T(\phi) = \sum_{0 \le |\alpha| \le m} T_{v_\alpha}(D^\alpha \phi) = L(\phi)$$

so that $L$ is clearly an extension of $T$. Moreover, we have, following (6),

$$\|L; (W^{m,p}(\Omega))'\| = \min\{\|v; L_N^{p'}\| : L \text{ extends } T \text{ given by (8)}\}.$$

The above remarks also hold for $L \in (W_0^{m,p}(\Omega))'$ since any such functional possesses a norm-preserving extension to $W^{m,p}(\Omega)$.

Now suppose $T$ is any element of $\mathscr{D}'(\Omega)$ having the form (8) for some $v \in L_N^{p'}$, where $1 \le p' \le \infty$. Then $T$ possesses possibly nonunique continuous extensions to $W^{m,p}(\Omega)$. We show, however, that $T$ does possess a unique such extension to $W_0^{m,p}(\Omega)$. If $u \in W_0^{m,p}(\Omega)$, let $\{\phi_n\}$ be a sequence in $C_0{}^\infty(\Omega) =$

$\mathscr{D}(\Omega)$ such that $\|\phi_n - u\|_{m,p} \to 0$ as $n \to \infty$. Then

$$\begin{aligned} |T(\phi_k) - T(\phi_n)| &\le \sum_{0 \le |\alpha| \le m} |T_{v_\alpha}(D^\alpha \phi_k - D^\alpha \phi_n)| \\ &\le \sum_{0 \le |\alpha| \le m} \|D^\alpha(\phi_k - \phi_n)\|_p \|v_\alpha\|_{p'} \\ &\le \|\phi_k - \phi_n\|_{m,p} \|v; L_N^{p'}\| \to 0 \qquad \text{as} \quad k, n \to \infty. \end{aligned}$$

Therefore $\{T(\phi_n)\}$ is a Cauchy sequence in $\mathbb{C}$ and so converges to a limit that we may denote by $L(u)$ since it is clear that if also $\{\psi_n\} \subset \mathscr{D}(\Omega)$ and $\|\psi_n - u\|_{m,p} \to 0$, then $T(\phi_n) - T(\psi_n) \to 0$ as $n \to \infty$. The functional $L$ thus defined is linear and belongs to $(W_0^{m,p}(\Omega))'$, for if $u = \lim_{n\to\infty} \phi_n$ as above, then

$$|L(u)| = \lim_{n\to\infty} |T(\phi_n)| \le \lim_{n\to\infty} \|\phi_n\|_{m,p} \|v; L_N^{p'}\| = \|u\|_{m,p} \|v; L_N^{p'}\|.$$

We have therefore proved the following theorem.

**3.10 THEOREM** Let $1 \le p < \infty$. The dual space $(W_0^{m,p}(\Omega))'$ is isometrically isomorphic to the Banach space consisting of those distributions $T \in \mathscr{D}'(\Omega)$ satisfying (8) for some $v \in L_N^{p'}$, normed by

$$\|T\| = \inf\{\|v; L_N^{p'}\| : v \text{ satisfies (8)}\}.$$

In general one cannot expect any such simple characterization of $(W^{m,p}(\Omega))'$ if $W_0^{m,p}(\Omega)$ is a proper subspace of $W^{m,p}(\Omega)$.

**3.11** If $m = 1, 2, \ldots$ and $1 \le p < \infty$, let $p'$ denote the exponent conjugate to $p$ and denote by $W^{-m,p'}(\Omega)$ the Banach space of distributions on $\Omega$ referred to in the above theorem. (The completeness of this space is a consequence of the isometric isomorphism asserted there.) Evidently $W^{-m,p'}(\Omega)$ is separable and reflexive if $1 < p < \infty$.

**3.12** Let $1 < p < \infty$. Each element $v \in L^{p'}(\Omega)$ determines an element $L_v$ of $(W_0^{m,p}(\Omega))'$ by means of $L_v(u) = \langle u, v \rangle$ for

$$|L_v(u)| = |\langle u, v \rangle| \le \|v\|_{p'} \|u\|_p \le \|v\|_{p'} \|u\|_{m,p}.$$

We define the $(-m, p')$-norm of $v \in L^{p'}(\Omega)$ to be the norm of $L_v$, that is,

$$\|v\|_{-m,p'} = \|L_v; (W_0^{m,p}(\Omega))'\| = \sup_{\substack{u \in W \\ \|u\|_{m,p} \le 1}} |\langle u, v \rangle|,$$

where we have written $W$ to represent $W_0^{m,p}(\Omega)$ on the right side, a practice we continue below, for simplicity. Clearly, for any $u \in W$ and $v \in L^{p'}(\Omega)$ we have

$$\|v\|_{-m,p'} \le \|v\|_{p'}$$

$$|\langle u, v \rangle| = \|u\|_{m,p} |\langle u/\|u\|_{m,p}, v \rangle| \le \|u\|_{m,p} \|v\|_{-m,p'}. \tag{9}$$

The latter formula is a generalized Hölder's inequality.

Let $V = \{L_v : v \in L^{p'}(\Omega)\}$. Thus $V$ is a vector subspace of $W' = (W_0^{m,p}(\Omega))'$. We show that $V$ is in fact dense in $W'$. This is easily seen to be equivalent to showing that if $F \in W''$ satisfies $F(L_v) = 0$ for every $L_v \in V$, then $F = 0$ in $W''$. Since $W$ is reflexive, there exists $f \in W$ corresponding to $F \in W''$ such that $\langle f, v \rangle = L_v(f) = F(L_v) = 0$ for every $v \in L^{p'}(\Omega)$. Since $f \in L^p(\Omega)$, it follows that $f(x) = 0$ a.e. in $\Omega$. Hence $f = 0$ in $W$ and $F = 0$ in $W''$.

Let $H^{-m,p'}(\Omega)$ denote the completion of $L^{p'}(\Omega)$ with respect to the norm $\|\cdot\|_{-m,p'}$. Then we have

$$H^{-m,p'}(\Omega) \cong (W_0^{m,p}(\Omega))' \cong W^{-m,p'}(\Omega).$$

In particular, corresponding to each $v \in H^{-m,p'}(\Omega)$, there exists $T_v \in W^{-m,p'}(\Omega)$ such that $T_v(\Omega) = \lim_{n\to\infty}\langle \phi, v_n \rangle$ for every $\phi \in \mathscr{D}(\Omega)$ and every sequence $\{v_n\} \subset L^{p'}(\Omega)$ such that $\lim_{n\to\infty}\|v_n - v\|_{-m,p'} = 0$ and conversely any $T \in H^{-m,p'}(\Omega)$ satisfies $T = T_v$ for some such $v$. Moreover, by (9), $|T_v(\phi)| \le \|\phi\|_{m,p}\|v\|_{-m,p'}$.

**3.13** By an argument similar to that in Section 3.12 the dual space $(W^{m,p}(\Omega))'$ can be characterized for $1 < p < \infty$ as the completion of $L^{p'}(\Omega)$ with respect to the norm

$$\|v\|_{-m,p'}^{\sim} = \sup_{\substack{u \in W^{m,p}(\Omega) \\ \|u\|_{m,p} \le 1}} |\langle u, v \rangle|.$$

## Approximation by Smooth Functions on $\Omega$

We wish to prove that $\{\phi \in C^\infty(\Omega) : \|\phi\|_{m,p} < \infty\}$ is dense in $W^{m,p}(\Omega)$. To this end we require the following standard existence theorem for infinitely differentiable *partitions of unity*.

**3.14 THEOREM** Let $A$ be an arbitrary subset of $\mathbb{R}^n$ and let $\mathscr{O}$ be a collection of open sets in $\mathbb{R}^n$ which cover $A$, that is, such that $A \subset \bigcup_{U \in \mathscr{O}} U$. Then there exists a collection $\Psi$ of functions $\psi \in C_0^\infty(\mathbb{R}^n)$ having the following properties:

(i) For every $\psi \in \Psi$ and every $x \in \mathbb{R}^n$, $0 \le \psi(x) \le 1$.

(ii) If $K \subset\subset A$, all but possibly finitely many $\psi \in \Psi$ vanish identically on $K$.

(iii) For every $\psi \in \Psi$ there exists $U \in \mathscr{O}$ such that $\operatorname{supp}\psi \subset U$.

(iv) For every $x \in A$, $\sum_{\psi\in\Psi}\psi(x) = 1$.

Such a collection $\Psi$ is called a *$C^\infty$-partition of unity for $A$ subordinate to $\mathscr{O}$.*

PROOF Since the proof can be found in many texts we give only an outline of it here, leaving the details to the reader. Suppose first that $A$ is compact so that $A \subset \bigcup_{j=1}^{N} U_j$, where $U_1, \ldots, U_N \in \mathcal{O}$. Compact sets $K_1 \subset U_1, \ldots, K_N \subset U_N$ can be constructed so that $A \subset \bigcup_{j=1}^{N} K_j$. For $1 \le j \le N$ there exists a non-negative-valued function $\phi_j \in C_0^{\infty}(U_j)$ such that $\phi_j(x) > 0$ for $x \in K_j$. A function $\phi$ can then be constructed so as to be infinitely differentiable and positive on $\mathbb{R}^n$ and to satisfy $\phi(x) = \sum_{j=1}^{N} \phi_j(x)$ for $x \in A$. Now $\Psi = \{\psi_j : \psi_j(x) = \phi_j(x)/\phi(x), 1 \le j \le N\}$ has the desired properties. Now suppose $A$ is open. Then $A = \bigcup_{j=1}^{N} A_j$, where

$$A_j = \{x \in A : |x| \le j \text{ and } \operatorname{dist}(x, \operatorname{bdry} A) \ge 1/j\}$$

is compact. For each $j$ the collection

$$\mathcal{O}_j = \{U \cap (\operatorname{interior} A_{j+1} \cap A_{j-2}^{c}) : U \in \mathcal{O}\}$$

covers $A_j$ and so there exists a finite $C^{\infty}$-partition of unity $\Psi_j$ for $A_j$ subordinate to $\mathcal{O}_j$. The sum $\sigma(x) = \sum_{j=1}^{\infty} \sum_{\phi \in \Psi_j} \phi(x)$ involves only finitely many nonzero terms at each point, and is positive at each $x \in A$. The collection $\Psi = \{\psi : \psi(x) = \phi(x)/\sigma(x)$ for some $\phi$ in some $\Psi_j$ if $x \in A$, $\psi(x) = 0$ if $x \notin A\}$ has the prescribed properties. Finally, if $A$ is arbitrary, then $A \subset B = \bigcup_{U \in \mathcal{O}} U$, where $B$ is open. Any partition of unity for $B$ will do for $A$ as well. ∎

**3.15 LEMMA** Let $J_\varepsilon$ be defined as in Section 2.17 and let $1 \le p < \infty$ and $u \in W^{m,p}(\Omega)$. If $\Omega' \subset\subset \Omega$, then $\lim_{\varepsilon \to 0+} J_\varepsilon * u = u$ in $W^{m,p}(\Omega')$.

PROOF Let $\varepsilon < \operatorname{dist}(\Omega', \operatorname{bdry}\Omega)$. For any $\phi \in \mathcal{D}(\Omega')$ we have

$$\begin{aligned}\int_{\Omega'} J_\varepsilon * u(x)\, D^\alpha \phi(x)\, dx &= \int_{\mathbb{R}^n} \int_{\mathbb{R}^n} \tilde{u}(x-y)\, J_\varepsilon(y)\, D^\alpha \phi(x)\, dx\, dy \\ &= (-1)^{|\alpha|} \int_{\mathbb{R}^n} \int_{\mathbb{R}^n} D_x^{\alpha} \tilde{u}(x-y)\, J_\varepsilon(y)\, \phi(x)\, dx\, dy \\ &= (-1)^{|\alpha|} \int_{\Omega'} J_\varepsilon * D^\alpha u(x)\, \phi(x)\, dx,\end{aligned}$$

where $\tilde{u}$ is the zero extension of $u$ outside $\Omega$. Thus $D^\alpha J_\varepsilon * u = J_\varepsilon * D^\alpha u$ in the distributional sense in $\Omega'$. Since $D^\alpha u \in L^p(\Omega)$ for $0 \le |\alpha| \le m$ we have by Lemma 2.18(c)

$$\lim_{\epsilon \to 0+} \|D^\alpha J_\varepsilon * u - D^\alpha u\|_{p,\Omega'} = \lim_{\epsilon \to 0+} \|J_\varepsilon * D^\alpha u - D^\alpha u\|_{p,\Omega'} = 0.$$

Thus $\lim_{\varepsilon \to 0+} \|J_\varepsilon * u - u\|_{m,p,\Omega'} = 0$. ∎

**3.16 THEOREM** (*Meyers and Serrin* [46]) If $1 \le p < \infty$, then

$$H^{m,p}(\Omega) = W^{m,p}(\Omega).$$

PROOF By virtue of Corollary 3.3 it is sufficient to show that $W^{m,p}(\Omega) \subset H^{m,p}(\Omega)$, that is, that $\{\phi \in C^m(\Omega) : \|\phi\|_{m,p} < \infty\}$ is dense in $W^{m,p}(\Omega)$. If

$u \in W^{m,p}(\Omega)$ and $\varepsilon > 0$, we in fact show that there exists $\phi \in C^\infty(\Omega)$ such that $\|u-\phi\|_{m,p} < \varepsilon$. For $k = 1, 2, \ldots$ let

$$\Omega_k = \{x \in \Omega : |x| < k \text{ and } \operatorname{dist}(x, \operatorname{bdry}\Omega) > 1/k\},$$

and let $\Omega_0 = \Omega_{-1} = \emptyset$, the empty set. Then

$$\mathcal{O} = \{U_k : U_k = \Omega_{k+1} \cap (\overline{\Omega_{k-1}})^c,\ k = 1, 2, \ldots\}$$

is a collection of open subsets of $\Omega$ that covers $\Omega$. Let $\Psi$ be a $C^\infty$-partition of unity for $\Omega$ subordinate to $\mathcal{O}$. Let $\psi_k$ denote the sum of the finitely many functions $\psi \in \Psi$ whose supports are contained in $U_k$. Then $\psi_k \in C_0^\infty(U_k)$ and $\sum_{k=1}^\infty \psi_k(x) = 1$ on $\Omega$.

If $0 < \varepsilon < 1/(k+1)(k+2)$, then $J_\varepsilon * (\psi_k u)$ has support in $\Omega_{k+2} \cap (\Omega_{k-2})^c = \mathscr{V}_k \subset\subset \Omega$. Since $\psi_k u \in W^{m,p}(\Omega)$ we may choose $\varepsilon_k$, $0 < \varepsilon_k < 1/(k+1)(k+2)$, such that

$$\|J_{\varepsilon_k} * (\psi_k u) - \psi_k u\|_{m,p,\Omega} = \|J_{\varepsilon_k} * (\psi_k u) - \psi_k u\|_{m,p,\mathscr{V}_k} < \varepsilon/2^k.$$

Let $\phi = \sum_{k=1}^\infty J_{\varepsilon_k} * (\psi_k u)$. On any $\Omega' \subset\subset \Omega$ only finitely many terms in the sum can fail to vanish. Thus $\phi \in C^\infty(\Omega)$. For $x \in \Omega_k$ we have

$$u(x) = \sum_{j=1}^{k+2} \psi_j(x) u(x), \qquad \phi(x) = \sum_{j=1}^{k+2} J_{\varepsilon_j} * (\psi_j u)(x).$$

Thus

$$\|u - \phi\|_{m,p,\Omega_k} \le \sum_{j=1}^{k+2} \|J_{\varepsilon_j} * (\psi_j u) - \psi_j u\|_{m,p,\Omega} < \varepsilon.$$

By the monotone convergence theorem 1.43, $\|u-\phi\|_{m,p,\Omega} < \varepsilon$. ∎

We remark that the theorem does not extend to the case $p = \infty$. For instance, if $\Omega = \{x \in \mathbb{R} : -1 < x < 1\}$ and $u(x) = |x|$, then $u \in W^{1,\infty}(\Omega)$ but $u \notin H^{1,\infty}(\Omega)$; in fact, if $\varepsilon < \frac{1}{2}$, there exists no function $\phi \in C^1(\Omega)$ such that $\|\phi' - u'\|_\infty < \varepsilon$.

## Approximation by Smooth Functions on $\mathbb{R}^n$

**3.17** Having shown that an element of $W^{m,p}(\Omega)$ can always be approximated by functions smooth on $\Omega$ we now ask whether or not the approximation can in fact be done with bounded functions having bounded derivatives of all orders, or, say, of all orders up to $m$. That is, we are asking whether for any values of $k \ge m$ the space $C^k(\overline{\Omega})$ is dense in $W^{m,p}(\Omega)$. The answer may be negative as the following example shows:

Let $\Omega = \{(x, y) \in \mathbb{R}^2 : 0 < |x| < 1,\ 0 < y < 1\}$. Then the function $u$ specified by

$$u(x,y) = \begin{cases} 1 & \text{if} \quad x > 0 \\ 0 & \text{if} \quad x < 0 \end{cases}$$

evidently belongs to $W^{1,p}(\Omega)$. The reader may verify, however, that for sufficiently small $\varepsilon > 0$ no function $\phi \in C^1(\overline{\Omega})$ can satisfy $\|u-\phi\|_{1,p} < \varepsilon$. The difficulty with this domain is that it lies on both sides of part of its boundary (the segment $x = 0$, $0 < y < 1$).

We shall say that a domain $\Omega$ has the *segment property* if for every $x \in \operatorname{bdry}\Omega$ there exists an open set $U_x$ and a nonzero vector $y_x$ such that $x \in U_x$ and if $z \in \overline{\Omega} \cap U_x$, then $z + ty_x \in \Omega$ for $0 < t < 1$. A domain having this property must have $(n-1)$-dimensional boundary and cannot simultaneously lie on both sides of any given part of its boundary.

The following theorem shows that this property is sufficient to guarantee that $C_0^\infty(\mathbb{R}^n)$ is dense in $W^{m,p}(\Omega)$, and hence in particular that $C^k(\overline{\Omega})$ is dense in $W^{m,p}(\Omega)$ for any $m$.

**3.18 THEOREM** If $\Omega$ has the segment property, then the set of restrictions to $\Omega$ of functions in $C_0^\infty(\mathbb{R}^n)$ is dense in $W^{m,p}(\Omega)$ for $1 \le p < \infty$.

PROOF Let $f$ be a fixed function in $C_0^\infty(\mathbb{R}^n)$ satisfying

(i) $f(x) = 1$ if $|x| \le 1$,
(ii) $f(x) = 0$ if $|x| \ge 2$,
(iii) $|D^\alpha f(x)| \le M$ (constant) for all $x$ and $0 \le |\alpha| \le m$.

Let $f_\varepsilon(x) = f(\varepsilon x)$ for $\varepsilon > 0$. Then $f_\varepsilon(x) = 1$ if $|x| \le 1/\varepsilon$ and $|D^\alpha f_\varepsilon(x)| \le M\varepsilon^{|\alpha|} \le M$ if $\varepsilon \le 1$. If $u \in W^{m,p}(\Omega)$, then $u_\varepsilon = f_\varepsilon \cdot u$ belongs to $W^{m,p}(\Omega)$ and has bounded support. Since, for $0 < \varepsilon \le 1$ and $|\alpha| \le m$,

$$|D^\alpha u_\varepsilon(x)| = \left|\sum_{\beta \le \alpha} \binom{\alpha}{\beta} D^\beta u(x)\, D^{\alpha-\beta} f_\varepsilon(x)\right| \le M \sum_{\beta \le \alpha} \binom{\alpha}{\beta} |D^\beta u(x)|,$$

we have, setting $\Omega_\varepsilon = \{x \in \Omega : |x| > 1/\varepsilon\}$,

$$\begin{aligned} \|u - u_\varepsilon\|_{m,p,\Omega} &= \|u - u_\varepsilon\|_{m,p,\Omega_\varepsilon} \\ &\le \|u\|_{m,p,\Omega_\varepsilon} + \|u_\varepsilon\|_{m,p,\Omega_\varepsilon} \le \text{const}\, \|u\|_{m,p,\Omega_\varepsilon}. \end{aligned}$$

The right side tends to zero as $\varepsilon$ tends to 0. Thus any $u \in W^{m,p}(\Omega)$ can be approximated in that space by functions with bounded supports.

We may now, therefore, assume $K = \{x \in \Omega : u(x) \ne 0\}$ is bounded. The set $F = \overline{K} \sim (\bigcup_{x \in \operatorname{bdry}\Omega} U_x)$ is thus compact and contained in $\Omega$, $\{U_x\}$ being the collection of open sets referred to in the definition of the segment property.

There exists an open set $U_0$ such that $F \subset\subset U_0 \subset\subset \Omega$. Since $\bar{K}$ is compact, there exist finitely many of the sets $U_x$; let us rename them $U_1, \ldots, U_k$, such that $\bar{K} \subset U_0 \cup U_1 \cup \cdots \cup U_k$. Moreover, we may find other open sets $\tilde{U}_0, \tilde{U}_1, \ldots, \tilde{U}_k$ such that $\tilde{U}_j \subset\subset U_j$, $0 \le j \le k$, but still $\bar{K} \subset \tilde{U}_0 \cup \tilde{U}_1 \cup \cdots \cup \tilde{U}_k$.

Let $\Psi$ be a $C^\infty$-partition of unity subordinate to $\{\tilde{U}_j : 0 \le j \le k\}$, and let $\psi_j$ be the sum of the finitely many functions $\psi \in \Psi$ whose supports lie in $\tilde{U}_j$. Let $u_j = \psi_j u$. Suppose that for each $j$ we can find $\phi_j \in C_0^\infty(\mathbb{R}^n)$ such that

$$\|u_j - \phi_j\|_{m,p,\Omega} < \varepsilon/(k+1). \tag{10}$$

Then putting $\phi = \sum_{j=0}^k \phi_j$, we would obtain

$$\|\phi - u\|_{m,p,\Omega} \le \sum_{j=0}^{k} \|\phi_j - u_j\|_{m,p,\Omega} < \varepsilon.$$

A function $\phi_0 \in C_0^\infty(\mathbb{R}^n)$ satisfying (10) for $j = 0$ can be found via Lemma 3.15 since $\operatorname{supp} u_0 \subset \tilde{U}_0 \subset\subset \Omega$. It remains, therefore, to find $\phi_j$ satisfying (10) for $1 \le j \le k$. For fixed such $j$ we extend $u_j$ to be identically zero outside $\Omega$. Thus $u_j \in W^{m,p}(\mathbb{R}^n \sim \Gamma)$, where $\Gamma = \bar{\tilde{U}}_j \cap \operatorname{bdry} \Omega$. Let $y$ be the nonzero vector associated with the set $U_j$ in the definition of the segment property (Fig. 1). Let $\Gamma_t = \Gamma - ty$, where $t$ is so chosen that

$$0 < t < \min(1, \operatorname{dist}(\tilde{U}_j, \mathbb{R}^n \sim U_j)/|y|).$$

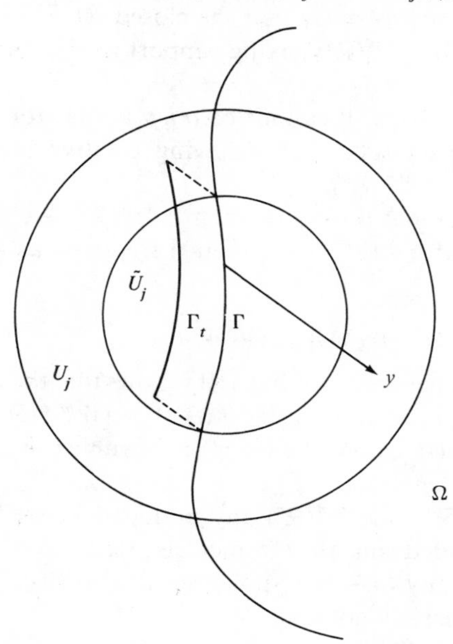


FIG. 1

Then $\Gamma_t \subset U_j$ and $\Gamma_t \cap \overline{\Omega}$ is empty by the segment property. Let $U_{j,t}(x) = u_j(x+ty)$. Then $u_{j,t} \in W^{m,p}(\mathbb{R}^n \sim \Gamma_t)$. Translation is continuous in $L^p(\Omega)$ so $D^\alpha u_{j,t} \to D^\alpha u_j$ in $L^p(\Omega)$ as $t \to 0+$, $|\alpha| \le m$. Thus $u_{j,t} \to u_j$ in $W^{m,p}(\Omega)$ as $t \to 0+$ and so it is sufficient to find $\phi_j \in C_0^\infty(\mathbb{R}^n)$ such that $\|u_{j,t} - \phi_j\|_{m,p}$ is sufficiently small. However, $\Omega \cap U_j \subset\subset \mathbb{R}^n \sim \Gamma_t$ and so by Lemma 3.15 we may take $\phi_j = J_\delta * u_{j,t}$ for suitably small $\delta > 0$. This completes the proof. ▌

**3.19 COROLLARY** $W_0^{m,p}(\mathbb{R}^n) = W^{m,p}(\mathbb{R}^n)$.

## Approximation by Functions in $C_0^\infty(\Omega)$; $(m,p')$-Polar Sets

Corollary 3.19 suggests the question: For what domains $\Omega$ is it true that $W^{m,p}(\Omega) = W_0^{m,p}(\Omega)$, that is, when is $C_0^\infty(\Omega)$ dense in $W^{m,p}(\Omega)$? A partial answer to this problem can be formulated in terms of the nature of the distributions belonging to $W^{-m,p'}(\mathbb{R}^n)$. The approach given below is due to Lions [39].

**3.20** Throughout the following discussion we assume that $1 < p < \infty$ and $p'$ is conjugate to $p$. Let $F$ be a closed subset of $\mathbb{R}^n$. A distribution $T \in \mathscr{D}'(\mathbb{R}^n)$ has support in $F$ ($\operatorname{supp} T \subset F$) provided that $T(\phi) = 0$ for every $\phi \in \mathscr{D}(\mathbb{R}^n)$ vanishing identically on $F$. We say that the closed set $F$ is *$(m,p')$-polar* if the only distribution $T$ in $W^{-m,p'}(\mathbb{R}^n)$ having support in $F$ is the zero distribution $T = 0$.

If $F$ has positive measure, it cannot be $(m,p')$-polar for the characteristic function of any compact subset of $F$ having positive measure belongs to $L^{p'}(\mathbb{R}^n)$ and hence to $W^{-m,p'}(\mathbb{R}^n)$.

We shall show later that if $mp > n$, then $W^{m,p}(\mathbb{R}^n) \to C(\mathbb{R}^n)$ (see Theorem 5.4) in the sense that if $u \in W^{m,p}(\mathbb{R}^n)$, then there exists $u_0 \in C(\mathbb{R}^n)$ such that $u(x) = u_0(x)$ a.e. and

$$|u_0(x)| \le \text{const}\, \|u\|_{m,p},$$

the constant being independent of $x$ and $u$. It follows that the Dirac distribution $\delta_x$ given by $\delta_x(\phi) = \phi(x)$ belongs to $(W^{m,p}(\mathbb{R}^n))' = (W_0^{m,p}(\mathbb{R}^n))' \cong W^{-m,p'}(\mathbb{R}^n)$. Hence, if $mp > n$, a set $F$ cannot be $(m,p')$-polar unless it is empty.

**3.21** Since $W^{m+1,p}(\mathbb{R}^n) \to W^{m,p}(\mathbb{R}^n)$ any bounded linear functional on the latter space is bounded on the former as well, that is, $W^{-m,p'}(\Omega) \subset W^{-m-1,p'}(\Omega)$. Hence any $(m+1,p')$-polar set is also $(m,p')$-polar. The converse is, of course, generally not true.

Let the mapping $u \to \tilde{u}$ denote zero extension of $u$ outside a domain

$\Omega \subset \mathbb{R}^n$:

$$\tilde{u}(x) = \begin{cases} u(x) & \text{if } x \in \Omega \\ 0 & \text{if } x \in \Omega^c. \end{cases} \tag{11}$$

The following lemma shows that this mapping takes $W_0^{m,p}(\Omega)$ (isometrically) into $W^{m,p}(\mathbb{R}^n)$.

**3.22 LEMMA** Let $u \in W_0^{m,p}(\Omega)$. If $|\alpha| \le m$, then $D^\alpha \tilde{u} = (D^\alpha u)^\sim$ in the distributional sense in $\mathbb{R}^n$. Hence $\tilde{u} \in W^{m,p}(\mathbb{R}^n)$.

PROOF Let $\{\phi_n\}$ be a sequence in $C_0^\infty(\Omega)$ converging to $u$ in the space $W_0^{m,p}(\Omega)$. If $\psi \in \mathscr{D}(\mathbb{R}^n)$, we have for $|\alpha| \le m$

$$\begin{aligned} (-1)^{|\alpha|} \int_{\mathbb{R}^n} \tilde{u}(x)\, D^\alpha \psi(x)\, dx &= (-1)^{|\alpha|} \int_\Omega u(x)\, D^\alpha \psi(x)\, dx \\ &= \lim_{n\to\infty} (-1)^{|\alpha|} \int_\Omega \phi_n(x)\, D^\alpha \psi(x)\, dx \\ &= \lim_{n\to\infty} \int_\Omega D^\alpha \phi_n(x)\, \psi(x)\, dx \\ &= \int_{\mathbb{R}^n} (D^\alpha u)^\sim(x)\, \psi(x)\, dx. \end{aligned}$$

Thus $D^\alpha \tilde{u} = (D^\alpha u)^\sim$ in the distributional sense on $\mathbb{R}^n$ and hence these locally integrable functions are equal a.e. in $\mathbb{R}^n$. It follows that $\|\tilde{u}\|_{m,p,\mathbb{R}^n} = \|u\|_{m,p,\Omega}$. ∎

We now give a necessary and sufficient condition on $\Omega$ that mapping (11) carry $W_0^{m,p}(\Omega)$ isometrically *onto* $W^{m,p}(\mathbb{R}^n)$.

**3.23 THEOREM** $C_0^\infty(\Omega)$ is dense in $W^{m,p}(\mathbb{R}^n)$ if and only if the complement $\Omega^c$ of $\Omega$ is $(m,p')$-polar.

PROOF First suppose $C_0^\infty(\Omega)$ is dense in $W^{m,p}(\mathbb{R}^n)$. Let $T \in W^{-m,p'}(\mathbb{R}^n)$ have support contained in $\Omega^c$. If $u \in W^{m,p}(\mathbb{R}^n)$, then there exists a sequence $\{\phi_n\} \subset C_0^\infty(\Omega)$ converging in $W^{m,p}(\mathbb{R}^n)$ to $u$. Hence $T(u) = \lim_{n\to\infty} T(\phi_n) = 0$ and so $T = 0$ and $\Omega^c$ is $(m,p')$-polar.

Conversely, if $C_0^\infty(\Omega)$ is not dense in $W^{m,p}(\mathbb{R}^n)$, then there exists $u \in W^{m,p}(\mathbb{R}^n)$ such that $\|u - \phi\|_{m,p,\mathbb{R}^n} \ge k > 0$ for every $\phi \in C_0^\infty(\Omega)$, $k$ being independent of $\phi$. By the Hahn–Banach extension theorem there exists $T \in W^{-m,p'}(\mathbb{R}^n)$ such that $T(\phi) = 0$ for all $\phi \in C_0^\infty(\Omega)$ but $T(u) \ne 0$. Since $\operatorname{supp} T \subset \Omega^c$ but $T \ne 0$, $\Omega^c$ cannot be $(m,p')$-polar. ∎

As a final preparation for our investigation of the possible identity of $W_0^{m,p}(\Omega)$ and $W^{m,p}(\Omega)$ we establish a distributional analog of the fact, obvious for differentiable functions, that identical vanishing of first derivatives over a rectangle implies constancy on that rectangle. We extend this first to distributions and then to locally integrable functions.

**3.24 LEMMA** Let $B = (a_1, b_1) \times (a_2, b_2) \times \cdots \times (a_n, b_n)$ be an open rectangular box in $\mathbb{R}^n$. Let $\phi \in \mathscr{D}(B)$. If $\int_B \phi(x)\,dx = 0$, then $\phi(x) = \sum_{j=1}^n \phi_j(x)$, where $\phi_j \in \mathscr{D}(B)$ and

$$\int_{a_j}^{b_j} \phi_j(x_1, \ldots, x_j, \ldots, x_n)\,dx_j = 0 \tag{12}$$

for every fixed $(x_1, \ldots, x_{j-1}, x_{j+1}, \ldots, x_n) \in \mathbb{R}^{n-1}$.

PROOF For $1 \le j \le n$ select functions $u_j \in C_0^\infty(a_j, b_j)$ such that $\int_{a_j}^{b_j} u_j(t)\,dt = 1$. Let

$$B_j = (a_j, b_j) \times (a_{j+1}, b_{j+1}) \times \cdots \times (a_n, b_n),$$

$$\psi_j(x_j, \ldots, x_n) = \int_{a_1}^{b_1} dt_1 \int_{a_2}^{b_2} dt_2 \cdots \int_{a_{j-1}}^{b_{j-1}} \phi(t_1, \ldots, t_{j-1}, x_j, \ldots, x_n)\,dt_{j-1},$$

$$\omega_j(x) = u_1(x_1) \cdots u_{j-1}(x_{j-1})\,\psi_j(x_j, \ldots, x_n).$$

Then $\psi_j \in \mathscr{D}(B_j)$ and $\omega_j \in \mathscr{D}(B)$. Moreover,

$$\int_{B_j} \psi_j(x_j, \ldots, x_n)\,dx_j \cdots dx_n = \int_B \phi(x)\,dx = 0.$$

Let $\phi_1 = \phi - \omega_2$, $\phi_j = \omega_j - \omega_{j+1}$ $(2 \le j \le n-1)$, $\phi_n = \omega_n$. Clearly, $\phi_j \in \mathscr{D}(B)$ for $1 \le j \le n$, and $\phi = \sum_{j=1}^n \phi_j$. Finally,

$$\int_{a_1}^{b_1} \phi_1(x_1, \ldots, x_n)\,dx_1$$

$$= \int_{a_1}^{b_1} \phi(x_1, \ldots, x_n)\,dx_1 - \psi_2(x_2, \ldots, x_n) \int_{a_1}^{b_1} u_1(x_1)\,dx_1 = 0,$$

$$\int_{a_j}^{b_j} \phi_j(x_1, \ldots, x_n)\,dx_j$$

$$= u_1(x_1) \cdots u_{j-1}(x_{j-1})$$

$$\times \left( \int_{a_j}^{b_j} \psi_j(x_j, \ldots, x_n)\,dx_j - \psi_{j+1}(x_{j+1}, \ldots, x_n) \int_{a_j}^{b_j} u_j(x_j)\,dx_j \right)$$

$$= 0, \qquad 2 \le j \le n-1,$$

$$\int_{a_n}^{b_n} \phi_n(x_1,\ldots,x_n)\,dx_n = u_1(x_1)\cdots u_{n-1}(x_{n-1})\int_{a_n}^{b_n} \psi_n(x_n)\,dx_n$$

$$= u_1(x_1)\cdots u_{n-1}(x_{n-1})\int_B \phi(x)\,dx = 0. \quad \blacksquare$$

**3.25 COROLLARY** If $T \in \mathscr{D}'(B)$ and $D_j T = 0$ for $1 \le j \le n$, then there exists a constant $k$ such that for all $\phi \in \mathscr{D}(B)$

$$T(\phi) = k\int_B \phi(x)\,dx.$$

PROOF First note that if $\int_B \phi(x)\,dx = 0$, then $T(\phi) = 0$, for, by the above lemma we may write $\phi = \sum_{j=1}^n \phi_j$, where $\phi_j \in \mathscr{D}(B)$ satisfies (12), and hence $\phi_j = D_j\theta_j$, where $\theta_j \in \mathscr{D}(B)$ is defined by

$$\theta_j(x) = \int_{a_j}^{x_j} \phi_j(x_1,\ldots,x_{j-1},t,x_{j+1},\ldots,x_n)\,dt.$$

Thus $T(\phi) = \sum_{j=1}^n T(D_j\theta_j) = -\sum_{j=1}^n D_j T(\theta_j) = 0$.

Now suppose $T \ne 0$. Then there exists $\phi_0 \in \mathscr{D}(B)$ such that $T(\phi_0) = k_1 \ne 0$. Hence $\int_B \phi_0(x)\,dx = k_2 \ne 0$ and $T(\phi_0) = k\int_B \phi_0(x)\,dx$, where $k = k_1/k_2$. If $\phi \in \mathscr{D}(B)$ is arbitrary, let $K(\phi) = \int_B \phi(x)\,dx$. Then

$$\int_B \left(\phi(x) - \frac{K(\phi)}{k_2}\phi_0(x)\right)dx = 0$$

and so $T(\phi - [K(\phi)/k_2]\,\phi_0) = 0$. It follows that

$$T(\phi) = \frac{K(\phi)}{k_2}T(\phi_0) = kK(\phi) = k\int_B \phi(x)\,dx. \quad \blacksquare$$

It should be remarked that this corollary can be extended to any open, connected set $B$ in $\mathbb{R}^n$ via a partition of unity for $\Omega$ subordinate to some open cover of $\Omega$ by open rectangular boxes that are contained in $\Omega$. We shall not, however, require this extension.

The following lemma shows that different locally integrable functions on an open set $\Omega$ determine different distributions on $\Omega$.

**3.26 LEMMA** Let $u \in L^1_{\mathrm{loc}}(\Omega)$ satisfy $\int_\Omega u(x)\phi(x)\,dx = 0$ for every $\phi \in \mathscr{D}(\Omega)$. Then $u(x) = 0$ a.e. in $\Omega$.

PROOF If $\psi \in C_0(\Omega)$, then for sufficiently small positive $\varepsilon$, the mollifier $J_\varepsilon * \psi$ belongs to $\mathscr{D}(\Omega)$. By Lemma 2.18, $J_\varepsilon * \psi \to \psi$ uniformly on $\Omega$ as $\varepsilon \to 0+$. Hence $\int_\Omega u(x)\psi(x)\,dx = 0$ for every $\psi \in C_0(\Omega)$.

Let $K \subset\subset \Omega$ and let $\varepsilon > 0$. Let $\chi_K$ be the characteristic function of $K$. Then $\int_K |u(x)|\,dx < \infty$. There exists $\delta > 0$ such that for any measurable set $A \subset K$ with $\mu(A) < \delta$ we have $\int_A |u(x)|\,dx < \varepsilon/2$ (see, e.g., the book by Munroe [48, p. 136]).

By Lusin's theorem 1.37(f) there exists $\psi \in C_0(\Omega)$ with $\operatorname{supp}\psi \subset K$ and $|\psi(x)| \le 1$ for all $x$, such that

$$\mu(\{x \in \Omega : \psi(x) \ne \chi_K(x)\operatorname{sgn}\overline{u(x)}\}) < \delta.$$

Here

$$\operatorname{sgn} v(x) = \begin{cases} v(x)/|v(x)| & \text{if } v(x) \ne 0 \\ 0 & \text{if } v(x) = 0. \end{cases}$$

Hence

$$\begin{aligned}\int_K |u(x)|\,dx &= \int_\Omega u(x)\chi_K(x)\operatorname{sgn}\overline{u(x)}\,dx \\ &\int_\Omega u(x)\psi(x)\,dx + \int_\Omega u(x)[\chi_K(x)\operatorname{sgn}\overline{u(x)} - \psi(x)]\,dx \\ &\le 2\int_{\{x\in\Omega:\psi(x)\ne\chi_K(x)\operatorname{sgn}\overline{u(x)}\}} |u(x)|\,dx < \varepsilon.\end{aligned}$$

Since $\varepsilon$ is arbitrary, $u(x) = 0$ a.e. in $K$ and hence a.e. in $\Omega$. ∎

**3.27 COROLLARY** If $B$ is a rectangular box as in Lemma 3.24 and $u \in L^1_{\mathrm{loc}}(B)$ possesses weak derivatives $D_j u = 0$ for $1 \le j \le n$, then for some constant $k$, $u(x) = k$ a.e. in $B$.

PROOF By Corollary 3.25 we have, since $D_j T_u = 0$, $1 \le j \le n$,

$$\int_B u(x)\phi(x)\,dx = T_u(\phi) = k\int_B \phi(x)\,dx.$$

Hence $u(x) - k = 0$ a.e. in $B$. ∎

**3.28 THEOREM** (1) If $W^{m,p}(\Omega) = W_0^{m,p}(\Omega)$, then $\Omega^c$ is $(m,p')$-polar.

(2) If $\Omega^c$ is both $(1,p)$-polar and $(m,p')$-polar, then $W^{m,p}(\Omega) = W_0^{m,p}(\Omega)$.

PROOF (1) Assume $W^{m,p}(\Omega) = W_0^{m,p}(\Omega)$. We deduce first that $\Omega^c$ must have measure zero. If not, there would exist some open rectangle $B \subset \mathbb{R}^n$ which intersects both $\Omega$ and $\Omega^c$ in sets of positive measure. Let $u$ be the restriction to $\Omega$ of a function in $C_0^\infty(\mathbb{R}^n)$ which is identically one on $B \cap \Omega$. Then $u \in W^{m,p}(\Omega)$ and so $u \in W_0^{m,p}(\Omega)$. By Lemma 3.22, $\tilde{u} \in W^{m,p}(\mathbb{R}^n)$ and $D_j\tilde{u} = (D_j u)^\sim$ in the distributional sense in $\mathbb{R}^n$, for $1 \le j \le n$. Now $(D_j u)^\sim$ vanishes identically on $B$ whence so does $D_j\tilde{u}$ as a distribution on $B$. By Corollary 3.27, $\tilde{u}$ must have a constant value a.e. in $B$. Since $\tilde{u}(x) = 1$ on

$B \cap \Omega$ and $\tilde{u}(x) = 0$ on $B \cap \Omega^c$, we have a contradiction. Thus $\Omega^c$ has measure zero.

Now if $v \in W^{m,p}(\mathbb{R}^n)$ and $u$ is the restriction of $v$ to $\Omega$, then $u$ belongs to $W^{m,p}(\Omega)$ and hence, by assumption, to $W_0^{m,p}(\Omega)$. By Lemma 3.22, $\tilde{u} \in W^{m,p}(\mathbb{R}^n)$ and can be approximated by elements of $C_0^\infty(\Omega)$. But $v(x) = \tilde{u}(x)$ on $\Omega$, that is, a.e. in $\mathbb{R}^n$. Hence $v$ and $\tilde{u}$ have the same distributional derivatives and so coincide in $W^{m,p}(\mathbb{R}^n)$. Therefore $C_0^\infty(\Omega)$ is dense in $W^{m,p}(\mathbb{R}^n)$ and $\Omega^c$ is $(m,p')$-polar by Theorem 3.23.

(2) Now assume $\Omega^c$ is $(1,p)$-polar and $(m,p')$-polar. Let $u \in W^{m,p}(\Omega)$. We show that $u \in W_0^{m,p}(\Omega)$. Since $\tilde{u} \in L^p(\mathbb{R}^n)$, the distribution $T_{D_j\tilde{u}}$, corresponding to $D_j\tilde{u}$, belongs to $W^{-1,p}(\mathbb{R}^n)$. Since $(D_j u)^\sim \in L^p(\mathbb{R}^n) \subset H^{-1,p}(\mathbb{R}^n)$, therefore $T_{(D_j u)^\sim} \in W^{-1,p}(\mathbb{R}^n)$. Hence $T_{D_j\tilde{u}-(D_j u)^\sim} \in W^{-1,p}(\mathbb{R}^n)$. But $D_j\tilde{u}-(D_j u)^\sim$ vanishes on $\Omega$ so $\operatorname{supp} T_{D_j\tilde{u}-(D_j u)^\sim} \subset \Omega^c$. Since $\Omega_c$ is $(1,p)$-polar $D_j\tilde{u} = (D_j u)^\sim$ in the distributional sense on $\mathbb{R}^n$. By induction on $|\alpha|$ we can show similarly that $D^\alpha\tilde{u} = (D^\alpha u)^\sim$ in the distributional sense, for $|\alpha| \le m$. Therefore $\tilde{u} \in W^{m,p}(\mathbb{R}^n)$ whence, by Theorem 3.23, $u$, the restriction of $\tilde{u}$ to $\Omega$, belongs to $W_0^{m,p}(\Omega)$, $\Omega^c$ being $(m,p')$-polar. ∎

If $(m,p')$-polarity implies $(1,p)$-polarity, then Theorem 3.28 amounts to the assertion that $(m,p')$-polarity of $\Omega^c$ is necessary and sufficient for the equality of $W^{m,p}(\Omega)$ and $W_0^{m,p}(\Omega)$. We now examine this possibility, establishing first two lemmas containing important properties of polarity. The first of these shows that $(m,p')$-polarity is a local property.

**3.29 LEMMA** $F \subset \mathbb{R}^n$ is $(m,p')$-polar if and only if $F \cap K$ is $(m,p')$-polar for every compact set $K \subset \mathbb{R}^n$.

PROOF Clearly the $(m,p')$-polarity of $F$ implies that of $F \cap K$ for every compact $K$. We prove the converse. Let $T \in W^{-m,p'}(\mathbb{R}^n)$ be given by (8) and have support in $F$. We must show that $T = 0$. Let $f \in C_0^\infty(\mathbb{R}^n)$ satisfy $f(x) = 1$ if $|x| \le 1$ and $f(x) = 0$ if $|x| \ge 2$. For $\varepsilon > 0$ let $f_\varepsilon(x) = f(\varepsilon x)$ so that $D^\alpha f_\varepsilon(x) = \varepsilon^{|\alpha|} D^\alpha f(\varepsilon x) \to 0$ uniformly in $x$ as $\varepsilon \to 0+$. Then $f_\varepsilon T \in W^{-m,p'}(\mathbb{R}^n)$ and for any $\phi \in \mathscr{D}(\mathbb{R}^n)$ we have

$$
\begin{aligned}
|T(\phi)-f_\varepsilon T(\phi)| &= |T(\phi)-T(f_\varepsilon \phi)| \\
&= \left| \sum_{0 \le |\alpha| \le m} \int_{\mathbb{R}^n} v_\alpha(x)\, D^\alpha[\phi(x)(1-f_\varepsilon(x))]\, dx \right| \\
&= \left| \sum_{0 \le |\alpha| \le m} \sum_{\beta \le \alpha} \binom{\alpha}{\beta} \int_{\mathbb{R}^n} v_\alpha(x)\, D^\beta\phi(x)\, D^{\alpha-\beta}(1-f_\varepsilon(x))\, dx \right| \\
&\le \sum_{0 \le |\beta| \le m} \int_{\mathbb{R}^n} |w_\beta(x)\, D^\beta\phi(x)|\, dx \le \|\phi\|_{m,p} \|w; L_N^{p'}\|
\end{aligned}
$$

where

$$\begin{aligned} w_\beta(x) &= \sum_{\substack{|\alpha|\le m\\ \alpha\ge\beta}} \binom{\alpha}{\beta} v_\alpha(x)\, D^{\alpha-\beta}(1-f_\varepsilon(x)) \\ &= v_\beta(x)\,(1-f_\varepsilon(x)) - \sum_{\substack{|\alpha|\le m\\ \alpha\ge\beta,\,\alpha\ne\beta}} \binom{\alpha}{\beta} v_\alpha(x)\, D^{\alpha-\beta} f_\varepsilon(x). \end{aligned}$$

Since $f_\varepsilon(x) = 1$ for $|x| \le 1/\varepsilon$, we have $\lim_{\varepsilon\to 0+} \|w_\beta\|_p = 0$. Thus $f_\varepsilon T \to T$ in $W^{-m,p'}(\mathbb{R}^n)$ as $\varepsilon \to 0+$. Since $f_\varepsilon T$ has compact support in $K$, it vanishes by assumption. Thus $T = 0$. ∎

**3.30 LEMMA** If $p' \le q'$ and $F \subset \mathbb{R}^n$ is $(m,p')$-polar, then $F$ is also $(m,q')$-polar.

PROOF Let $K \subset \mathbb{R}^n$ be compact. By Lemma 3.29 it is sufficient to show that $F \cap K$ is $(m,q')$-polar. Let $G$ be an open, bounded set in $\mathbb{R}^n$ containing $K$. By Lemma 2.8, $W_0^{m,p}(G) \to W_0^{m,q}(G)$ so that $W^{-m,q'}(G) \subset W^{-m,p'}(G)$. Any distribution $T \in W^{-m,q'}(\mathbb{R}^n)$ having support in $K \cap F$ also belongs to $W^{-m,q'}(G)$ and hence to $W^{-m,p'}(G)$. Since $K \cap F$ is $(m,p')$-polar, $T = 0$. Thus $K \cap F$ is $(m,q')$-polar. ∎

**3.31 THEOREM** Let $p \ge 2$. Then $W^{m,p}(\Omega) = W_0^{m,p}(\Omega)$ if and only if $\Omega^c$ is $(m,p')$-polar.

PROOF Since $p' \le p$, $\Omega$ is $(m,p)$-polar, and hence $(1,p)$-polar, if it is $(m,p')$-polar. The result now follows by Theorem 3.28. ∎

**3.32** The Sobolev imbedding theorem (Theorem 5.4) can be used to extend Theorem 3.31 to cover certain values of $p < 2$. If $(m-1)p < n$, the imbedding theorem gives

$$W^{m,p}(\mathbb{R}^n) \to W^{1,q}(\mathbb{R}^n), \qquad q = np/[n-(m-1)p],$$

which in turn implies that $W^{-1,q'}(\mathbb{R}^n) \subset W^{-1,p'}(\mathbb{R}^n)$. If also $p \le 2n/(n+m-1)$, then $q' \le p$, and so by Lemma 3.30, $\Omega^c$ is $(1,p)$-polar if it is $(m,p')$-polar. Note that $2n/(n+m-1) < 2$ provided $m > 1$. If, on the other hand, $(m-1)p \ge n$, then $mp > n$ and, as pointed out in Section 3.20, $\Omega^c$ cannot be $(m,p')$-polar unless it is empty, in which case it is $(1,p)$-polar trivially.

Thus, the only values of $p$ for which the $(m,p')$-polarity of $\Omega^c$ is not known to imply $(1,p)$-polarity and hence be equivalent to the identity of $W^{m,p}(\Omega)$ and $W_0^{m,p}(\Omega)$ are given by $1 \le p \le \min(n/(m-1),\, 2n/(n+m-1))$.

**3.33** Whenever $W_0^{m,p}(\Omega) \neq W^{m,p}(\Omega)$, the former space is a closed subspace of the latter. In the Hilbert space case, $p = 2$, we may consider the space $W_0{}^{\perp}$ consisting of all $v \in W^{m,p}(\Omega)$ such that $(v, \phi)_m = 0$ for every $\phi \in C_0{}^{\infty}(\Omega)$. Every $u \in W^{m,p}(\Omega)$ can be uniquely decomposed in the form $u = u_0 + v$, where $u_0 \in W_0^{m,p}(\Omega)$ and $v \in W_0{}^{\perp}$. Integration by parts shows that any $v \in W_0{}^{\perp}$ must satisfy

$$\sum_{0 \leq |\alpha| \leq m} (-1)^{|\alpha|} D^{2\alpha} v(x) = 0$$

in the weak sense, and hence a.e. in $\Omega$.

## Transformation of Coordinates

**3.34** Let $\Phi$ be a one-to-one transformation of a domain $\Omega \subset \mathbb{R}^n$ onto a domain $G \subset \mathbb{R}^n$, having inverse $\Psi = \Phi^{-1}$. We call $\Phi$ *m*-smooth if, writing $y = \Phi(x)$ and

$$\begin{aligned} y_1 &= \phi_1(x_1, \ldots, x_n), & x_1 &= \psi_1(y_1, \ldots, y_n) \\ y_2 &= \phi_2(x_1, \ldots, x_n), & x_2 &= \psi_2(y_1, \ldots, y_n) \\ &\vdots & &\vdots \\ y_n &= \phi_n(x_1, \ldots, x_n), & x_n &= \psi_n(y_1, \ldots, y_n), \end{aligned}$$

the functions $\phi_1, \ldots, \phi_n$ belong to $C^m(\overline{\Omega})$ and the functions $\psi_1, \ldots, \psi_n$ belong to $C^m(\overline{G})$.

If $u$ is a measurable function defined on $\Omega$, we can define a measurable function on $G$ by

$$Au(y) = u(\Psi(y)). \tag{13}$$

Suppose that $\Phi$ is 1-smooth so that for all $x \in \Omega$

$$c \leq |\det \Phi'(x)| \leq C \tag{14}$$

for certain constants $c, C, 0 < c \leq C$. [Here, of course, $\Phi'(x)$ denotes the Jacobian matrix $\partial(y_1, \ldots, y_n)/\partial(x_1, \ldots, x_n)$.] It is readily seen that the operator $A$ defined by (13) transforms $L^p(\Omega)$ boundedly onto $L^p(G)$ and has a bounded inverse; in fact (for $1 \leq p < \infty$),

$$c^{1/p} \|u\|_{p,\Omega} \leq \|Au\|_{p,G} \leq C^{1/p} \|u\|_{p,\Omega}.$$

We establish a similar result for Sobolev spaces.

**3.35 THEOREM** Let $\Phi$ be *m*-smooth, where $m \geq 1$. Then $A$ transforms $W^{m,p}(\Omega)$ boundedly onto $W^{m,p}(G)$ and has a bounded inverse.

PROOF We show that the inequality $\|Au\|_{m,p,G} \le \text{const}\|u\|_{m,p,\Omega}$ holds for any $u \in W^{m,p}(\Omega)$, the constant depending only on the transformation $\Phi$. The reverse inequality $\|Au\|_{m,p,G} \ge \text{const}\|u\|_{m,p,\Omega}$ can be established in a similar manner using the operator $A^{-1}$ taking functions defined on $G$ into functions defined on $\Omega$.

By Theorem 3.16 there exists for any $u \in W^{m,p}(\Omega)$ a sequence $\{u_n\}$ of functions in $C^\infty(\Omega)$ converging to $u$ in $W^{m,p}(\Omega)$-norm. For such smooth $u_n$ it is easily checked by induction that

$$D^\alpha(Au_n)(y) = \sum_{|\beta| \le |\alpha|} M_{\alpha\beta}(y)[A(D^\beta u_n)](y), \tag{15}$$

where $M_{\alpha\beta}$ is a polynomial of degree not exceeding $|\beta|$ in derivatives, of orders not exceeding $|\alpha|$, of the various components of $\Psi$. If $\phi \in \mathscr{D}(G)$, we obtain from (15) and integration by parts

$$(-1)^{|\alpha|}\int_G (Au_n)(y)\, D^\alpha\phi(y)\, dy = \sum_{|\beta| \le |\alpha|} \int_G [A(D^\beta u_n)](y)\, M_{\alpha\beta}(y)\, dy, \tag{16}$$

or, replacing $y$ by $\Phi(x)$ and expressing the integrals over $\Omega$,

$$(-1)^{|\alpha|}\int_\Omega u_n(x)\,(D^\alpha\phi)(\Phi(x))\,|\det \Phi'(x)|\, dx$$

$$= \sum_{|\beta| \le |\alpha|} \int_\Omega D^\beta u_n(x)\, M_{\alpha\beta}(\Phi(x))\,|\det \Phi'(x)|\, dx. \tag{17}$$

Since $D^\beta u_n \to u$ in $L^p(\Omega)$ for $|\beta| \le m$, we may take the limit through (17) as $n \to \infty$ and hence obtain (16) with $u$ replacing $u_n$. Thus (15) holds in the weak sense for any $u \in W^{m,p}(\Omega)$. We now obtain from (15) and (14)

$$\int_G |D^\alpha(Au)(y)|^p\, dy$$

$$\le \left(\sum_{|\beta| \le |\alpha|} 1\right)^p \max_{|\beta| \le |\alpha|} \left(\sup_{y \in G}|M_{\alpha\beta}(y)|^p \int_G |(D^\beta u)(\Psi(y))|^p\, dy\right)$$

$$\le \text{const} \max_{|\beta| \le |\alpha|} \int_\Omega |D^\beta u(x)|^p\, dx$$

whence it follows that $\|Au\|_{m,p,G} \le \text{const}\|u\|_{m,p,\Omega}$. ∎

Of special importance in later chapters is the case of the above theorem corresponding to nonsingular linear transformations $\Phi$ or, more generally, affine transformations (compositions of nonsingular linear transformations and translations). For such transformations $\det \Phi'(x)$ is a nonzero constant.

# IV

# Interpolation and Extension Theorems

## Geometrical Properties of Domains

**4.1** Many properties of Sobolev spaces defined on a domain $\Omega$, and in particular the imbedding properties of these spaces, depend on regularity properties of $\Omega$. Such regularity is normally expressed in terms of geometrical conditions that may or may not be satisfied by any given domain. We specify below five such geometric conditions, including the segment property already encountered in Section 3.17, and consider their interrelationships. First, however, we standardize some geometrical concepts and notations that will prove useful.

Given a point $x \in \mathbb{R}^n$, an open ball $B_1$ with center $x$, and an open ball $B_2$ not containing $x$, the set $C_x = B_1 \cap \{x+\lambda(y-x): y \in B_2, \lambda > 0\}$ is called a *finite cone* in $\mathbb{R}^n$ having vertex at $x$. We also denote by $x + C_0 = \{x+y : y \in C_0\}$ the finite cone with vertex at $x$ obtained by parallel translation of a finite cone $C_0$ with vertex at 0.

Given linearly independent vectors $y_1, y_2, \ldots, y_n \in \mathbb{R}^n$, the set $P = \{\sum_{j=1}^{n} \lambda_j y_j : 0 < \lambda_j < 1,\ 1 \le j \le n\}$ is a *parallelepiped* with one vertex at the origin. Similarly, $x+P$ is a parallel translate of $P$ having one vertex at $x$. By the center of $x+P$ we mean, of course, the point $c(x+P) = x+\frac{1}{2}(y_1 + \cdots + y_n)$. Every parallelepiped with one vertex at $x$ contains a finite cone with vertex at $x$, and conversely is also contained in such a cone.

An open cover $\mathcal{O}$ of a set $S \subset \mathbb{R}^n$ is said to be *locally finite* if any compact set in $\mathbb{R}^n$ can intersect at most finitely many elements of $\mathcal{O}$. Such locally finite collections of sets must be countable, so their elements can be listed in sequence. If $S$ is closed, then any open cover of $S$ possesses a locally finite subcover.

We now define five regularity properties which an open domain $\Omega \subset \mathbb{R}^n$ may possess.

**4.2** $\Omega$ has the *segment property* if there exists a locally finite open cover $\{U_j\}$ of bdry $\Omega$ and a corresponding sequence $\{y_j\}$ of nonzero vectors such that if $x \in \overline{\Omega} \cap U_j$ for some $j$, then $x + ty_j \in \Omega$ for $0 < t < 1$.

**4.3** $\Omega$ has the *cone property* if there exists a finite cone $C$ such that each point $x \in \Omega$ is the vertex of a finite cone $C_x$ contained in $\Omega$ and congruent to $C$. (Note that $C_x$ need not be obtained from $C$ by parallel translation, just by rigid motion.)

**4.4** $\Omega$ has the *uniform cone property* if there exists a locally finite open cover $\{U_j\}$ of bdry $\Omega$, and a corresponding sequence $\{C_j\}$ of finite cones, each congruent to some fixed finite cone $C$, such that:

(i) For some finite $M$, every $U_j$ has diameter less than $M$.
(ii) For some $\delta > 0$, $\bigcup_{j=1}^{\infty} U_j \supset \Omega_\delta \equiv \{x \in \Omega : \operatorname{dist}(x, \operatorname{bdry}\Omega) < \delta\}$.
(iii) For every $j$, $\bigcup_{x \in \Omega \cap U_j} (x + C_j) \equiv Q_j \subset \Omega$.
(iv) For some finite $R$, every collection of $R+1$ of the sets $Q_j$ has empty intersection.

**4.5** $\Omega$ has the *strong local Lipschitz property* provided there exist positive numbers $\delta$ and $M$, a locally finite open cover $\{U_j\}$ of bdry $\Omega$, and for each $U_j$ a real-valued function $f_j$ of $n-1$ real variables, such that the following conditions hold:

(i) For some finite $R$, every collection of $R+1$ of the sets $U_j$ has empty intersection.

(ii) For every pair of points $x, y \in \Omega_\delta = \{x \in \Omega : \operatorname{dist}(x, \operatorname{bdry}\Omega) < \delta\}$ such that $|x-y| < \delta$ there exists $j$ such that

$$x, y \in \mathscr{V}_j = \{x \in U_j : \operatorname{dist}(x, \operatorname{bdry} U_j) > \delta\}.$$

(iii) Each function $f_j$ satisfies a Lipschitz condition with constant $M$:

$$|f(\xi_1, \ldots, \xi_{n-1}) - f(\eta_1, \ldots, \eta_{n-1})| \le M|(\xi_1 - \eta_1, \ldots, \xi_{n-1} - \eta_{n-1})|.$$

(iv) For some Cartesian coordinate system $(\xi_{j,1}, \ldots, \xi_{j,n})$ in $U_j$ the set $\Omega \cap U_j$ is represented by the inequality

$$\xi_{j,n} < f_j(\xi_{j,1}, \ldots, \xi_{j,n-1}).$$

We remark that if $\Omega$ is bounded, the rather complicated conditions above reduce to the simple condition that $\Omega$ have a locally Lipschitz boundary, that is, that each point $x$ on the boundary of $\Omega$ should have a neighborhood $U_x$ such that $\text{bdry}\,\Omega \cap U_x$ is the graph of a Lipschitz continuous function.

**4.6** $\Omega$ has the *uniform $C^m$-regularity property* if there exists a locally finite open cover $\{U_j\}$ of $\text{bdry}\,\Omega$, and a corresponding sequence $\{\Phi_j\}$ of $m$-smooth one-to-one transformations (see Section 3.34) with $\Phi_j$ taking $U_j$ onto $B = \{y \in \mathbb{R}^n : |y| < 1\}$, such that:

(i) For some $\delta > 0$, $\bigcup_{j=1}^{\infty} \Psi_j(\{y \in \mathbb{R}^n : |y| < \frac{1}{2}\}) \supset \Omega_\delta$, where $\Psi_j = \Phi_j^{-1}$.

(ii) For some finite $R$, every collection of $R+1$ of the sets $U_j$ has empty intersection.

(iii) For each $j$, $\Phi_j(U_j \cap \Omega) = \{y \in B : y_n > 0\}$.

(iv) If $(\phi_{j,1}, \ldots, \phi_{j,n})$ and $(\psi_{j,1}, \ldots, \psi_{j,n})$ denote the components of $\Phi_j$ and $\Psi_j$, respectively, then there exists a finite $M$ such that for all $\alpha$, $|\alpha| \le m$, for every $i$, $1 \le i \le n$, and for every $j$, we have

$$|D^\alpha \phi_{j,i}(x)| \le M, \qquad x \in U_j$$

$$|D^\alpha \psi_{j,i}(y)| \le M, \qquad y \in B.$$

**4.7** With the exception of the cone property, all the other properties above require $\Omega$ to lie on only one side of its boundary. The two-dimensional domain $\Omega$ mentioned in Section 3.17, that is,

$$\Omega = \{(x, y) \in \mathbb{R}^2 : 0 < |x| < 1,\ 0 < y < 1\}$$

has the cone property but none of the other four. The reader may wish to convince himself that

$$\begin{aligned} &\text{uniform } C^m\text{-regularity property } (m \ge 1) \\ &\Rightarrow \text{strong local Lipschitz property} \\ &\Rightarrow \text{uniform cone property} \\ &\Rightarrow \text{segment property} \end{aligned}$$

for any domain $\Omega$.

Most of the important imbedding results of Chapter V require only the cone property though one requires the strong local Lipschitz property. Although the cone property implies none of the other above properties it "almost" implies the strong local Lipschitz property for bounded domains, in a sense made precise in the following theorem of Gagliardo [24].

**4.8 THEOREM** (*Gagliardo* [24]) Let $\Omega$ be a bounded domain in $\mathbb{R}^n$ having the cone property. For each $\rho > 0$ there exists a finite collection $\{\Omega_1, \Omega_2, \dots, \Omega_m\}$ of open subsets of $\Omega$ such that $\Omega = \bigcup_{j=1}^m \Omega_j$, and such that to each $\Omega_j$ there corresponds a subset $A_j$ of $\overline{\Omega}_j$ having diameter not exceeding $\rho$, and an open parallelepiped $P_j$ with one vertex at 0, such that $\Omega_j = \bigcup_{x \in A_j}(x + P_j)$. Moreover, if $\rho$ is sufficiently small, then each $\Omega_j$ has the strong local Lipschitz property.

PROOF Let $C_0$ be a finite cone with vertex at 0 such that any $x \in \Omega$ is the vertex of a finite cone $C_x \subset \Omega$ congruent to $C_0$. It is clearly possible to select a finite number of finite cones $C_1, \dots, C_k$ each having vertex at 0 (and each having aperture angle smaller than that of $C_0$) such that any finite cone congruent to $C_0$ and having vertex at 0 must contain one of the cones $C_j$, $1 \le j \le k$. For each $C_j$ let $P_j$ be an open parallelepiped with one vertex at the origin and such that $P_j \subset C_j$. Then for each $x \in \Omega$ there exists $j$, $1 \le j \le k$, such that

$$x + P_j \subset x + C_j \subset C_x \subset \Omega.$$

Since $\Omega$ is open and $\overline{x + P_j}$ is compact, it follows that $y + P_j \subset \Omega$ for all $y$ sufficiently close to $x$. Hence for every $x \in \Omega$ we can find $y \in \Omega$ such that $x \in y + P_j \subset \Omega$ for some $j$, $1 \le j \le k$. (Any domain with the cone property can therefore be expressed as a union of translates of finitely many parallelepipeds.)

Let $\tilde{A}_j = \{x \in \overline{\Omega} : x + P_j \subset \Omega\}$. If $\operatorname{diam} \tilde{A}_j \le \rho$ for each $j$, we take m $= k$, set $A_j = \tilde{A}_j$ and $\Omega_j = \bigcup_{x \in A_j}(x + P_j)$, and note that the first part of the theorem is proved. Otherwise we decompose $\tilde{A}_j$ into a finite union of sets $A_{ji}$ such that $\operatorname{diam} A_{ji} \le \rho$, set corresponding $P_{ji} = P_j$, rearrange the totality of sets $A_{ji}$ into a finite sequence $A_1, \dots, A_m$, rename the corresponding $P_{ji}$'s as $P_1, \dots, P_m$, and finally set $\Omega_j = \bigcup_{x \in A_j}(x + P_j)$ to achieve the same end. (Figure 2 attempts to illustrate these notions for the case

$$\Omega = \{(x, y) \in \mathbb{R}^2 : 0 < |x| < 1,\ 0 < y < 1\},$$

$$C_0 = \{(x, y) \in \mathbb{R}^2 : x > 0,\ y > 0,\ x^2 + y^2 < \tfrac{1}{4}\},$$

$$\rho = 13/16.$$

In this case $\Omega$ can be covered by as few as four open subsets $\Omega_j$ corresponding to only two distinct parallelepipeds.)

It remains to be shown that if $\rho$ is sufficiently small, then each $\Omega_j$ has the strong local Lipschitz property. For simplicity of notation we suppose, therefore, that $\Omega = \bigcup_{x \in A}(x + P)$, where $\operatorname{diam} A \le \rho$ and $P$ is a fixed parallelepiped, and we show that $\Omega$ has the strong local Lipschitz property.

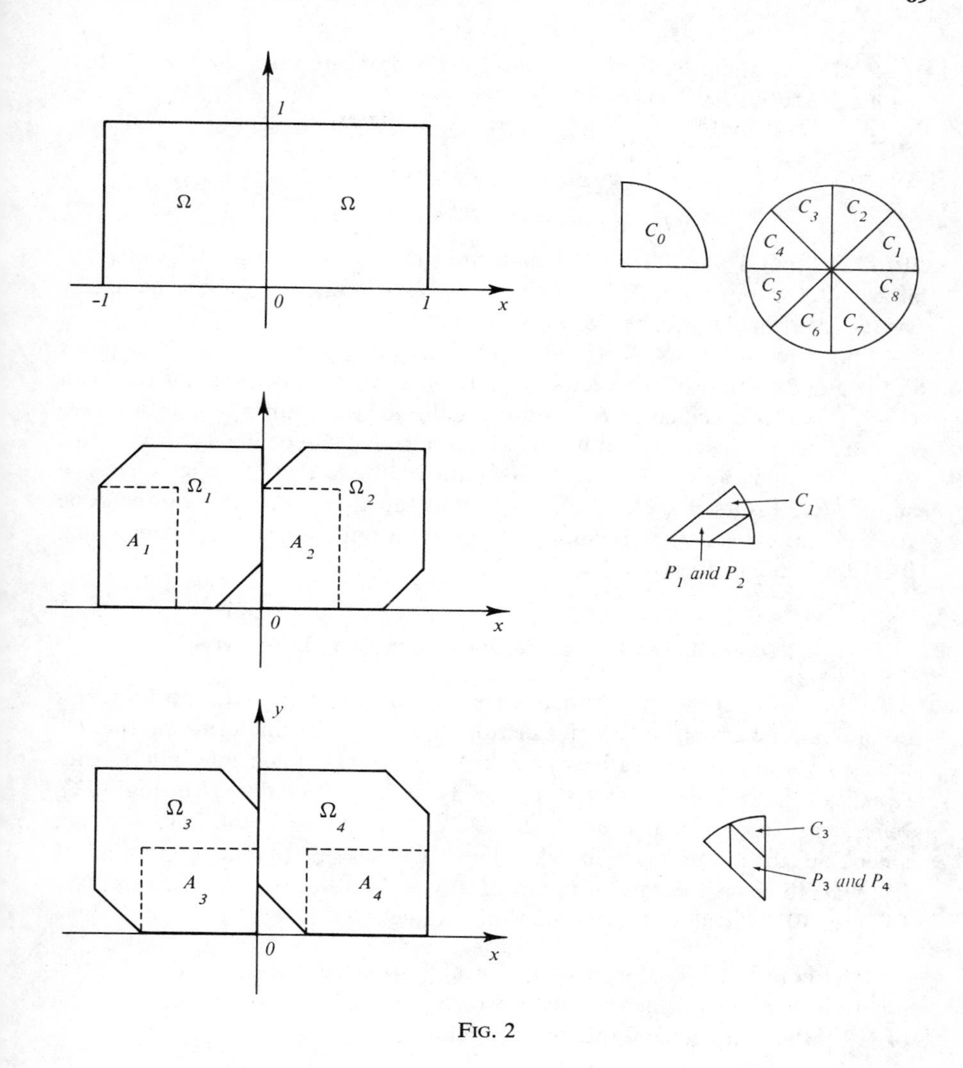


FIG. 2

For each vertex $v_j$ of $P$ let $Q_j = \{y = v_j + \lambda(x - v_j) : x \in P, \lambda > 0\}$ be the infinite pyramid with vertex $v_j$ generated by $P$. Then $P = \bigcap Q_j$, the intersection being taken over all $2^n$ vertices of $P$. Let $\Omega_{(j)} = \bigcup_{x \in A}(x + Q_j)$. Let $\delta = \operatorname{dist}(\text{center of } P, \text{bdry } P)$ and let $B$ be an arbitrary ball of radius $\sigma = \delta/2$. For any fixed $x \in \Omega$, $B$ cannot intersect opposite faces of $x + P$ so we may pick a vertex $v_j$ of $P$ with the property that $x + v_j$ is common to all faces of $x + P$ that meet $B$, if any such faces exist. Then $B \cap (x + P) = B \cap (x + Q_j)$. Now let $x, y \in A$ and suppose $B$ could intersect relatively opposite faces of $x + P$ and

$y+P$, that is, there exist points $a$ and $b$ on opposite faces of $P$ such that $x+a \in B$ and $y+b \in B$. Then

$$\begin{aligned}\rho \geq \operatorname{dist}(x, y) &= \operatorname{dist}(x+b, y+b)\\ &\geq \operatorname{dist}(x+b, x+a) - \operatorname{dist}(x+a, y+b)\\ &\geq 2\delta - 2\sigma = \delta.\end{aligned}$$

It follows that if $\rho < \delta$, then $B$ cannot meet relatively opposite faces of $x+P$ and $y+P$ for any $x, y \in A$. Thus $B \cap (x+P) = B \cap (x+Q_j)$ for some fixed $j$ independent of $x \in A$, whence $B \cap \Omega = B \cap \Omega_{(j)}$.

Choose coordinates $\xi = (\xi', \xi_n) = (\xi_1, \ldots, \xi_{n-1}, \xi_n)$ in $B$ so that the $\xi_n$ axis lies in the direction of the vector from the center of $P$ to the point $v_j$. Then $(x+Q_j) \cap B$ is specified in $B$ by an inequality of the form $\xi_n < f_x(\xi')$, where $f_x$ satisfies a Lipschitz condition with constant independent of $x$. Thus $\Omega_{(j)} \cap B$, and hence $\Omega \cap B$, is specified by $\xi_n < f(\xi')$ where $f(\xi') = \sup_{x \in A} f_x(\xi')$ is itself a Lipschitz continuous function. Since this can be done for a neighborhood $B$ of any point on bdry $\Omega$ it follows that $\Omega$ has the strong local Lipschitz property. ▮

## Interpolation Inequalities for Intermediate Derivatives

**4.9** We consider the problem of determining upper bounds for $L^p$-norms of derivatives $D^\beta u$, $|\beta| \leq m$, of functions $u \in W^{m,p}(\Omega)$, in terms of the $L^p$-norms of $u$ and its derivatives $D^\alpha u$ of order $|\alpha| = m$. Such interpolation inequalities have been obtained by many writers including Ehrling [23], Nirenberg [53, 54], Browder [11, 12], and Gagliardo [24, 25], and are amenable to numerous generalizations. Extensions of the definition of $W^{m,p}(\Omega)$ to cover the case of nonintegral values of $m$ can be carried out (see Chapter VII) via suitable interpolation arguments.

It is convenient to begin with a straightforward one-dimensional interpolation inequality which nevertheless typifies and provides a basis for the proof of the more general theorems which follow.

**4.10 LEMMA** Let $-\infty \leq a < b \leq \infty$, let $1 \leq p < \infty$, and let $0 < \varepsilon_0 < \infty$. There exists a finite constant $K = K(\varepsilon_0, p, b-a)$, depending continuously on $b-a$ for $0 < b-a \leq \infty$, such that for every $\varepsilon$ satisfying $0 < \varepsilon \leq \varepsilon_0$, and for every function $f$ twice continuously differentiable on the open interval $(a, b)$

$$\int_a^b |f'(t)|^p\, dt \leq K\varepsilon \int_a^b |f''(t)|^p\, dt + K\varepsilon^{-1} \int_a^b |f(t)|^p\, dt. \tag{1}$$

Moreover, if $b-a = \infty$, then $K = K(p)$ can be found so that (1) holds for every positive number $\varepsilon$.

PROOF It is sufficient to prove (1) for real-valued functions $f$, for, assuming this done and writing arbitrary $f$ in the form $f = u+iv$ with $u, v$ real valued, we obtain

$$\int_a^b |f'(t)|^p\,dt = \int_a^b [u'(t)^2 + v'(t)^2]^{p/2}\,dt$$

$$\leq \max(1, 2^{(p-2)/2}) \int_a^b [|u'(t)|^p + |v'(t)|^p]\,dt$$

$$\leq 2K \max(1, 2^{(p-2)/2}) \left\{ \varepsilon \int_a^b |f''(t)|^p\,dt + \varepsilon^{-1} \int_a^b |f(t)|^p\,dt \right\}.$$

We may also assume, without loss of generality, that $\varepsilon_0 = 1$, for, assuming the lemma proved in this case, we obtain from (1), since $0 < \varepsilon/\varepsilon_0 \leq 1$,

$$\int_a^b |f'(t)|^p\,dt \leq K\cdot(\varepsilon/\varepsilon_0) \int_a^b |f''(t)|^p\,dt + K\cdot(\varepsilon_0/\varepsilon) \int_a^b |f(t)|^p\,dt.$$

This, in turn, implies (1) with $K = K(\varepsilon_0, p, b-a) = K(1, p, b-a) \max(\varepsilon_0, \varepsilon_0^{-1})$.

We assume, therefore, that $f$ is real valued and $\varepsilon_0 = 1$. For the moment we suppose also that $a = 0$ and $b = 1$. If $0 < \xi < \frac{1}{3}$ and $\frac{2}{3} < \eta < 1$, then there exists $\lambda \in (\xi, \eta)$ such that

$$|f'(\lambda)| = \left| \frac{f(\eta) - f(\xi)}{\eta - \xi} \right| \leq 3\,|f(\xi)| + 3\,|f(\eta)|.$$

It follows that for any $x \in (0, 1)$

$$|f'(x)| = |f'(\lambda) + \int_\lambda^x f''(t)\,dt\,|$$

$$\leq 3\,|f(\xi)| + 3\,|f(\eta)| + \int_0^1 |f''(t)|\,dt.$$

Integrating the above inequality with respect to $\xi$ over $(0, \frac{1}{3})$ and with respect to $\eta$ over $(\frac{2}{3}, 1)$, we obtain

$$\tfrac{1}{9}\,|f'(x)| \leq \int_0^{1/3} |f(\xi)|\,d\xi + \int_{2/3}^1 |f(\eta)|\,d\eta + \tfrac{1}{9} \int_0^1 |f''(t)|\,dt$$

$$\leq \int_0^1 |f(t)|\,dt + \tfrac{1}{9} \int_0^1 |f''(t)|\,dt,$$

whence, by Hölder's inequality,

$$|f'(x)|^p \leq 2^{p-1} \cdot 9^p \int_0^1 |f(t)|^p\,dt + 2^{p-1} \int_0^1 |f''(t)|^p\,dt.$$

Hence,

$$\int_0^1 |f'(t)|^p\,dt \le K_p \int_0^1 |f''(t)|^p\,dt + K_p \int_0^1 |f(t)|^p\,dt,$$

where $K_p = 2^{p-1}\cdot 9^p$. It follows by a change of variable that for any finite interval $(a,b)$

$$\int_a^b |f'(t)|^p\,dt \le K_p(b-a)^p \int_a^b |f''(t)|^p\,dt + K_p(b-a)^{-p} \int_a^b |f(t)|^p\,dt. \qquad (2)$$

Since $0 < \varepsilon \le 1$ there exists a positive integer $n$ such that

$$\tfrac{1}{2}\varepsilon^{1/p} \le 1/n \le \varepsilon^{1/p}.$$

Setting $a_j = a+(b-a)j/n$ for $j = 0, 1, \dots, n$, we obtain from (2), noting that $a_j - a_{j-1} = (b-a)/n$,

$$\begin{aligned}\int_a^b |f'(t)|^p\,dt &= \sum_{j=1}^n \int_{a_{j-1}}^{a_j} |f'(t)|^p\,dt \\ &\le K_p \sum_{j=1}^n \left\{\left(\frac{b-a}{n}\right)^p \int_{a_{j-1}}^{a_j} |f''(t)|^p\,dt + \left(\frac{n}{b-a}\right)^p \int_{a_{j-1}}^{a_j} |f(t)|^p\,dt\right\} \\ &\le \tilde{K}(p, b-a)\left\{\varepsilon \int_a^b |f''(t)|^p\,dt + \varepsilon^{-1} \int_a^b |f(t)|^p\,dt\right\}, \end{aligned} \qquad (3)$$

where $\tilde{K}(p, b-a) = K_p \max[(b-a)^p, 2^p(b-a)^{-p}]$.

Now let

$$K(1, p, b-a) = \begin{cases} \max\limits_{1 \le s \le 2} \tilde{K}(p, s) & \text{if } \ b-a \ge 1 \\ \max\limits_{b-a \le s \le 2} \tilde{K}(p, s) & \text{if } \ 0 < b-a < 1. \end{cases}$$

Then $K(1, p, b-a)$ is finite for $0 < b-a \le \infty$ and depends continuously on $b-a$. For $b-a < 1$, (1) follows directly from (3). For $1 \le b-a \le \infty$, the interval $(a,b)$ may be partitioned into (possibly infinitely many) subintervals each of length between 1 and 2, whence (1) follows upon summing (3) applied to each of these subintervals.

Finally, suppose that $b-a = \infty$. To be specific we assume $a$ is finite and $b = \infty$, the other possibilities being similar. For given $\varepsilon > 0$ let $a_j = a + j\varepsilon^{1/p}$, $j = 0, 1, 2, \dots$. Then $a_j - a_{j-1} = \varepsilon^{1/p}$ and we have, using (2),

$$\begin{aligned}\int_a^\infty |f'(t)|^p\,dt &= \sum_{j=1}^\infty \int_{a_{j-1}}^{a_j} |f'(t)|^p\,dt \\ &\le K_p\varepsilon \sum_{j=1}^\infty \int_{a_{j-1}}^{a_j} |f''(t)|^p\,dt + K_p\varepsilon^{-1} \sum_{j=1}^\infty \int_{a_{j-1}}^{a_j} |f(t)|^p\,dt\end{aligned}$$

which is (1) with $K = K_p$ depending only on $p$. ∎

**4.11** For $1 \le p < \infty$ and for integers $j$, $0 \le j \le m$, we introduce functionals $|\cdot|_{j,p}$ on $W^{m,p}(\Omega)$ as follows:

$$|u|_{j,p} = |u|_{j,p,\Omega} = \left\{ \sum_{|\alpha|=j} \int_\Omega |D^\alpha u(x)|^p\, dx \right\}^{1/p}.$$

Clearly, $|u|_{0,p} = \|u\|_{0,p} = \|u\|_p$ is the norm of $u$ in $L^p(\Omega)$ and

$$\|u\|_{m,p} = \left\{ \sum_{0 \le j \le m} |u|_{j,p}^p \right\}^{1/p}.$$

If $j \ge 1$, $|\cdot|_{j,p}$ is a seminorm—it has all the properties of a norm except that $|u|_{j,p} = 0$ does not imply that $u$ vanishes in $W^{m,p}(\Omega)$; for instance, $u$ may be a nonzero constant on a domain $\Omega$ having finite volume. Under certain circumstances which we investigate later, $|\cdot|_{m,p}$ is an equivalent norm for the space $W_0^{m,p}(\Omega)$. In particular this is so if $\Omega$ is bounded.

At the moment we are concerned with establishing interpolation inequalities of the form

$$|u|_{j,p} \le K\varepsilon |u|_{m,p} + K\varepsilon^{-j/(m-j)} |u|_{0,p}, \tag{4}$$

where $0 \le j \le m-1$. The following lemma shows that in general we need only establish (4) for the special case $j = 1$, $m = 2$, a reduction that will be used in the three interpolation theorems that follow.

**4.12 LEMMA** Let $0 < \delta_0 < \infty$, let $m \ge 2$, and let

$$\varepsilon_0 = \min(\delta_0, \delta_0^2, \ldots, \delta_0^{m-1}).$$

Suppose that for given $p$, $1 \le p < \infty$, and given $\Omega \subset \mathbb{R}^n$ there exists a constant $K = K(\delta_0, p, \Omega)$ such that for every finite $\delta$, $0 < \delta \le \delta_0$, and for every $u \in W^{2,p}(\Omega)$, we have

$$|u|_{1,p} \le K\delta |u|_{2,p} + K\delta^{-1} |u|_{0,p}. \tag{5}$$

Then there exists a constant $K = K(\varepsilon_0, m, p, \Omega)$ such that for every finite $\varepsilon$, $0 < \varepsilon \le \varepsilon_0$, every integer $j$, $0 \le j \le m-1$, and every $u \in W^{m,p}(\Omega)$, we have

$$|u|_{j,p} \le K\varepsilon |u|_{m,p} + K\varepsilon^{-j/(m-j)} |u|_{0,p}. \tag{6}$$

PROOF Since (6) is obvious for $j = 0$ we consider only the case $1 \le j \le m-1$. The proof is accomplished by double induction on $m$ and $j$. The constants $K_1, K_2, \ldots$ appearing in the argument may depend on $\delta_0$ (or $\varepsilon_0$), $m$, $p$, and $\Omega$. We first prove (6) for $j = m-1$ by induction on $m$, so that (5) is the special case $m = 2$. Assume, therefore, that for some $k$, $2 \le k \le m-1$,

$$|u|_{k-1,p} \le K_1 \delta |u|_{k,p} + K_1 \delta^{-(k-1)} |u|_{0,p} \tag{7}$$

holds for all $\delta$, $0 < \delta \le \delta_0$, and all $u \in W^{k,p}(\Omega)$. If $u \in W^{k+1,p}(\Omega)$, we prove that (7) also holds with $k+1$ replacing $k$ (and a different constant $K_1$). If $|\alpha| = k-1$, we obtain from (5)

$$|D^\alpha u|_{1,p} \le K_2 \delta |D^\alpha u|_{2,p} + K_2 \delta^{-1} |D^\alpha u|_{0,p}.$$

Combining this inequality with (7), we obtain, for $0 < \eta \le \delta_0$,

$$\begin{aligned} |u|_{k,p} &\le K_3 \sum_{|\alpha| = k-1} |D^\alpha u|_{1,p} \\ &\le K_4 \delta |u|_{k+1,p} + K_4 \delta^{-1} |u|_{k-1,p} \\ &\le K_4 \delta |u|_{k+1,p} + K_4 K_1 \delta^{-1} \eta |u|_{k,p} + K_4 K_1 \delta^{-1} \eta^{1-k} |u|_{0,p}. \end{aligned}$$

We may assume without prejudice that $2K_1 K_4 \ge 1$. Hence we may take $\eta = \delta/2K_1 K_4$ and so obtain

$$\begin{aligned} |u|_{k,p} &\le 2K_4 \delta |u|_{k+1,p} + (\delta/2K_1 K_4)^{-k} |u|_{0,p} \\ &\le K_5 \delta |u|_{k+1,p} + K_5 \delta^{-k} |u|_{0,p}. \end{aligned}$$

This completes the induction establishing (7) for $0 < \delta \le \delta_0$ and hence (6) for $j = m-1$ and $0 < \varepsilon \le \delta_0$.

We now prove by downward induction on $j$ that

$$|u|_{j,p} \le K_6 \delta^{m-j} |u|_{m,p} + K_6 \delta^{-j} |u|_{0,p} \tag{8}$$

holds for $1 \le j \le m-1$ and $0 < \delta \le \delta_0$. Note that (7) with $k = m$ is the special case $j = m-1$ of (8). Assume, therefore, that (8) holds for some $j$, $2 \le j \le m-1$. We prove that it also holds with $j$ replaced by $j-1$ (and a different constant $K_6$). From (7) and (8) we obtain

$$\begin{aligned} |u|_{j-1,p} &\le K_7 \delta |u|_{j,p} + K_7 \delta^{1-j} |u|_{0,p} \\ &\le K_7 \delta \{K_6 \delta^{m-j} |u|_{m,p} + K_6 \delta^{-j} |u|_{0,p}\} + K_7 \delta^{1-j} |u|_{0,p} \\ &\le K_8 \delta^{m-(j-1)} |u|_{m,p} + K_8 \delta^{-(j-1)} |u|_{0,p}. \end{aligned}$$

Thus (8) holds, and (6) follows by setting $\delta = \varepsilon^{1/(m-j)}$ in (8) and noting that $\varepsilon \le \varepsilon_0$ if $\delta \le \delta_0$. ∎

**4.13 THEOREM** There exists a constant $K = K(m, p, n)$ such that for any $\Omega \subset \mathbb{R}^n$, any $\varepsilon > 0$, any integer $j$, $0 \le j \le m-1$, and any $u \in W_0^{m,p}(\Omega)$,

$$|u|_{j,p} \le K\varepsilon |u|_{m,p} + K\varepsilon^{-j/(m-j)} |u|_{0,p}. \tag{9}$$

PROOF By Lemma 3.22 the operator of zero extension outside $\Omega$ is an isometric isomorphism of $W_0^{m,p}(\Omega)$ into $W^{m,p}(\mathbb{R}^n)$. Thus it is sufficient to establish (9) for $\Omega = \mathbb{R}^n$. Also, by Lemma 4.12 we need consider only the case $j = 1$, $m = 2$. (The case $j = 0$ $m = 1$ is trivial.) For any $\varepsilon > 0$ and any $\phi \in C_0^\infty(\mathbb{R}^n)$

we obtain from Lemma 4.10

$$\int_{-\infty}^{\infty} \left| \frac{\partial}{\partial x_j} \phi(x) \right|^p dx_j \le K\varepsilon^p \int_{-\infty}^{\infty} \left| \frac{\partial^2}{\partial x_j^2} \phi(x) \right|^p dx_j + K\varepsilon^{-p} \int_{-\infty}^{\infty} |\phi(x)|^p \, dx_j.$$

Integrating the remaining components of $x$, we are led to

$$\|D_j \phi\|_p^p \le K\varepsilon^p \|D_j^2 \phi\|_p^p + K\varepsilon^{-p} \|\phi\|_p^p,$$

whence

$$|\phi|_{1,p}^p \le K\varepsilon^p \sum_{j=1}^{n} \|D_j^2 \phi\|_p^p + nK\varepsilon^{-p} |\phi|_{0,p}^p \le K\varepsilon^p |\phi|_{2,p}^p + nK\varepsilon^{-p} |\phi|_{0,p}^p.$$

The case $j = 1$, $m = 2$ of (9) now follows by taking $p$th roots and noting that $C_0^\infty(\mathbb{R}^n)$ is dense in $W^{m,p}(\mathbb{R}^n)$. ∎

**4.14 THEOREM** (*Ehrling* [23], *Nirenberg* [53], *Gagliardo* [24]) Let $\Omega \subset \mathbb{R}^n$ have the uniform cone property (Section 4.4), and let $\varepsilon_0$ be a finite, positive number. Then there exists a constant $K = K(\varepsilon_0, m, p, \Omega)$ such that for any $\varepsilon$, $0 < \varepsilon \le \varepsilon_0$, any integer $j$, $0 \le j \le m-1$, and any $u \in W^{m,p}(\Omega)$

$$|u|_{j,p} \le K\varepsilon |u|_{m,p} + K\varepsilon^{-j/(m-j)} |u|_{0,p}. \tag{10}$$

PROOF The case $m = 1$ is trivial; again Lemma 4.12 shows that it is sufficient to establish (10) for $j = 1$, $m = 2$. In addition, the argument used in the second paragraph of the proof of Lemma 4.10 shows that we may assume $\varepsilon_0 = 1$.

In this proof we make constant use of the notations of Section 4.4 describing the uniform cone property possessed by $\Omega$. If $\delta$ is the constant of condition (ii) of that section and if $\lambda = (\lambda_1, \ldots, \lambda_n)$ is an $n$-tuple of integers, we consider the cube

$$H_\lambda = \{x \in \mathbb{R}^n : \lambda_k \delta/2\sqrt{n} \le x_k \le (\lambda_k + 1)\delta/2\sqrt{n}\}.$$

Then $\mathbb{R}^n = \bigcup_\lambda H_\lambda$ and $\operatorname{diam} H_\lambda = \delta/2$. Let $\Omega_0 = \bigcup_{H_\lambda \subset \Omega} H_\lambda$. Thus $\Omega \sim \Omega_\delta \subset \Omega_0 \subset \Omega$. If the sets $U_1, U_2, \ldots$ and $Q_1, Q_2, \ldots$ are as in Section 4.4, then

$$\Omega = \bigcup_{j=1}^{\infty} (U_j \cap \Omega) \cup \Omega_0 = \bigcup_{j=1}^{\infty} Q_j \cup \Omega_0.$$

We shall prove that for any $u \in W^{2,p}(\Omega)$

$$|u|_{1,p,\Omega_0}^p \le K_1 \varepsilon^p |u|_{2,p,\Omega_0}^p + K_1 \varepsilon^{-p} |u|_{0,p,\Omega_0}^p \tag{11}$$

and for $j = 1, 2, 3, \ldots$

$$|u|_{1,p,U_j \cap \Omega}^p \le K_2 \varepsilon^p |u|_{2,p,Q_j}^p + K_2 \varepsilon^{-p} |u|_{0,p,Q_j}^p, \tag{12}$$

where $K_2$ is independent of $j$. Since any $R+2$ of the sets $\Omega_0, Q_1, Q_2, \ldots$ have empty intersection, (11) and (12) imply that

$$
\begin{aligned}
|u|_{1,p,\Omega}^p &\le |u|_{1,p,\Omega_0}^p + \sum_{j=1}^{\infty} |u|_{1,p,U_j\cap\Omega}^p \\
&\le \max(K_1, K_2)\left\{\varepsilon^p |u|_{2,p,\Omega_0}^p + \varepsilon^p \sum_{j=1}^{\infty} |u|_{2,p,Q_j}^p \right. \\
&\qquad \left. + \varepsilon^{-p} |u|_{0,p,\Omega_0}^p + \varepsilon^{-p} \sum_{j=1}^{\infty} |u|_{0,p,Q_j}^p \right\} \\
&\le (R+1)\max(K_1, K_2)\{\varepsilon^p |u|_{2,p,\Omega}^p + \varepsilon^{-p} |u|_{0,p,\Omega}^p\}
\end{aligned}
$$

and this inequality yields (10) (case $j = 1$, $m = 2$) on taking $p$th roots. It remains therefore to verify the validity of (11) and (12).

If $u \in C^\infty(\Omega) \cap W^{2,p}(\Omega)$, we apply Lemma 4.10 to $u$ considered as a function of $x_k$ on the interval from $\lambda_k \delta/2\sqrt{n}$ to $(\lambda_k+1)\delta/2\sqrt{n}$, and then integrate the remaining variables over similar intervals to obtain, for any $H_\lambda \subset \Omega$,

$$\int_{H_\lambda} |D_k u(x)|^p \, dx \le K_3 \varepsilon^p \int_{H_\lambda} |D_k{}^2 u(x)|^p \, dx + K_3 \varepsilon^{-p} \int_{H_\lambda} |u(x)|^p \, dx, \tag{13}$$

where $K_3$ depends only on $p$ and the length of a side of $H_\lambda$ (which in turn depends on $\Omega$ via $\delta$ and $n$). Summing (13) for $1 \le k \le n$, we obtain

$$|u|_{1,p,H_\lambda}^p \le K_3 \varepsilon^p |u|_{2,p,H_\lambda}^p + nK_3 \varepsilon^{-p} |u|_{0,p,H_\lambda}^p. \tag{14}$$

Since the cubes $H$ do not overlap, we sum (14) for all cubes $H_\lambda \subset \Omega$ and obtain (11) with $K_1 = nK_3$. Since $C^\infty(\Omega) \cap W^{2,p}(\Omega)$ is dense in $W^{2,p}(\Omega)$, (11) holds for all $u \in W^{2,p}(\Omega)$.

The constant $K_2$ in (12) will turn out to depend only on $p$, $M$ and the dimensions of the cone $C_j$ (see Section 4.4). Anticipating this, and noting that these dimensions are specified by the single cone $C$ to which all cones $C_j$ are congruent, we drop, for simplicity, all subscripts $j$ in considering (12). Let $\xi$ be a unit vector in a direction in $C$, and let $\Omega_\xi = \{y + t\xi : y \in \Omega \cap U, 0 \le t \le h\}$, where $h$ is the height of the cone $C$ (Fig. 3). Thus $(\Omega \cap U) \subset \Omega_\xi \subset Q$ by condition (iii) of the uniform cone property. Any line $L$ parallel to $\xi$ either has empty intersection with $\Omega_\xi$ or else intersects $\Omega_\xi$ in an interval of length $\rho$, where $h \le \rho \le h + \operatorname{diam} U \le h + M$ by condition (i) of the uniform cone property. By Lemma 4.10, if $u \in C^\infty(\Omega) \cap W^{2,p}(\Omega)$,

$$\int_{L\cap\Omega_\xi} |D_\xi u|^p \, ds \le K_4 \varepsilon^p \int_{L\cap\Omega_\xi} |D_\xi{}^2 u|^p \, ds + K_4 \varepsilon^{-p} \int_{L\cap\Omega_\xi} |u|^p \, ds, \tag{15}$$

where $D_\xi$ denotes differentiation in the direction of $\xi$ and where $K_4$ can be chosen to depend only on $p$, $h$, and $M$, that is, on $p$ and $\Omega$. We now integrate

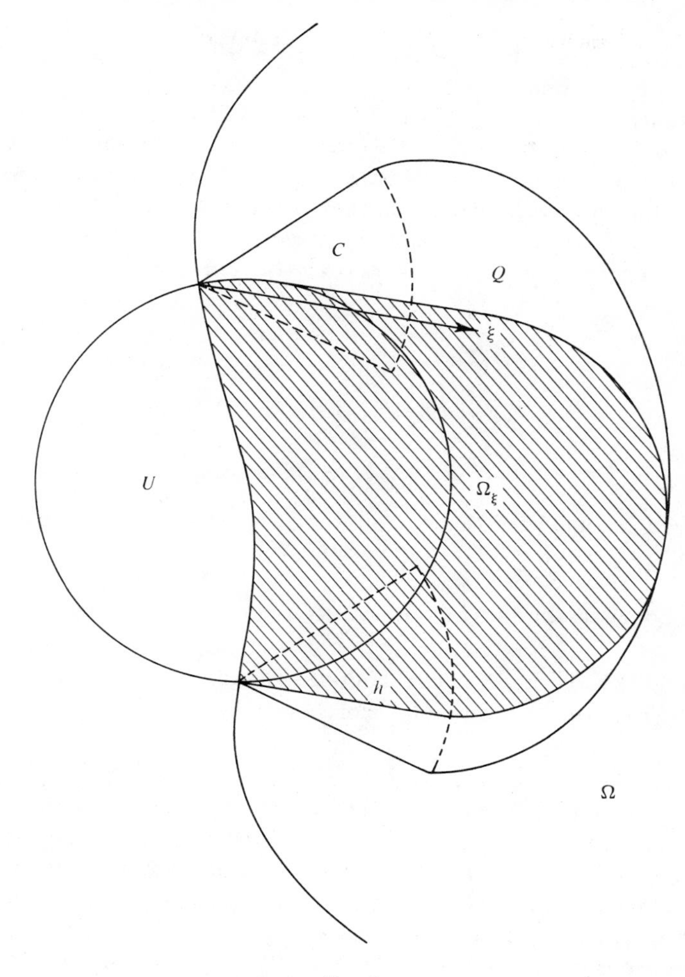


FIG. 3

(15) over the projection of $\Omega_\xi$ on a hyperplane perpendicular to $\xi$ and so obtain

$$\begin{aligned}\int_{\Omega \cap U} |D_\xi u(x)|^p \, dx &\leq \int_{\Omega_\xi} |D_\xi u(x)|^p \, dx \\ &\leq K_4 \varepsilon^p \int_{\Omega_\xi} |D_\xi{}^2 u(x)|^p \, dx + K_4 \varepsilon^{-p} \int_{\Omega_\xi} |u(x)|^p \, dx \\ &\leq K_4 \varepsilon^p \int_Q |D_\xi{}^2 u(x)|^p \, dx + K_4 \varepsilon^{-p} \int_Q |u(x)|^p \, dx. \qquad (16)\end{aligned}$$

Now let $\xi_1, \ldots, \xi_n$ be a basis of unit vectors in $\mathbb{R}^n$ each lying in a direction contained in the cone $C$. For $1 \le k \le n$, $D_k u(x) = \sum_{j=1}^{n} a_j D_{\xi_j} u(x)$ where the constants $a_j$ satisfy $|a_j| \le 1/V$, $1 \le j \le n$, $V$ being the volume of the parallelepiped spanned by $\xi_1, \ldots, \xi_n$. (The reader might verify this assertion. It is a simple exercise in linear algebra.) A lower bound for $V$ can be specified in terms of the aperture angle of the cone $C$—that is to say, the basis $\xi_1, \ldots, \xi_n$, which varies with the covering patch $U$, may always be chosen in such a way that $V$ is independent of $U$. It now follows from (16) that

$$\begin{aligned}\int_{\Omega \cap U} |D_k u(x)|^p \, dx &\le K_5 \sum_{j=1}^{n} \int_{\Omega \cap U} |D_{\xi_j} u(x)|^p \, dx \\ &\le K_5 \sum_{j=1}^{n} \left\{ K_4 \varepsilon^p \int_Q |D_{\xi_j}^2 u(x)|^p \, dx + K_4 \varepsilon^{-p} \int_Q |u(x)|^p \, dx \right\} \\ &\le K_6 \varepsilon^p |u|_{2,p,Q}^p + K_6 \varepsilon^{-p} |u|_{0,p,Q}^p.\end{aligned}$$

The desired inequality (12) now follows by summing on $k$ and using the density of $C^\infty(\Omega) \cap W^{2,p}(\Omega)$ in $W^{2,p}(\Omega)$. ∎

If $\Omega$ is bounded, the above theorem can be proved under weaker hypotheses.

**4.15 THEOREM** Let $\Omega$ be a bounded domain in $\mathbb{R}^n$ having the cone property. Then the conclusion of Theorem 4.14 holds for $\Omega$.

PROOF By Theorem 4.8 there exists a finite collection $\{\Omega_1, \ldots, \Omega_k\}$ of open subsets of $\Omega$ such that $\Omega = \bigcup_{j=1}^{k} \Omega_j$ and such that each $\Omega_j$ is a union of translates of some open parallelepiped. It is clearly sufficient to prove an inequality analogous to (11) for each $\Omega_j$. We may therefore assume without loss of generality that $\Omega = \bigcup_{x \in A} (x + P)$, where $A$ is a bounded set in $\mathbb{R}^n$ and $P$ is an open parallelepiped with one vertex at the origin. Let $\xi_1, \ldots, \xi_n$ be unit vectors in the directions of the $n$ edges of $P$ that concur in the origin, and let $l$ be the minimum length of these edges. Then the intersection with $\Omega$ of any line $L$ parallel to one of the vectors $\xi_j$ is either empty or a finite collection of segments each having length between $l$ and $\operatorname{diam} \Omega$. It follows as in (16) that

$$\int_\Omega |D_{\xi_j} u(x)|^p \, dx \le K_1 \varepsilon^p \int_\Omega |D_{\xi_j}^2 u(x)|^p \, dx + K \varepsilon^{-p} \int_\Omega |u(x)|^p \, dx$$

for smooth functions $u$. Since $\{\xi_1, \xi_2, \ldots, \xi_n\}$ is a basis for $\mathbb{R}^n$ one can now show, by an argument similar to that following (16), that

$$|u|_{1,p,\Omega}^p \le K_2 \varepsilon^p |u|_{2,p,\Omega}^p + K_2 \varepsilon^{-p} |u|_{0,p,\Omega}^p$$

holds for all $u \in W^{2,p}(\Omega)$. ∎

**4.16 COROLLARY** The functional $((\cdot))_{m,p,\Omega}$ defined by

$$((u))_{m,p,\Omega} = \{|u|^p_{m,p,\Omega} + |u|^p_{0,p,\Omega}\}^{1/p}$$

is a norm, equivalent to the usual norm $\|\cdot\|_{m,p,\Omega}$, on each of the following spaces:

(i) $W_0^{m,p}(\Omega)$ for any domain $\Omega$,
(ii) $W^{m,p}(\Omega)$ for any domain $\Omega$ having the uniform cone property,
(iii) $W^{m,p}(\Omega)$ for any bounded domain $\Omega$ having the cone property.

**4.17 THEOREM** (*Ehrling* [23], *Browder* [12]) If $\Omega \subset \mathbb{R}^n$ has the uniform cone property or if it is bounded and has the cone property, and if $1 \le p < \infty$, then there exists a constant $K = K(m, p, \Omega)$ such that for $0 \le j \le m$ and any $u \in W^{m,p}(\Omega)$,

$$\|u\|_{j,p} \le K \|u\|^{j/m}_{m,p} \|u\|^{(m-j)/m}_{0,p}. \tag{17}$$

In addition, (17) is valid for all $u \in W_0^{m,p}(\Omega)$ with a constant $K = K(m, p, n)$ independent of $\Omega$.

PROOF Inequality (17) is obvious if either $j = 0$ or $j = m$. For $0 < j < m$ we can obtain from successive applications of (10) that

$$\|u\|_{j,p} \le K_1 \varepsilon \|u\|_{m,p} + K_1 \varepsilon^{-j/(m-j)} \|u\|_{0,p} \tag{18}$$

holds for all $\varepsilon$, $0 < \varepsilon \le 1$, and all $u \in W^{m,p}(\Omega)$ with $K_1$ depending only on $m$, $p$, and $\Omega$. (By Theorem 4.13 the same inequality holds for all $u \in W_0^{m,p}(\Omega)$ with $K_1$ depending only on $m$, $p$, and $n$.) Inequality (17) now follows for $u \ne 0$ if we set $\varepsilon = (\|u\|_{0,p}/\|u\|_{m,p})^{(m-j)/m}$ in (18). ∎

We remark that (17) also implies (18) algebraically: specifically, set $p = m/j$, $p' = m/(m-j)$, $a = (\varepsilon \|u\|_{m,p})^{j/m}$, and $b = \varepsilon^{-j/m} \|u\|^{(m-j)/m}_{0,p}$ in inequality (4) of Chapter II,

$$ab \le (a^p/p) + (b^{p'}/p'), \qquad (1/p) + (1/p') = 1,$$

to show that the right side of (17) does not exceed the right side of (18).

## Interpolation Inequalities Involving Compact Subdomains

**4.18** Upper bounds for the $L^p(\Omega)$-norms of intermediate derivatives $D^\beta u$, $|\beta| \le m-1$, of a function $u \in W^{m,p}(\Omega)$ can be expressed in terms of the seminorm $|u|_{m,p,\Omega}$ and the $L^p$-norm of $u$ over a suitable subdomain whose closure is a compact subset of the bounded domain $\Omega$. We establish some hybrid interpolation inequalities of this sort by methods that follow somewhat the same lines as those used for the interpolation inequalities derived above. (See, for instance, the work of Agmon [6].)

**4.19 LEMMA** Let $(a, b)$ be a finite open interval in $\mathbb{R}$ and let $1 \leq p < \infty$. There exists a finite constant $K = K(p, b-a)$ and, for every positive number $\varepsilon$, a number $\delta = \delta(\varepsilon, b-a)$ satisfying $0 < 2\delta < b-a$ such that every continuously differentiable function $f$ on $(a, b)$ satisfies

$$\int_a^b |f(t)|^p \, dt \leq K\varepsilon \int_a^b |f'(t)|^p \, dt + K \int_{a+\delta}^{b-\delta} |f(t)|^p \, dt. \tag{19}$$

Moreover, fixed values of $K$ and $\delta$, independent of $b-a$, can be chosen so that (19) holds for all intervals $(a, b)$ whose lengths lie between fixed positive bounds: $0 < l_1 \leq b-a \leq l_2 < \infty$.

PROOF The proof is similar to that of Lemma 4.10. Suppose for the moment that $a = 0$ and $b = 1$, and let $\frac{1}{3} < \eta < \frac{2}{3}$. If $0 < x < 1$, then

$$|f(x)| = \left| f(\eta) + \int_\eta^x f'(t) \, dt \right| \leq |f(\eta)| + \int_0^1 |f'(t)| \, dt.$$

Integrating $\eta$ over $(\frac{1}{3}, \frac{2}{3})$, we are led to

$$|f(x)| \leq 3 \int_{1/3}^{2/3} |f(\eta)| \, d\eta + \int_0^1 |f'(t)| \, dt,$$

so that by Hölder's inequality if $p > 1$,

$$|f(x)|^p \leq 3 \cdot 2^{p-1} \int_{1/3}^{2/3} |f(t)|^p \, dt + 2^{p-1} \int_0^1 |f'(t)|^p \, dt.$$

Integrating $x$ over $(0, 1)$, we obtain

$$\int_0^1 |f(t)|^p \, dt \leq K_p \int_0^1 |f'(t)|^p \, dt + K_p \int_{1/3}^{2/3} |f(t)|^p \, dt,$$

where $K_p = 3 \cdot 2^{p-1}$. The change of variable $a + t(b-a) \to t$ now yields, for any finite interval $(a, b)$,

$$\int_a^b |f(t)|^p \, dt \leq K_p (b-a)^p \int_a^b |f'(t)|^p \, dt + K_p \int_{a+(b-a)/3}^{b-(b-a)/3} |f(t)|^p \, dt.$$

For given $\varepsilon > 0$ pick a positive integer $n$ such that $n^{-p} \leq \varepsilon$. Let $a_j = a + (b-a)j/n$ for $j = 0, 1, \ldots, n$ and pick $\delta$ so that $0 < \delta \leq (b-a)/3n$. Then

$$\int_a^b |f(t)|^p \, dt = \sum_{j=1}^{n} \int_{a_{j-1}}^{a_j} |f(t)|^p \, dt$$

*equation continues*

$$\leq K_p \sum_{j=1}^{n} \left\{ \left( \frac{b-a}{n} \right)^p \int_{a_{j-1}}^{a_j} |f'(t)|^p \, dt + \int_{a_{j-1}+\delta}^{a_j-\delta} |f(t)|^p \, dt \right\}$$

$$\leq K_p \max(1, (b-a)^p) \left\{ \varepsilon \int_a^b |f'(t)|^p \, dt + \int_{a+\delta}^{b-\delta} |f(t)|^p \, dt \right\}$$

which is the desired inequality. ∎

The reader may convince himself that Lemma 4.19, unlike Lemma 4.10, does not extend to infinite intervals $(a, b)$ if the intention is that the second integral on the right side of (19) should be taken over a compact subinterval.

**4.20 THEOREM** Let $\Omega$ be a bounded domain in $\mathbb{R}^n$ that has the segment property. Then there exists a constant $K = K(p, \Omega)$ such that to any positive number $\varepsilon$ there corresponds a domain $\Omega_\varepsilon \subset\subset \Omega$ such that

$$|u|_{0,p,\Omega} \leq K\varepsilon |u|_{1,p,\Omega} + K|u|_{0,p,\Omega_\varepsilon} \tag{20}$$

holds for every $u \in W^{1,p}(\Omega)$.

PROOF The proof is similar to that of Theorem 4.14. Since $\Omega$ is bounded the locally finite open cover $\{U_j\}$ of bdry $\Omega$ and the corresponding set $\{y_j\}$ of nonzero vectors referred to in the description of the segment property (Section 4.2) are both finite sets. Therefore open sets $\mathscr{V}_j \subset\subset U_j$ can be found so that bdry $\Omega \subset \bigcup_j \mathscr{V}_j$. (See the first part of the proof of Theorem 3.14.) Moreover, for some $\delta > 0$, $\Omega_\delta = \{x \in \Omega : \operatorname{dist}(x, \operatorname{bdry}\Omega) < \delta\} \subset \bigcup_j \mathscr{V}_j$ so that we may write $\Omega = \bigcup_j (\mathscr{V}_j \cap \Omega) \cup \tilde{\Omega}$, where $\tilde{\Omega} \subset\subset \Omega$. It is thus sufficient to prove that for each $j$

$$|u|^p_{0,p,\mathscr{V}_j \cap \Omega} \leq K_1 \varepsilon^p |u|^p_{1,p,\Omega} + K_1 |u|^p_{0,p,\Omega_{\varepsilon,j}}$$

for some $\Omega_{\varepsilon,j} \subset\subset \Omega$. For simplicity we drop all subscripts $j$.

Consider the sets $Q, Q_\eta, 0 \leq \eta < 1$, defined by

$$Q = \{x + ty : x \in U \cap \Omega,\ 0 < t < 1\},$$
$$Q_\eta = \{x + ty : x \in \mathscr{V} \cap \Omega,\ \eta < t < 1\}.$$

If $\eta > 0$, $Q_\eta \subset\subset Q$. By the segment property, $Q \subset \Omega$ and any line $L$ parallel to $y$ and passing through a point of $\mathscr{V} \cap \Omega$ intersects $Q_0$ in one or more intervals each having length between $|y|$ and diam $\Omega$. By Lemma 4.19 there exists $\eta > 0$ and a constant $K$ such that for any $u \in C^\infty(\Omega)$ and any such line $L$

$$\int_{L \cap Q_0} |u(x)|^p \, ds \leq K_1 \varepsilon^p \int_{L \cap Q_0} |D_y u(x)|^p \, ds + K_1 \int_{L \cap Q_\eta} |u(x)|^p \, ds,$$

$D_y$ denoting differentiation in the direction of $y$. We integrate this inequality over the projection of $Q_0$ on a hyperplane perpendicular to $y$ and so obtain

$$|u|_{0,p,\Omega\cap\mathscr{V}}^p \le |u|_{0,p,Q_0}^p \le K_1 \varepsilon^p |u|_{1,p,Q_0}^p + K_1 |u|_{0,p,Q_\eta}^p$$
$$\le K_1 \varepsilon^p |u|_{1,p,\Omega}^p + K_1 |u|_{0,p,\Omega_\varepsilon}^p,$$

where $\Omega_\varepsilon = \Omega_\eta \subset\subset \Omega$. By density this inequality holds for any $u \in W^{1,p}(\Omega)$. ∎

**4.21 COROLLARY** The conclusion of Theorem 4.20 is also valid if $\Omega$ is bounded and has the cone property.

PROOF As remarked earlier, a domain $\Omega$ with the cone property need not have the segment property. By Theorem 4.8, however, $\Omega$ is a finite union of domains having the strong local Lipschitz property. We leave it to the reader to show that a bounded domain with the strong local Lipschitz property has the segment property, and thus complete the proof. ∎

**4.22 LEMMA** Let $\Omega_0, \Omega$ be domains in $\mathbb{R}^n$ with $\Omega_0 \subset\subset \Omega$. Then there exists a domain $\Omega'$ having the cone property such that $\Omega_0 \subset \Omega' \subset\subset \Omega$.

PROOF Since $\overline{\Omega}_0$ is a compact subset of $\Omega$ there exists $\delta > 0$ such that $\operatorname{dist}(\overline{\Omega}_0, \operatorname{bdry}\Omega) > \delta$. The domain $\Omega' = \{y \in \mathbb{R}^n : |y-x| < \delta \text{ for some } x \in \Omega_0\}$ clearly has the desired properties. ∎

**4.23 THEROEM** Let $\Omega$ be a bounded domain in $\mathbb{R}^n$ having either the segment property or the cone property. Let $0 < \varepsilon_0 < \infty$, let $1 \le p < \infty$, and let $j$ and $m$ be integers with $0 \le j \le m-1$. Then there exists a constant $K = K(\varepsilon_0, m, p, \Omega)$, and for each $\varepsilon$, $0 < \varepsilon \le \varepsilon_0$, a domain $\Omega_\varepsilon$ such that $\Omega_\varepsilon \subset\subset \Omega$ and such that for every $u \in W^{m,p}(\Omega)$

$$|u|_{j,p,\Omega} \le K\varepsilon |u|_{m,p,\Omega} + K\varepsilon^{-j/(m-j)} |u|_{0,p,\Omega_\varepsilon}. \tag{21}$$

PROOF We apply Theorem 4.20 or its corollary to derivatives $D^\beta u$, $|\beta| = m-1$, to obtain

$$|u|_{m-1,p,\Omega} \le K_1 \varepsilon |u|_{m,p,\Omega} + K_1 |u|_{m-1,p,\Omega_\varepsilon}, \tag{22}$$

where $\Omega_\varepsilon \subset\subset \Omega$. By Lemma 4.22 we may assume that $\Omega_\varepsilon$ has the cone property. For $0 < \varepsilon \le \varepsilon_0$, we have by Theorem 4.15

$$|u|_{m-1,p,\Omega_\varepsilon} \le K_2 \varepsilon |u|_{m,p,\Omega_\varepsilon} + K_2 \varepsilon^{-(m-1)} |u|_{0,p,\Omega_\varepsilon}. \tag{23}$$

Combining (22) and (23), we get the case $j = m-1$ of (21). We complete the proof by downward induction on $j$. Assuming (21) holds for some $j \ge 1$ and replacing $\varepsilon$ by $\varepsilon^{m-j}$ (with consequent alteration of $K$ and $\Omega_\varepsilon$), we obtain

$$|u|_{j,p,\Omega} \le K_3 \varepsilon^{m-j} |u|_{m,p,\Omega} + K_3 \varepsilon^{-j} |u|_{0,p,\Omega_\varepsilon'}. \tag{24}$$

Also by (21) with $j$ and $m$ replaced by $j-1$ and $j$, respectively (the case already proved), we have

$$|u|_{j-1,p,\Omega} \le K_4 \varepsilon |u|_{j,p,\Omega} + K_4 \varepsilon^{-(j-1)} |u|_{0,p,\Omega_\varepsilon''}. \tag{25}$$

Combining (24) and (25), we get

$$|u|_{j-1,p,\Omega} \le K_5 \varepsilon^{m-(j-1)} |u|_{m,p,\Omega} + K_5 \varepsilon^{-(j-1)} |u|_{0,p,\Omega_\varepsilon},$$

where $K_5 = K_4(K_3+1)$ and $\Omega_\varepsilon = \Omega_\varepsilon' \cup \Omega_\varepsilon''$. Replacing $\varepsilon$ by $\varepsilon^{1/(m-j+1)}$, we complete the induction. ∎

## Extension Theorems

**4.24** Let $\Omega$ be a domain in $\mathbb{R}^n$. For given $m$ and $p$ a linear operator $E$ mapping $W^{m,p}(\Omega)$ into $W^{m,p}(\mathbb{R}^n)$ is called a *simple $(m,p)$-extension operator for $\Omega$* provided there exists a constant $K = K(m,p)$ such that for every $u \in W^{m,p}(\Omega)$ the following conditions hold:

(i) $Eu(x) = u(x)$ a.e. in $\Omega$,
(ii) $\|Eu\|_{m,p,\mathbb{R}^n} \le K \|u\|_{m,p,\Omega}$.

$E$ is called a *strong $m$-extension operator* for $\Omega$ if $E$ is a linear operator mapping functions defined a.e. in $\Omega$ into functions defined a.e. in $\mathbb{R}^n$ and if for every $p$, $1 \le p < \infty$, and every $k$, $0 \le k \le m$, the restriction of $E$ to $W^{k,p}(\Omega)$ is a simple $(k,p)$-extension operator for $\Omega$. Finally, $E$ is called a *total extension operator* for $\Omega$ provided $E$ is a strong $m$-extension operator for $\Omega$, for every $m$.

**4.25** The existence of even a simple $(m,p)$-extension operator for a domain $\Omega$ guarantees that $W^{m,p}(\Omega)$ inherits many properties possessed by $W^{m,p}(\mathbb{R}^n)$. For instance, if the imbedding $W^{m,p}(\mathbb{R}^n) \to L^q(\mathbb{R}^n)$ is known to hold, then the imbedding $W^{m,p}(\Omega) \to L^q(\Omega)$ follows via the chain of inequalities

$$\|u\|_{0,q,\Omega} \le \|Eu\|_{0,q,\mathbb{R}^n} \le K_1 \|Eu\|_{m,p,\mathbb{R}^n} \le K_1 K \|u\|_{m,p,\Omega}.$$

We shall not, however, use this technique to prove the Sobolev imbedding theorem in Chapter V as we shall prove that theorem under rather weaker hypothesis on $\Omega$ than are needed to guarantee the existence of an $(m,p)$-extension operator.

We shall construct extension operators of each of the three types defined above. The method used in two of these constructions is based on successive reflections in smooth boundaries. It is attributed to Lichenstein [35] and later Hestenes [31] and Seeley [61]. The third construction, due to Calderón [14] involves the use of the Calderón–Zygmund theory of singular integrals. It is less transparent than the reflections method and yields a weaker result, but requires much less regularity of the domain $\Omega$. Except for very simple domains all of the constructions require the use of partitions of unity subordinate to open covers of bdry $\Omega$ chosen in such a way that the functions in the partition have uniformly bounded derivatives. Because of this, domains with bounded boundaries (both exterior domains and bounded domains) are more likely to be easily seen to satisfy the conditions of our extension theorems. Exceptions are half-spaces, quadrants, etc., and smooth images of these.

**4.26 THEOREM** Let $\Omega$ be either

(i) a half-space in $\mathbb{R}^n$, or
(ii) a domain in $\mathbb{R}^n$ having the uniform $C^m$-regularity property, and also having a bounded boundary.

For any positive integer $m$ there exists a strong $m$-extension operator $E$ for $\Omega$. Moreover, if $\alpha$ and $\gamma$ are multi-indices with $|\gamma| \le |\alpha| \le m$, there exists a linear operator $E_{\alpha\gamma}$ continuous from $W^{j,p}(\Omega)$ into $W^{j,p}(\mathbb{R}^n)$ for $1 \le j \le m - |\alpha|$ such that if $u \in W^{|\alpha|,p}(\Omega)$, then

$$D^\alpha Eu(x) = \sum_{|\gamma| \le |\alpha|} E_{\alpha\gamma} D^\gamma u(x) \qquad \text{a.e. in } \mathbb{R}^n. \tag{26}$$

PROOF First let $\Omega$ be the half-space $\mathbb{R}_+{}^n = \{x \in \mathbb{R}^n : x_n > 0\}$. For functions $u$ defined a.e. on $\mathbb{R}_+{}^n$ we define extensions $Eu$ and $E_\alpha u$, $|\alpha| \le m$, a.e. on $\mathbb{R}^n$ via

$$\begin{aligned} Eu(x) &= \begin{cases} u(x) & x_n > 0 \\ \sum_{j=1}^{m+1} \lambda_j u(x_1, \ldots, x_{n-1}, -jx_n) & \text{if} \quad x_n \le 0, \end{cases} \\ E_\alpha u(x) &= \begin{cases} u(x) & \text{if} \quad x_n > 0 \\ \sum_{j=1}^{m+1} (-j)^{\alpha_n} \lambda_j u(x_1, \ldots, x_{n-1}, -jx_n) & \text{if} \quad x_n \le 0, \end{cases} \end{aligned} \tag{27}$$

where the coefficients $\lambda_1, \ldots, \lambda_{m+1}$ are the unique solutions of the $(m+1) \times (m+1)$ system of linear equations

$$\sum_{j=1}^{m+1} (-j)^k \lambda_j = 1, \qquad k = 0, 1, \ldots, m.$$

If $u \in C^m(\overline{\mathbb{R}_+{}^n})$, then it is readily checked that $Eu \in C^m(\mathbb{R}^n)$ and

$$D^\alpha Eu(x) = E_\alpha D^\alpha u(x), \qquad |\alpha| \le m.$$

Thus

$$\int_{\mathbb{R}^n} |D^\alpha Eu(x)|^p \, dx = \int_{\mathbb{R}_+{}^n} |D^\alpha u(x)|^p \, dx$$
$$+ \int_{\mathbb{R}_-{}^n} \left| \sum_{j=1}^{m+1} (-j)^{\alpha_n} \lambda_j \, D^\alpha u(x_1, \ldots, x_{n-1}, -jx_n) \right|^p dx$$
$$\leq K(m, p, \alpha) \int_{\mathbb{R}_+{}^n} |D^\alpha u(x)|^p \, dx.$$

The above inequality extends, by virtue of Theorem 3.18, to functions $u \in W^{k,p}(\mathbb{R}_+{}^n)$, $m \geq k \geq |\alpha|$. Hence $E$ is a strong $m$-extension operator for $\mathbb{R}_+{}^n$. Since $D^\beta E_\alpha u(x) = E_{\alpha+\beta} u(x)$, a similar calculation shows that $E_\alpha$ is a strong $(m - |\alpha|)$-extension. Thus the theorem is proved for half-spaces (with $E_{\alpha\alpha} = E_\alpha$, $E_{\alpha\gamma} = 0$ otherwise).

Now suppose $\Omega$ is uniformly $C^m$-regular and has bounded boundary. Then the open cover $\{U_j\}$ of $\operatorname{bdry}\Omega$, and the corresponding $m$-smooth maps $\Phi_j$ from $U_j$ onto $B$, referred to in Section 4.6, are finite collections, say $1 \leq j \leq N$. Let $Q = \{y \in \mathbb{R}^n : |y'| = (\sum_{j=1}^{n-1} y_j{}^2)^{1/2} < \frac{1}{2}, |y_n| < \sqrt{3}/2\}$. Then

$$\{y \in \mathbb{R}^n : |y| < \tfrac{1}{2}\} \subset Q \subset B = \{y \in \mathbb{R}^n : |y| < 1\}.$$

By condition (i) of Section 4.6 the open sets $\mathscr{V}_j = \Psi_j(Q)$, $1 \leq j \leq N$, form an open cover of $\Omega_\delta = \{x \in \Omega : \operatorname{dist}(x, \operatorname{bdry}\Omega) < \delta\}$ for some $\delta > 0$. There exists an open set $\mathscr{V}_0$ of $\Omega$, bounded away from $\operatorname{bdry}\Omega$, such that $\Omega \subset \bigcup_{j=0}^N \mathscr{V}_j$. By Theorem 3.14 we may find infinitely differentiable functions $\omega_0, \omega_1, \ldots, \omega_N$ such that $\operatorname{supp}\omega_j \subset \mathscr{V}_j$ and $\sum_{j=0}^N \omega_j(x) = 1$ for all $x \in \Omega$. (Note that $\operatorname{supp}\omega_0$ need not be compact if $\Omega$ is unbounded.)

Since $\Omega$ is uniformly $C^m$-regular it has the segment property and so restrictions to $\Omega$ of functions in $C_0{}^\infty(\mathbb{R}^n)$ are dense in $W^{k,p}(\Omega)$. If $\phi \in C_0{}^\infty(\mathbb{R}^n)$, then $\phi$ agrees on $\Omega$ with the function $\sum_{j=0}^N \phi_j$, where $\phi_j = \omega_j \cdot \phi \in C_0{}^\infty(\mathscr{V}_j)$.

For $j \geq 1$ and $y \in B$ let $\psi_j(y) = \phi_j(\Psi_j(y))$. Then $\psi_j \in C_0{}^\infty(Q)$. We extend $\psi_j$ to be identically zero outside $Q$. With $E$ (and $E_\alpha$) defined as in (27), we have $E\psi_j \in C_0{}^m(Q)$, $E\psi_j = \psi_j$ on $Q_+ = \{y \in Q : y_n > 0\}$, and

$$\|E\psi_j\|_{k,p,Q} \leq K_1 \|\psi_j\|_{k,p,Q_+}, \qquad 0 \leq k \leq m,$$

where $K_1$ depends on $k$, $m$, and $p$. If $\theta_j(x) = E\psi_j(\Phi_j(x))$, then $\theta_j \in C_0{}^m(\mathscr{V}_j)$ and $\theta_j(x) = \phi_j(x)$ if $x \in \Omega$. It may be checked by induction that if $|\alpha| \leq m$, then

$$D^\alpha \theta_j(x) = \sum_{|\beta| \leq |\alpha|} \sum_{|\gamma| \leq |\alpha|} a_{j;\alpha\beta}(x) [E_\beta(b_{j;\beta\gamma} \cdot (D^\gamma \phi_j \circ \Psi_j))](\Phi_j(x)),$$

where $a_{j;\alpha\beta} \in C^{m-|\alpha|}(\overline{U}_j)$ and $b_{j;\beta\gamma} \in C^{m-|\beta|}(\overline{B})$ depend on the transformations $\Phi_j$ and $\Psi_j = \Phi_j^{-1}$ and satisfy

$$\sum_{|\beta| \leq |\alpha|} a_{j;\alpha\beta}(x) \, b_{j;\beta\gamma}(\Phi_j(x)) = \begin{cases} 1 & \text{if} \quad \gamma = \alpha \\ 0 & \text{otherwise.} \end{cases}$$

By Theorem 3.35 we have for $k \le m$,

$$\|\theta_j\|_{k,p,\mathbb{R}^n} \le K_2 \|E\psi_j\|_{k,p,Q} \le K_1 K_2 \|\psi_j\|_{k,p,Q_+} \le K_3 \|\psi_j\|_{k,p,\Omega},$$

where $K_3$ may be chosen to be independent of $j$. The operator $\tilde{E}$ defined by

$$\tilde{E}\phi(x) = \phi_0(x) + \sum_{j=1}^{N} \theta_j(x)$$

clearly satisfies $\tilde{E}\phi(x) = \phi(x)$ if $x \in \Omega$, and

$$\begin{aligned} \|\tilde{E}\phi\|_{k,p,\mathbb{R}^n} &\le \|\phi_0\|_{k,p,\Omega} + K_3 \sum_{j=1}^{N} \|\phi_j\|_{k,p,\Omega} \\ &\le K_4(1+NK_3)\|\phi\|_{k,p,\Omega}, \end{aligned} \tag{28}$$

where

$$K_4 = \max_{0\le j\le N} \max_{|\alpha|\le m} \sup_{x\in\mathbb{R}^n} |D^\alpha \omega_j(x)| < \infty.$$

Thus $\tilde{E}$ is a strong $m$-extension operator for $\Omega$. Also

$$D^\alpha \tilde{E}\phi(x) = \sum_{|\gamma|\le|\alpha|} (E_{\alpha\gamma} D^\gamma \phi)(x),$$

where

$$E_{\alpha\gamma} v(x) = \sum_{j=1}^{N} \sum_{|\beta|\le|\alpha|} a_{j;\alpha\beta}(x)[E_\beta(b_{j;\beta\gamma}\cdot(v\cdot\omega_j)\circ\Psi_j)](\Phi_j(x))$$

if $\alpha \ne \gamma$, and

$$E_{\alpha\alpha} v(x) = (v\cdot\omega_0)(x) + \sum_{j=1}^{N} \sum_{|\beta|\le|\alpha|} a_{j;\alpha\beta}(x)[E_\beta(b_{j;\beta\alpha}\cdot(v\cdot\omega_j)\circ\Psi_j)](\Phi_j(x)).$$

We note that if $x \in \Omega$, $E_{\alpha\gamma}v(x) = 0$ for $\alpha \ne \gamma$ and $E_{\alpha\alpha}v(x) = v(x)$. Clearly $E_{\alpha\gamma}$ is a linear operator. By the differentiability properties of $a_{j;\alpha\beta}$ and $b_{j;\beta\gamma}$, $E_{\alpha\gamma}$ is continuous on $W^{j,p}(\Omega)$ into $W^{j,p}(\mathbb{R}^n)$ for $1 \le j \le m - |\alpha|$. This completes the proof. ▌

The representation (26) for derivatives of extended functions was included in the above theorem because it will be needed when we study fractional-order spaces in Chapter VII.

The reflection technique used in the above proof can be modified to yield a total extension operator for smoothly bounded domains. The proof, due to Seeley [61], is based on the following lemma.

**4.27 LEMMA** There exists a real sequence $\{a_k\}_{k=0}^{\infty}$ such that for every nonnegative integer $n$ we have

$$\sum_{k=0}^{\infty} 2^{nk} a_k = (-1)^n \tag{29}$$

and

$$\sum_{k=0}^{\infty} 2^{nk} |a_k| < \infty. \tag{30}$$

Proof For fixed $N$, let $a_{k,N}$, $k = 0, 1, 2, \ldots, N$, be the solution of the system of linear equations

$$\sum_{k=0}^{N} 2^{nk} a_{k,N} = (-1)^n, \qquad n = 0, 1, 2, \ldots, N. \tag{31}$$

In terms of the Vandermonde determinant

$$V(x_0, x_1, \ldots, x_N) = \begin{vmatrix} 1 & 1 & \cdots & 1 \\ x_0 & x_1 & \cdots & x_N \\ {x_0}^2 & {x_1}^2 & \cdots & {x_N}^2 \\ \vdots & \vdots & & \vdots \\ {x_0}^N & {x_1}^N & \cdots & {x_N}^N \end{vmatrix} = \prod_{\substack{i,j=0 \\ i<j}}^{N} (x_j - x_i)$$

the solution of (31), given by Cramer's rule, is

$$\begin{aligned} a_{k,N} &= \frac{V(1, 2, \ldots, 2^{k-1}, -1, 2^{k+1}, \ldots, 2^N)}{V(1, 2, \ldots, 2^N)} \\ &= \left\{ \prod_{\substack{i,j=0 \\ i,j \neq k \\ i<j}}^{N} (2^j - 2^i) \prod_{i=0}^{k-1} (-1-2^i) \prod_{j=k+1}^{N} (2^j + 1) \right\} \left\{ \prod_{\substack{i,j=0 \\ i<j}}^{N} (2^j - 2^i) \right\}^{-1} \\ &= A_k B_{k,N} \end{aligned}$$

where

$$A_k = \prod_{i=0}^{k-1} \frac{1+2^i}{2^i - 2^k}, \qquad B_{k,N} = \prod_{j=k+1}^{N} \frac{1+2^j}{2^j - 2^k},$$

it being understood that $\prod_{i=l}^{m} P_i = 1$ if $l > m$. Now

$$|A_k| \leq \prod_{i=1}^{k-1} \frac{2^{i+1}}{2^{k-1}} = 2^{(3k-k^2)/2}.$$

Also

$$\begin{aligned} \log B_{k,N} &= \sum_{j=k+1}^{N} \log\left(1 + \frac{1+2^k}{2^j - 2^k}\right) \\ &< \sum_{j=k+1}^{N} \frac{1+2^k}{2^j - 2^k} < (1+2^k) \sum_{j=k+1}^{N} \frac{1}{2^{j-1}} < 4, \end{aligned}$$

where we have used the inequality $\log(1+x) < x$ valid for $x > 0$. It follows that the increasing sequence $\{B_{k,N}\}_{N=0}^{\infty}$ converges to a limit $B_k \leq e^4$. Let

$a_k = A_k B_k$ so that

$$|a_k| \le e^4 2^{(3k-k^2)/2}.$$

Then for any $n$

$$\sum_{k=0}^{\infty} 2^{nk} |a_k| \le e^4 \sum_{k=0}^{\infty} 2^{(2nk+3k-k^2)/2} < \infty.$$

Letting $N$ tend to infinity in (31), we complete the proof. ∎

**4.28 THEOREM** Let $\Omega$ be either

(i) a half-space in $\mathbb{R}^n$, or

(ii) a domain in $\mathbb{R}^n$ having the uniform $C^m$-regularity property for every $m$, and also having a bounded boundary.

Then there exists a total extension operator for $\Omega$.

PROOF It is sufficient to prove the theorem for the half-space $\mathbb{R}_+{}^n$; the proof for $\Omega$ satisfying (ii) then follows just as in Theorem 4.26.

The restrictions to $\mathbb{R}_+{}^n$ of functions $\phi \in C_0{}^\infty(\mathbb{R}^n)$ being dense in $W^{m,p}(\mathbb{R}_+{}^n)$ for any $m$ and $p$, we define the extension operator only on such functions. Let $f$ be a real-valued function, infinitely differentiable on $[0, \infty)$ and satisfying $f(t) = 1$ if $0 \le t \le \frac{1}{2}$, $f(t) = 0$ if $t \ge 1$. If $\phi \in C_0{}^\infty(\mathbb{R}^n)$, let

$$E\phi(x) = \begin{cases} \phi(x) & \text{if } x \in \overline{\mathbb{R}_+{}^n} \\ \sum_{k=0}^{\infty} a_k f(-2^k x_n)\,\phi(x', -2^k x_n) & \text{if } x \in \mathbb{R}_-{}^n, \end{cases} \tag{32}$$

where $\{a_k\}$ is the sequence constructed in the above lemma, and $x' = (x_1, \ldots, x_{n-1})$. Then clearly $E\phi$ is well defined on $\mathbb{R}^n$ since the sum in (32) has only finitely many nonvanishing terms for any particular $x \in \mathbb{R}_-{}^n = \{x \in \mathbb{R}^n : x_n < 0\}$. Moreover, $E\phi$ has compact support and belongs to $C^\infty(\overline{\mathbb{R}_+{}^n}) \cap C^\infty(\overline{\mathbb{R}_-{}^n})$. If $x \in \mathbb{R}_-{}^n$, we have

$$\begin{aligned} D^\alpha E\phi(x) &= \sum_{k=0}^{\infty} a_k \sum_{j=0}^{\alpha_n} \binom{\alpha_n}{j} (-2^k)^{\alpha_n} f^{(\alpha_n - j)}(-2^k x_n)\, D_n{}^j D^{\alpha'} \phi(x', -2^k x_n) \\ &= \sum_{k=0}^{\infty} \psi_k(x). \end{aligned}$$

Since $\psi_k(x) = 0$ when $-x_n > 1/2^{k-1}$ it follows from (30) that the above series converges absolutely and uniformly as $x_n$ tends to zero from below. Hence by (29)

$$\begin{aligned} \lim_{x_n \to 0-} D^\alpha E\phi(x) &= \sum_{k=0}^{\infty} (-2^k)^{\alpha_n} a_k\, D^\alpha \phi(x', 0+) \\ &= D^\alpha \phi(x', 0+) = \lim_{x_n \to 0+} D^\alpha E\phi(x) = D^\alpha E\phi(0). \end{aligned}$$

Thus $E\phi \in C_0^\infty(\mathbb{R}^n)$. Moreover, if $|\alpha| \leq m$,

$$|\psi_k(x)|^p \leq K_1{}^p |a_k|^p 2^{kmp} \sum_{|\beta| \leq m} |D^\beta \phi(x', -2^k x_n)|^p,$$

where $K_1$ depends on $m, p, n$, and $f$. Thus

$$\begin{aligned}\|\psi_k\|_{0,p,\mathbb{R}_-{}^n} &\leq K_1 |a_k| 2^{km} \left\{ \sum_{|\beta| \leq m} \int_{\mathbb{R}_-{}^n} |D^\beta \phi(x', -2^k x_n)|^p \, dx \right\}^{1/p} \\ &= K_1 |a_k| 2^{km} \left\{ (1/2^k) \sum_{|\beta| \leq m} \int_{\mathbb{R}_+{}^n} |D^\beta \phi(y)|^p \, dy \right\}^{1/p} \\ &\leq K_1 |a_k| 2^{km} \|\phi\|_{m,p,\mathbb{R}_+{}^n}.\end{aligned}$$

It follows by (30) that

$$\|D^\alpha E\phi\|_{0,p,\mathbb{R}_-{}^n} \leq K_1 \|\phi\|_{m,p,\mathbb{R}_+{}^n} \sum_{k=0}^{\infty} 2^{km} |a_k| \leq K_2 \|\phi\|_{m,p,\mathbb{R}_+{}^n}.$$

Combining this with a similar (trivial) inequality for $\|D^\alpha E\phi\|_{0,p,\mathbb{R}_+{}^n}$, we obtain

$$\|E\phi\|_{m,p,\mathbb{R}^n} \leq K_3 \|\phi\|_{m,p,\mathbb{R}_+{}^n}$$

with $K_3 = K_3(m, p, n)$. This completes the proof. ∎

**4.29** The restriction that bdry $\Omega$ be bounded was imposed in Theorem 4.26 (and similarly in Theorem 4.28) so that the cover $\{\mathscr{V}_j\}$ would be finite. This finiteness was used in two places in the proof, first in asserting the existence of the constant $K_4$, and secondly in obtaining the last inequality in (28). This latter use is, however, not essential for the proof, for, were the cover $\{\mathscr{V}_j\}$ not finite, (28) could still be obtained via the finite intersection property [Section 4.6, conidition (ii)]. Theorems 4.26 and 4.28 extend to any suitably regular domains for which there exists a partition of unity $\{\omega_j\}$ subordinate to the cover $\{\mathscr{V}_j\}$ with $D^\alpha \omega_j$ bounded on $\mathbb{R}^n$ uniformly in $j$ for any given $\alpha$. The reader may find it interesting to construct, by the above techniques, extension operators for domains not covered by the above theorems, for example, quadrants, strips, rectangular boxes, and smooth images of these.

It might be noted here that although the Calderón extension theorem (Theorem 4.32) is proved by methods quite different from the reflection approach used above, nevertheless the proof does make use of a partition of unity in the same way as does that of Theorem 4.26. Accordingly, the above considerations also apply to it. The theorem is proved under a strengthened form of uniform cone condition that reduces to the uniform cone condition of Section 4.4 if $\Omega$ has bounded boundary.

Before proceeding to Calderón's theorem we present two well-known results on convolution operators that will be needed for the proof. The first is a special case of a theorem of W. H. Young.

**4.30 THEOREM** (*Young*) Let $1 \le p < \infty$ and suppose that $u \in L^1(\mathbb{R}^n)$ and $v \in L^p(\mathbb{R}^n)$. Then the convolution products

$$u * v(x) = \int_{\mathbb{R}^n} u(x-y)\,v(y)\,dy, \qquad v * u(x) = \int_{\mathbb{R}^n} v(x-y)\,u(y)\,dy$$

are well defined and equal for almost all $x \in \mathbb{R}^n$. Moreover, $u * v \in L^p(\mathbb{R}^n)$ and

$$\|u * v\|_p \le \|u\|_1 \|v\|_p. \tag{33}$$

PROOF The proof is a simple consequence of Fubini's theorem if $p = 1$, so we assume $1 < p < \infty$. Let $w \in L^{p'}(\Omega)$. Then

$$\begin{aligned}
&\left|\int_{\mathbb{R}^n} w(x) \int_{\mathbb{R}^n} u(x-y)\,v(y)\,dy\,dx\right| \\
&\quad = \left|\int_{\mathbb{R}^n} w(x) \int_{\mathbb{R}^n} u(y)\,v(x-y)\,dy\,dx\right| \\
&\quad \le \int_{\mathbb{R}^n} |u(y)|\,dy \int_{\mathbb{R}^n} |v(x-y)|\,|w(x)|\,dx \\
&\quad \le \int_{\mathbb{R}^n} |u(y)|\,dy \left\{\int_{\mathbb{R}^n} |v(x-y)|^p\,dx\right\}^{1/p} \left\{\int_{\mathbb{R}^n} |w(x)|^{p'}\,dx\right\}^{1/p'} \\
&\quad = \|u\|_1 \|v\|_p \|w\|_{p'}.
\end{aligned}$$

Since $w$ may be chosen so as to vanish nowhere, it follows that $u * v(x)$ and $v * u(x)$ must be finite a.e. Moreover, the functional

$$F_{u*v}(w) = \int_{\mathbb{R}^n} u * v(x)\,w(x)\,dx$$

belongs to $[L^{p'}(\mathbb{R}^n)]'$ and so by Theorem 2.33 there exists $\lambda \in L^p(\mathbb{R}^n)$ with $\|\lambda\|_p \le \|u\|_1 \|v\|_p$ such that

$$\int_{\mathbb{R}^n} \lambda(x)\,w(x)\,dx = \int_{\mathbb{R}^n} u * v(x)\,w(x)\,dx$$

for every $w \in L^{p'}(\mathbb{R}^n)$. Hence $\lambda = u * v \in L^p(\mathbb{R}^n)$ and (33) is proved. The equality of $u * v$ and $v * u$ is elementary. ∎

The following theorem is a special case, suitable for our purposes, of a well-known inequality of Calderón and Zygmund [16] for convolutions involving kernels with nonintegrable singularities. The proof, which is rather lengthy and may be found in many sources (e.g., Stein and Weiss [65]), is omitted here. Neither the inequality nor the extension theorem based on it will be required hereafter in this monograph.

Let $B_R = \{x \in \mathbb{R}^n : |x| \le R\}$, $S_R = \{x \in \mathbb{R}^n : |x| = R\}$, and let $d\sigma_R$ be the area element [Lebesgue $(n-1)$-measure] on $S_R$. A function $g$ is said to be *homogeneous of degree* $\mu$ on $B_R \sim \{0\}$ if $g(tx) = t^\mu g(x)$ for all $x \in B_R \sim \{0\}$ and $0 < t \le 1$.

**4.31 THEOREM** (*The Calderón–Zygmund inequality*) Let

$$g(x) = G(x)|x|^{-n},$$

where

(i) $G$ is bounded on $\mathbb{R}^n \sim \{0\}$ and has compact support,
(ii) $G$ is homogeneous of degree 0 on $B_R \sim \{0\}$ for some $R > 0$, and
(iii) $\int_{S_R} G(x)\, d\sigma_R = 0$.

If $1 < p < \infty$ and $u \in L^p(\mathbb{R}^n)$, then the principal value convolution integral

$$u * g(x) = \lim_{\varepsilon \to 0+} \int_{\mathbb{R}^n \sim B_\varepsilon} u(x-y)\, g(y)\, dy$$

exists for almost all $x \in \mathbb{R}^n$, and there exists a constant $K = K(G, p)$ such that for all such $u$

$$\|u * g\|_p \le K \|u\|_p.$$

Conversely, if $G$ satisfies (i) and (ii) and if $u * g$ exists for all $u \in C_0^\infty(\mathbb{R}^n)$, then $G$ satisfies (iii).

**4.32 THEOREM** (*The Calderón extension theorem*) Let $\Omega$ be a domain in $\mathbb{R}^n$ having the uniform cone property (Section 4.4) modified as follows:

(i) the open cover $\{U_j\}$ of bdry $\Omega$ is required to be finite ,and
(ii) the sets $U_j$ are not required to be bounded.

Then for any $m \in \{1, 2, \ldots\}$ and any $p$, $1 < p < \infty$, there exists a simple $(m, p)$-extension operator $E = E(m, p)$ for $\Omega$.

PROOF Let $\{U_1, U_2, \ldots, U_N\}$ be the open cover of bdry $\Omega$ given by the uniform cone property, and let $U_0$ be an open subset of $\Omega$ bounded away from bdry $\Omega$ such that $\Omega \subset \bigcup_{j=0}^N U_j$. [Such $U_0$ exists by condition (ii), Section 4.4.] Let $\omega_0, \omega_1, \ldots, \omega_N$ be a $C^\infty$-partition of unity for $\Omega$ with $\operatorname{supp} \omega_j \subset U_j$. For $1 \le j \le N$ we shall define operators $E_j$ so that if $u \in W^{m,p}(\Omega)$, then $E_j u \in W^{m,p}(\mathbb{R}^n)$ and satisfies

$$E_j u = u \quad \text{in} \quad U_j \cap \Omega,$$

$$\|E_j u\|_{m,p,\mathbb{R}^n} \le K_{m,p,j} \|u\|_{m,p,\Omega}. \tag{34}$$

The desired extension operator is then clearly given by

$$Eu = \omega_0 u + \sum_{j=1}^{N} \omega_j E_j u.$$

We shall write $x \in \mathbb{R}^n$ in the polar coordinate form $x = \rho\sigma$ where $\rho \geq 0$ and $\sigma$ is a unit vector. Let $C_j$, the cone associated with $U_j$ in the description of the uniform cone property, have vertex at 0. Let $\phi_j$ be a function defined in $\mathbb{R}^n \sim \{0\}$ and satisfying

(i) $\phi_j(x) \geq 0$ for all $x \neq 0$,
(ii) $\operatorname{supp} \phi_j \subset -C_j \cup \{0\}$,
(iii) $\phi_j \in C^\infty(\mathbb{R}^n \sim \{0\})$,
(iv) for some $\varepsilon > 0$, $\phi_j$ is homogeneous of degree $m-n$ in $B_\varepsilon \sim \{0\}$.

Now $\rho^{n-1}\phi_j$ is homogeneous of degree $m-1 \geq 0$ on $B_\varepsilon \sim \{0\}$ and so the function $\psi_j(x) = (\partial/\partial p)^m[\rho^{n-1}\phi_j(x)]$ vanishes on $B_\varepsilon \sim \{0\}$. Hence $\psi_j$, extended to be zero at $x = 0$, belongs to $C_0^\infty(-C_j)$. Define

$$E_j u(y) = K_j\left\{(-1)^m \int_S \int_0^\infty \phi_j(\rho\sigma)\rho^{n-1}\left(\frac{\partial}{\partial\rho}\right)^m u(y-\rho\sigma)\, d\rho\, d\sigma - \int_S \int_0^\infty \psi_j(\rho\sigma) u(y-\rho\sigma)\, d\rho\, d\sigma\right\} \tag{35}$$

where $\int_S \cdot\, d\sigma$ denotes integration over the unit sphere, and the constant $K_j$ will be determined shortly. If $y \in U_j \cap \Omega$, then, assuming for the moment that $u \in C^\infty(\Omega)$, we have by condition (iii) of Section 4.4 that $u(y-\rho\sigma)$ is infinitely differentiable for $\rho\sigma \in \operatorname{supp} \phi_j$. Now integration by parts $m$ times yields

$$\begin{aligned}
&(-1)^m \int_0^\infty \rho^{n-1}\phi_j(\rho\sigma)\left(\frac{\partial}{\partial\rho}\right)^m u(y-\rho\sigma)\, d\rho \\
&= \sum_{k=0}^{m-1} (-1)^{m-k}\left(\frac{\partial}{\partial\rho}\right)^k [\rho^{n-1}\phi_j(\rho\sigma)]\left(\frac{\partial}{\partial\rho}\right)^{m-k-1} u(y-\rho\sigma)\Bigg|_{\rho=0}^{\rho=\infty} \\
&\quad + \int_0^\infty \left(\frac{\partial}{\partial\rho}\right)^m [\rho^{n-1}\phi_j(\rho\sigma)]\, u(y-\rho\sigma)\, d\rho \\
&= \left(\frac{\partial}{\partial\rho}\right)^{m-1} [\rho^{n-1}\phi_j(\rho\sigma)]\Bigg|_{\rho=0} u(y) + \int_0^\infty \psi_j(\rho\sigma) u(y-\rho\sigma)\, d\rho.
\end{aligned}$$

Hence

$$E_j u(y) = K_j u(y) \int_S \left(\frac{\partial}{\partial\rho}\right)^{m-1} [\rho^{n-1}\phi_j(\rho\sigma)]\Bigg|_{\rho=0} d\sigma.$$

Since $(\partial/\partial\rho)^{m-1}[\rho^{n-1}\phi_j(\rho\sigma)]$ is homogeneous of degree zero near 0, the above integral does not vanish if $\phi_j$ is not identically zero. Hence $K_j$ can be chosen so that $E_j u(y) = u(y)$ for $y \in U_j \cap \Omega$ and all $u \in C^\infty(\Omega)$. Since $C^\infty(\Omega)$ is dense in $W^{m,p}(\Omega)$ we have $E_j u(y) = u(y)$ a.e. in $U_j \cap \Omega$ for every $u \in W^{m,p}(\Omega)$. It remains, therefore, to show that (34) holds, that is, that

$$\|D^\alpha E_j u\|_{0,p,\mathbb{R}^n} \le K_\alpha \|u\|_{m,p,\Omega}$$

for any $\alpha$, $|\alpha| \le m$.

The last integral in (35) is of the form $\theta_j * u(y)$, where $\theta_j(x) = \psi_j(x)|x|^{1-n}$. Since $\theta_j \in L^1(\mathbb{R}^n)$ and has compact support we obtain via Young's theorem 4.30 and a suitable approximation of $u$ by smooth functions,

$$\|D^\alpha(\theta_j * u)\|_{0,p,\mathbb{R}^n} = \|\theta_j * (D^\alpha u)\|_{0,p,\mathbb{R}^n} \le \|\theta_j\|_{0,1,\mathbb{R}^n} \|D^\alpha u\|_{0,p,\Omega}.$$

It now remains to be shown that the first integral in (35) defines a bounded map from $W^{m,p}(\Omega)$ into $W^{m,p}(\mathbb{R}^n)$. Since $(\partial/\partial\rho)^m = \sum_{|\alpha|=m}(m!/\alpha!)\sigma^\alpha D^\alpha$ we obtain

$$\int_S \int_0^\infty \phi_j(\rho\sigma)\,\rho^{n-1}\left(\frac{\partial}{\partial\rho}\right)^m u(y-\rho\sigma)\,d\rho\,d\sigma$$

$$= \sum_{|\alpha|=m} \frac{m!}{\alpha!} \int_{\mathbb{R}^n} \phi_j(x)\,D_x{}^\alpha u(y-x)\,\sigma^\alpha\,dx$$

$$= \sum_{|\alpha|=m} \xi_\alpha * D^\alpha u,$$

where $\xi_\alpha = (-1)^{|\alpha|}(m!/\alpha!)\sigma^\alpha\phi_j$ is homogeneous of degree $m-n$ in $B_\varepsilon \sim \{0\}$ and belongs to $C^\infty(\mathbb{R}^n \sim \{0\})$. It is now clearly sufficient to show that for any $\beta$, $|\beta| \le m$

$$\|D^\beta(\xi_\alpha * v)\|_{0,p,\mathbb{R}^n} \le K_{\alpha,\beta}\|v\|_{0,p,\Omega}. \tag{36}$$

If $|\beta| \le m-1$, then $D^\beta\xi_\alpha$ is homogeneous of degree not exceeding $1-n$ in $B_\varepsilon \sim \{0\}$ and so belong to $L^1(\mathbb{R}^n)$. Inequality (36) now follows by a Young's theorem argument. Thus we need consider only the case $|\beta| = m$, in which we write $D^\beta = (\partial/\partial x_i)D^\gamma$ for some $\gamma$, $|\gamma| = m-1$, and some $i$, $1 \le i \le n$. Suppose, for the moment, that $v \in C_0{}^\infty(\Omega)$. Then we may write

$$D^\beta(\xi_\alpha * v)(x) = [D^\gamma\xi_\alpha] * \left[\left(\frac{\partial}{\partial x_i}\right)v\right](x) = \int_{\mathbb{R}^n} D_i v(x-y)\,D^\gamma\xi_\alpha(y)\,dy$$

$$= \lim_{\delta\to 0+} \int_{\mathbb{R}^n \sim B_\delta} D_i v(x-y)\,D^\gamma\xi_\alpha(y)\,dy.$$

We now integrate by parts in the last integral to free $v$ and obtain $D^\beta\xi_\alpha$ under the integral. The integrated term is a surface integral over the sphere $S_\delta$ of the product of $v(x-\cdot)$ and a function homogeneous of degree $1-n$ near zero.

This surface integral must therefore tend to $Kv(x)$ as $\delta \to 0+$, for some constant $K$. Noting that $D_i v(x-y) = -(\partial/\partial y_i)\, v(x-y)$, we now have

$$D^\beta(\xi_\alpha * v)(x) = \lim_{\delta\to 0+} \int_{\mathbb{R}^n} v(x-y)\, D^\beta \xi_\alpha(y)\, dy + Kv(x).$$

Now $D^\beta \xi_\alpha$ is homogeneous of degree $-n$ near the origin and so, by the last assertion of Theorem 4.31, $D^\beta \xi_\alpha$ satisfies all the conditions for the singular kernel $g$ of that theorem. Since $p > 1$ we have for any $v \in L^p(\Omega)$ (regarded as being identically zero outside $\Omega$)

$$\|D^\beta \xi_\alpha * v\|_{0,p,\mathbb{R}^n} \le K_{\alpha,\beta} \|v\|_{0,p,\Omega}.$$

This completes the proof. ∎

# V

# Imbeddings of $W^{m,p}(\Omega)$

## The Sobolev Imbedding Theorem

**5.1** It is primarily the imbedding characteristics of Sobolev spaces that render these spaces so useful in analysis, especially in the study of differential and integral operators. The most important of the imbedding properties of the spaces $W^{m,p}(\Omega)$ are usually lumped together in a single theorem referred to as the *Sobolev imbedding theorem*. The core results are due to Sobolev [63] but our statement (Theorem 5.4) includes refinements due to others, in particular to Morrey [47] and Gagliardo [24].

Most of the imbedding results hold for domains $\Omega$ in $\mathbb{R}^n$ having the cone property but otherwise unrestricted; some imbeddings however require the strong local Lipschitz property. Specifically no imbedding of $W^{m,p}(\Omega)$ into a space of uniformly continuous functions on $\Omega$ is possible under only the cone property, as can be seen by considering the example given in the second paragraph of Section 3.17.

**5.2** The Sobolev imbedding theorem asserts the existence of imbeddings of $W^{m,p}(\Omega)$ into spaces of the following types:

(i) $W^{j,q}(\Omega)$, $j \le m$, and in particular $L^q(\Omega)$.

(ii) $C_B{}^j(\Omega) = \{u \in C^j(\Omega) : D^\alpha u \text{ is bounded on } \Omega \text{ for } |\alpha| \le j\}$. This space is larger than $C^j(\overline{\Omega})$ in that its elements need not be uniformly continuous on

$\Omega$. However, $C_B{}^j(\Omega)$ is a Banach space under the norm

$$\|u; C_B{}^j(\Omega)\| = \max_{0 \le |\alpha| \le j} \sup_{x \in \Omega} |D^\alpha u(x)|.$$

(iii) $C^{j,\lambda}(\overline{\Omega})$ (see Section 1.27) and in particular $C^j(\overline{\Omega})$.

(iv) $W^{j,q}(\Omega^k)$, and in particular $L^q(\Omega^k)$. Here $\Omega^k$ denotes the intersection of $\Omega$ with a $k$-dimensional plane in $\mathbb{R}^n$, considered as a domain in $\mathbb{R}^k$.

Since elements of $W^{m,p}(\Omega)$ are, strictly speaking, not functions defined everywhere on $\Omega$ but rather equivalence classes of such functions defined and equal up to sets of measure zero, we must clarify what is meant by an imbedding of $W^{m,p}(\Omega)$ into a space of type (ii)–(iv). In the case of (ii) or (iii) what is intended is that the "equivalence class" $u \in W^{m,p}(\Omega)$ should contain an element belonging to the continuous function space that is target of the imbedding, and bounded in that space by a constant times $\|u\|_{m,p,\Omega}$. Hence, for example, $W^{m,p}(\Omega) \to C^j(\overline{\Omega})$ means that each $u \in W^{m,p}(\Omega)$ can, when considered as a function, be redefined on a set of zero measure in $\Omega$ in such a way that the modified function $\tilde{u}$ [which equals $u$ in $W^{m,p}(\Omega)$] belongs to $C^j(\overline{\Omega})$ and satisfies $\|\tilde{u}; C^j(\overline{\Omega})\| \le K\|u\|_{m,p,\Omega}$ with $K$ independent of $u$.

Even more care is necessary in interpreting the imbedding $W^{m,p}(\Omega) \to W^{j,q}(\Omega^k)$ where $k < n$. Each element $u \in W^{m,p}(\Omega)$ is, by Theorem 3.16, a limit in that space of a sequence $\{u_n\}$ of functions in $C^\infty(\Omega)$. The functions $u_n$ have traces on $\Omega^k$ belonging to $C^\infty(\Omega^k)$. The above imbedding signifies that these traces converge in $W^{j,q}(\Omega^k)$ to a function $\tilde{u}$ satisfying $\|\tilde{u}\|_{j,q,\Omega^k} \le K\|u\|_{m,p,\Omega}$ with $K$ independent of $u$.

Let us note as a point of interest (though of no use to us later) that the imbedding $W^{m,p}(\Omega) \to W^{j,q}(\Omega)$ is equivalent to the simple containment $W^{m,p}(\Omega) \subset W^{j,q}(\Omega)$. Certainly the former implies the latter. To see the converse suppose $W^{m,p}(\Omega) \subset W^{j,q}(\Omega)$ and let $I$ be the linear operator defined on $W^{m,p}(\Omega)$ into $W^{j,q}(\Omega)$ by $Iu = u$. If $u_n \to u$ in $W^{m,p}(\Omega)$ [and hence in $L^p(\Omega)$] and $Iu_n \to v$ in $W^{j,q}(\Omega)$ [and hence in $L^q(\Omega)$], then, passing to a subsequence if necessary, we have by Corollary 2.11 that $u_n(x) \to u(x)$ a.e. in $\Omega$ and $u_n(x) = Iu_n(x) \to v(x)$ a.e. in $\Omega$. Thus $u(x) = v(x)$ a.e. in $\Omega$, that is, $Iu = v$, and $I$ is continuous by the closed graph theorem of functional analysis.

**5.3** Let $\Omega$ be a domain in $\mathbb{R}^n$ having the cone property specified by a certain finite cone $C$ (see Section 4.3). $C$ may be regarded as the intersection of an infinite cone $C^*$ having the same vertex as $C$ with a ball $B$ centered at that vertex. By the *height* of $C$ we mean the radius of $B$. By the *opening* of $C$ we mean the surface area [$(n-1)$-measure] of the intersection of $C^*$ with the sphere of unit radius having center at the vertex of $C$. These geometric parameters are clearly invariant under rigid transformations of $C$.

In asserting that an imbedding of the form

$$W^{m,p}(\Omega) \to X \tag{1}$$

(where $X$ is a Banach space of functions defined over $\Omega$) holds for $\Omega$ having the cone property, it is intended that an *imbedding constant* for (1), that is, a constant $K$ for which the inequality

$$\|u; X\| \le K\|u\|_{m,p,\Omega}$$

is satisfied for all $u \in W^{m,p}(\Omega)$, can be chosen to depend on $\Omega$ only through the dimension $n$ and various such parameters of the cone $C$ which are invariant under rigid motions of $C$.

**5.4 THEOREM** (*The Sobolev imbedding theorem*) Let $\Omega$ be a domain in $\mathbb{R}^n$ and let $\Omega^k$ be the $k$-dimensional domain obtained by intersecting $\Omega$ with a $k$-dimensional plane in $\mathbb{R}^n$, $1 \le k \le n$. (Thus $\Omega^n \equiv \Omega$.) Let $j$ and $m$ be non-negative integers and let $p$ satisfy $1 \le p < \infty$.

**PART I** If $\Omega$ has the cone property, then there exist the following imbeddings:

CASE A Suppose $mp < n$ and $n - mp < k \le n$. Then

$$W^{j+m,p}(\Omega) \to W^{j,q}(\Omega^k), \qquad p \le q \le kp/(n-mp), \tag{2}$$

and in particular,

$$W^{j+m,p}(\Omega) \to W^{j,q}(\Omega), \qquad p \le q \le np/(n-mp), \tag{3}$$

or

$$W^{m,p}(\Omega) \to L^q(\Omega), \qquad p \le q \le np/(n-mp). \tag{4}$$

Moreover, if $p = 1$, so that $m < n$, imbedding (2) also exists for $k = n - m$.

CASE B Suppose $mp = n$. Then for each $k$, $1 \le k \le n$,

$$W^{j+m,p}(\Omega) \to W^{j,q}(\Omega^k), \qquad p \le q < \infty, \tag{5}$$

so that in particular

$$W^{m,p}(\Omega) \to L^q(\Omega), \qquad p \le q < \infty. \tag{6}$$

Moreover, if $p = 1$ so that $m = n$, imbeddings (5) and (6) exist with $q = \infty$ as well; in fact,

$$W^{j+n,1}(\Omega) \to C_B^j(\Omega). \tag{7}$$

CASE C Suppose $mp > n$. Then

$$W^{j+m,p}(\Omega) \to C_B^j(\Omega). \tag{8}$$

**PART II** If $\Omega$ has the strong local Lipschitz property, then Case C of Part I can be refined as follows:

CASE C′ Suppose $mp > n > (m-1)p$. Then

$$W^{j+m,p}(\Omega) \to C^{j,\lambda}(\overline{\Omega}), \qquad 0 < \lambda \le m - (n/p). \tag{9}$$

CASE C″ Suppose $n = (m-1)p$. Then

$$W^{j+m,p}(\Omega) \to C^{j,\lambda}(\overline{\Omega}), \qquad 0 < \lambda < 1. \tag{10}$$

Also, if $n = m-1$ and $p = 1$, then (10) holds for $\lambda = 1$ as well.

**PART III** All the conclusions of Parts I and II are valid for *arbitrary* domains provided the $W$-spaces undergoing imbedding are replaced with the corresponding $W_0$-spaces.

**5.5 REMARKS** (1) Imbeddings (2)–(8) are essentially due to Sobolev [62, 63] whose original proof did not, however, cover the cases $q = kp/(n-mp)$ in (2), or $q = np/(n-mp)$ in (3) and (4). Imbeddings (9) and (10) find their origins in the work of Morrey [47].

(2) Imbeddings of type (2) and (5) involving traces of functions on planes of lower dimension can be extended in a reasonable manner to apply to traces on more general smooth manifolds. For example, see Theorem 5.22.

(3) Part III of the theorem is an immediate consequence of Parts I and II applied to $\mathbb{R}^n$ because, by Lemma 3.22, the operator of zero extension of functions outside $\Omega$ maps $W_0^{m,p}(\Omega)$ isometrically into $W^{m,p}(\mathbb{R}^n)$.

(4) Suppose that all the conclusions of the imbedding theorem have been proven for $\Omega = \mathbb{R}^n$. It then follows that they must also hold for any domain $\Omega$ satisfying the requirements of the Calderón extension theorem 4.32. For example, if $W^{m,p}(\mathbb{R}^n) \to L^q(\mathbb{R}^n)$, and if $E$ is an $(m,p)$-extension operator for $\Omega$, then for any $u \in W^{m,p}(\Omega)$ we have

$$\|u\|_{0,q,\Omega} \le \|Eu\|_{0,q,\mathbb{R}^n} \le K_1 \|Eu\|_{m,p,\mathbb{R}^n} \le K_1 K_2 \|u\|_{m,p,\Omega}$$

with $K_1$ and $K_2$ independent of $u$. We shall not, however, prove the imbedding theorem by such extension arguments.

(5) It is sufficient to establish each of the imbeddings (2), (3), (5), (7)–(10) for the special case $j = 0$. For example, if $W^{m,p}(\Omega) \to L^q(\Omega)$ has been established, then for any $u \in W^{j+m,p}(\Omega)$ we have $D^\alpha u \in W^{m,p}(\Omega)$ for $|\alpha| \le j$, whence $D^\alpha u \in L^q(\Omega)$; thus $u \in W^{j,q}(\Omega)$; and

$$\|u\|_{j,q} = \left( \sum_{|\alpha| \le j} \|D^\alpha u\|_{0,q}^q \right)^{1/q}$$

$$\le K_1 \left( \sum_{|\alpha| \le j} \|D^\alpha u\|_{m,p}^p \right)^{1/p} \le K_2 \|u\|_{j+m,p}.$$

Accordingly, we will always specialize $j = 0$ in the proofs.

(6) If $\Omega^k$ (or $\Omega$) has finite volume, it follows by Theorem 2.8 that imbeddings (2)–(6) hold for $1 \le q < p$ in addition to the values of $q$ asserted in the theorem. It will be shown later (Section 6.38) that no imbedding of the form $W^{m,p}(\Omega) \to L^q(\Omega)$ where $q < p$ is possible unless $\Omega$ has finite volume.

## Proof of the Imbedding Theorem

**5.6** The proof given here is due to Gagliardo [24]. Though it is rather lengthy, the techniques involved are quite elementary, being based on little more than simple calculus combined with astute applications of Hölder's inequality. Moreover, Gagliardo's proof establishes the imbedding theorem in the greatest possible generality and is capable of generalization to produce imbedding results for some domains not having the cone property (see Theorems 5.35–5.37).

The proof is carried out in a chain of auxiliary lemmas. In each such lemma constants $K_1, K_2, \ldots$ appearing in the proof are allowed to depend on the same parameters as the constant $K$ referred to in the statement of the lemma.

**5.7 LEMMA** Let

$$R = \{x \in \mathbb{R}^n : a_i < x_i < b_i;\ 1 \le i \le n\}$$

and

$$R' = \{x' = (x_1, \ldots, x_{n-1}) \in \mathbb{R}^{n-1} : a_i < x_i < b_i;\ 1 \le i \le n-1\}$$

be bounded open rectangles in $\mathbb{R}^n$ and $\mathbb{R}^{n-1}$, respectively. If $a_n < \zeta < b_n$ and $p \ge 1$, then for every $u \in C^\infty(R) \cap W^{1,p}(R)$ we have

$$\|u(\cdot, \zeta)\|_{0,p,R'} \le K \|u\|_{1,p,R} \tag{11}$$

where $K = K(p, b_n - a_n)$. Thus the trace mapping $u \to u(\cdot, \zeta)$ extends to an imbedding of $W^{1,p}(R)$ into $L^p(R_\zeta^{n-1})$, where $R_\zeta^{n-1} = R \cap \{x \in \mathbb{R}^n : x_n = \zeta\}$.

PROOF By Theorem 3.18, $C^\infty(\bar{R})$ is dense in $W^{1,p}(R)$ so we may assume $u \in C^\infty(\bar{R})$. Thus $\int_{R'} |u(x', \cdot)|^p \, dx'$ belongs to $C^\infty([a_n, b_n])$ and by the mean value theorem for integrals we have

$$\|u\|_{0,p,R}^p = \int_{a_n}^{b_n} \left( \int_{R'} |u(x', x_n)|^p \, dx' \right) dx_n = (b_n - a_n) \int_{R'} |u(x', \sigma)|^p \, dx'$$

for some $\sigma \in [a_n, b_n]$. Now

$$\begin{aligned} |u(x', \zeta)|^p &= \left| u(x', \sigma) + \int_\sigma^\zeta D_n u(x', t) \, dt \right|^p \\ &\le 2^{p-1} \left[ |u(x', \sigma)|^p + |\zeta - \sigma|^{p-1} \int_\sigma^\zeta |D_n u(x', t)|^p \, dt \right] \end{aligned}$$

by Hölder's inequality. Integration over $R'$ leads to

$$\|u(\cdot,\zeta)\|_{0,p,R'}^p \leq 2^{p-1}[\|u(\cdot,\sigma)\|_{0,p,R'}^p + (b_n-a_n)^{p-1}\|D_n u\|_{0,p,R}^p]$$
$$\leq 2^{p-1}[(b_n-a_n)^{-1}\|u\|_{0,p,R}^p + (b_n-a_n)^{p-1}\|D_n u\|_{0,p,R}^p]$$

which yields (11) with $K=[2^{p-1}\max((b_n-a_n)^{-1},(b_n-a_n)^{p-1})]^{1/p}$. We note that $K$ depends continuously on $b_n-a_n$ but may tend to infinity if $b_n-a_n$ tends to zero or infinity. ∎

**5.8 LEMMA** Let $R$ be as in the previous lemma. Then

$$W^{n,1}(R) \to C(\bar{R}).$$

The imbedding constant depends only on $n$ and the dimensions of $R$.

PROOF Let $x$ be any point of $R$, and let $R'$ be as in the previous lemma. If $u \in C^\infty(\bar{R})$ and $|\alpha| \leq n-1$, we have by that lemma that

$$\|D^\alpha u(\cdot,x_n)\|_{0,1\,R'} \leq K_1\|D^\alpha u\|_{1,1,R}.$$

Thus

$$\|u(\cdot,x_n)\|_{n-1,1,R'} \leq K_2\|u\|_{n,1,R}$$

with $K_2$ depending on $b_n-a_n$. Iteration of this argument over successively lower-diemnsional rectangles leads to

$$\|u(\cdot,x_2,x_3,\dots,x_n)\|_{1,1,(a_1,b_1)} \leq K_3\|u\|_{n,1,R}$$

with $K_3$ depending on $b_j-a_j$, $2\leq j\leq n$. By the mean value theorem for integrals there exists $\sigma\in[a_1,b_1]$ such that

$$\|u(\cdot,x_2,\dots,x_n)\|_{0,1,(a_1,b_1)} = (b_1-a_1)|u(\sigma,x_2,\dots,x_n)|.$$

Hence

$$|u(x)| \leq |u(\sigma,x_2,\dots,x_n)| + \int_\sigma^{x_1}|D_1 u(t,x_2,\dots,x_n)|\,dt$$
$$\leq [1/(b_1-a_1)]\|u(\cdot,x_2,\dots,x_n)\|_{0,1,(a_1,b_1)}$$
$$+\|D_1 u(\cdot,x_2,\dots,x_n)\|_{0,1,(a_1,b_1)}$$
$$\leq K\|u\|_{n,1,R}. \tag{12}$$

Now suppose $u\in W^{n,1}(R)$. By Theorem 3.18, $u$ is the limit in $W^{n,1}(R)$ of a sequence of functions belonging to $C^\infty(\bar{R})$. It follows from (12) that this sequence converges uniformly on $\bar{R}$ to a function $\tilde{u}\in C(\bar{R})$. Since $\tilde{u}(x)=u(x)$ a.e. in $R$, the lemma is proved. ∎

We now turn our attention to more general domains. The following lemma of Gagliardo, which is essentially combinatorial in nature, is the foundation on which his proof of the imbedding theorem rests.

**5.9 LEMMA** Let $\Omega$ be a domain in $\mathbb{R}^n$ where $n \geq 2$. Let $k$ be an integer satisfying $1 \leq k \leq n$, and let $\kappa = (\kappa_1, \kappa_2, \ldots, \kappa_n)$ denote a $k$-tuple of integers satisfying $1 \leq \kappa_1 < \kappa_2 < \cdots < \kappa_k \leq n$. Let $S$ be the set of all $\binom{n}{k}$ such $k$-tuples. Also, given $x \in \mathbb{R}^n$, let $x_\kappa$ denote the point $(x_{\kappa_1}, \ldots, x_{\kappa_k}) \in \mathbb{R}^k$; $dx_\kappa = dx_{\kappa_1} \cdots dx_{\kappa_k}$.

For given $\kappa \in S$ let $E_\kappa$ be the $k$-dimensional plane in $\mathbb{R}^n$ spanned by the coordinate axes corresponding to the components of $x_\kappa$:

$$E_\kappa = \{x \in \mathbb{R}^n : x_i = 0 \text{ if } i \notin \kappa\};$$

and for any set $G \subset \mathbb{R}^n$ let $G_\kappa$ be the projection of $G$ onto $E_\kappa$; in particular

$$\Omega_\kappa = \{x \in E_\kappa : \exists y \in \Omega \text{ such that } y_\kappa = x_\kappa\}.$$

Let $F_\kappa$ be a function depending on the $k$ components of $x_\kappa$ and belonging to $L^\lambda(\Omega_\kappa)$, where $\lambda = \binom{n-1}{k-1}$. Then the function $F$ defined on $\Omega$ by

$$F(x) = \prod_{\kappa \in S} F_\kappa(x_\kappa)$$

belongs to $L^1(\Omega)$, and $\|F\|_{1,\Omega} \leq \prod_{\kappa \in S} \|F_\kappa\|_{\lambda, \Omega_\kappa}$, that is,

$$\left[\int_\Omega |F(x)|\, dx\right]^\lambda \leq \prod_{\kappa \in S} \int_{\Omega_\kappa} |F_\kappa(x_\kappa)|^\lambda\, dx_\kappa. \tag{13}$$

PROOF For $\kappa \in S$ and $\xi_\kappa \in \mathbb{R}^k$ let $\Omega(\xi_\kappa)$ denote the $k$-dimensional plane section of $\Omega$ by the plane $x_\kappa = \xi_\kappa$:

$$\Omega(\xi_\kappa) = \{x \in \Omega : x_\kappa = \xi_\kappa\}.$$

We establish (13) by induction on $n$, and so consider first the case $n = 2$. We may also suppose that $k = 1$ since, for any $n$, the subcase $k = n$ of (13) is trivial. For $n = 2$, $k = 1$, we have $\lambda = 1$ and $S$ has only two elements, $\kappa = 1$ and $\kappa = 2$. Hence

$$\begin{aligned}
\int_\Omega |F_1(x_1) F_2(x_2)|\, dx_1\, dx_2 &= \int_{\Omega_1} dx_1 \int_{\Omega(x_1)} |F_1(x_1) F_2(x_2)|\, dx_2 \\
&= \int_{\Omega_1} |F_1(x_1)|\, dx_1 \int_{(\Omega(x_1))_2} |F_2(x_2)|\, dx_2 \\
&\leq \int_{\Omega_1} |F_1(x_1)|\, dx_1 \int_{\Omega_2} |F_2(x_2)|\, dx_2
\end{aligned}$$

since clearly $(\Omega(x_1))_2 \subset \Omega_2$ for any $x_1$. This is (13) for the case being considered. (A similar calculation will yield (13) for arbitrary $n$ and $k = 1$.)

Now we assume that (13) has been established for $n = N-1$. We consider the case $n = N$ and, as noted above, may assume $2 \le k \le N-1$. Thus $\lambda = \binom{N-1}{k-1}$. Let $\mu = \binom{N-2}{k-1}$ and $\nu = \binom{N-2}{k-2}$. The integrand on the left side of (13) is a product of $\binom{N}{k}$ factors $|F_\kappa|$ each belonging to the corresponding space $L^\lambda(\Omega_\kappa)$. Exactly $\binom{N-1}{k}$ of these factors, say those corresponding to $\kappa \in A \subset S$, are independent of $x_n$. It follows from applying the induction hypothesis over the $(N-1)$-dimensional domain $\Omega(x_N)$ and noting that $(\Omega(x_N))_\kappa \subset \Omega_\kappa$ that

$$\int_{\Omega(x_N)} \prod_{\kappa \in A} |F_\kappa(x_\kappa)|^{\lambda/\mu}\, dx_1 \cdots dx_{N-1} \le \prod_{\kappa \in A} \left[ \int_{(\Omega(x_N))_\kappa} |F_\kappa(x_\kappa)|^\lambda\, dx_\kappa \right]^{1/\mu}$$
$$\le \prod_{\kappa \in A} \left[ \int_{\Omega_\kappa} |F_\kappa(x_\kappa)|^\lambda\, dx_\kappa \right]^{1/\mu}. \tag{14}$$

The remaining $\binom{N}{k} - \binom{N-1}{k} = \lambda$ factors $|F_\kappa|$ depend on $x_N$, and so when restricted to $\Omega(x_N)$ depend on only $k-1$ variables. Applying the induction hypothesis over $\Omega(x_N)$ again, but this time with $k-1$ in place of $k$, we obtain

$$\int_{\Omega(x_N)} \prod_{\kappa \in S \sim A} |F_\kappa(x_\kappa)|^{\lambda/\nu}\, dx_1 \cdots dx_{N-1}$$
$$\le \prod_{\kappa \in S \sim A} \left[ \int_{(\Omega(x_N))_\kappa} |F_\kappa(x_\kappa)|^\lambda\, dx_{\kappa_1} \cdots dx_{\kappa_{k-1}} \right]^{1/\nu}. \tag{15}$$

Now $\mu + \nu = \lambda$ and so by Hölder's inequality, and (14) and (15),

$$\int_{\Omega(x_N)} \prod_{\kappa \in S} |F_\kappa(x_\kappa)|\, dx_1 \cdots dx_{N-1}$$
$$\le \prod_{\kappa \in A} \left[ \int_{\Omega_\kappa} |F_\kappa(x_\kappa)|^\lambda\, dx_\kappa \right]^{1/\lambda}$$
$$\times \prod_{\kappa \in S \sim A} \left[ \int_{(\Omega(x_N))_\kappa} |F_\kappa(x_\kappa)|^\lambda\, dx_{\kappa_1} \cdots dx_{\kappa_{k-1}} \right]^{1/\lambda}. \tag{16}$$

Since $S \sim A$ contains $\lambda$ elements we obtain by (the several function form of) Hölder's inequality that

$$\int_{\Omega_N} \prod_{\kappa \in S \sim A} \left[ \int_{(\Omega(x_N))_\kappa} |F_\kappa(x_\kappa)|^\lambda\, dx_{\kappa_1} \cdots dx_{\kappa_{k-1}} \right]^{1/\lambda} dx_N$$
$$\le \prod_{\kappa \in S \sim A} \left[ \int_{\Omega_N} \int_{(\Omega(x_N))_\kappa} |F_\kappa(x_\kappa)|^\lambda\, dx_\kappa \right]^{1/\lambda}$$
$$\le \prod_{\kappa \in S \sim A} \left[ \int_{\Omega_\kappa} |F_\kappa(x_\kappa)|^\lambda\, dx_\kappa \right]^{1/\lambda}. \tag{17}$$

It follows by insertion of (17) into (16) that

$$\int_{\Omega} \prod_{\kappa \in S} |F_\kappa(x_\kappa)| \, dx = \int_{\Omega_N} dx_N \int_{\Omega(x_N)} \prod_{\kappa \in S} |F_\kappa(x_\kappa)| \, dx_1 \cdots dx_{N-1}$$
$$\leq \prod_{\kappa \in S} \left[ \int_{\Omega_\kappa} |F_\kappa(x_\kappa)|^\lambda \, dx_\kappa \right]^{1/\lambda},$$

which completes the induction and the proof of (13). ∎

**5.10 LEMMA** Let $\Omega$ be a bounded domain in $\mathbb{R}^n$ having the cone property. If $1 \leq p < n$, then $W^{1,p}(\Omega) \to L^q(\Omega)$, where $q = np/(n-p)$. The imbedding constant may be chosen to depend only on $m$, $p$, $n$, and the cone $C$ determining the cone property for $\Omega$.

PROOF We must show that for any $u \in W^{1,p}(\Omega)$,

$$\|u\|_{0,q,\Omega} \leq K \|u\|_{1,p,\Omega}, \tag{18}$$

with $K = K(m, p, n, C)$. By Theorem 4.8, $\Omega$ may be expressed as a union of finitely many subdomains each of which has the strong local Lipschitz property (and therefore the segment property), and each of which is itself a union of parallel translates of a corresponding parallelepiped. A review of the proof of that theorem shows that the number of subdomains and the dimensions of the corresponding parallelepipeds depend on $n$ and $C$. It is therefore sufficient to establish (18) for one of these subdomains.

By Theorem 3.35 and a suitable nonsingular linear transformation we may assume that the parallelepiped involved is, in fact, a cube $Q$ having edge length 2 units, and having edges parallel to the coordinate axes. Accordingly we assume hereafter that $\Omega = \bigcup_{x \in A}(x + Q)$ with $A \subset \Omega$, and that $\Omega$ has the segment property. By Theorem 3.18 it is sufficient to establish (18) for $u \in C^\infty(\overline{\Omega})$.

For $x \in \Omega$ let $w_i(x)$ denote the intersection of $\Omega$ with the straight line through $x$ parallel to the $x_i$ coordinate axis. Clearly, $w_i(x)$ contains a segment of unit length with one endpoint at $x$, say the segment $x + te_i$, $0 \leq t < 1$, where $e_i$ is a unit vector along the $x_i$-axis.

Let $\gamma = (np - p)/(n - p)$ so that $\gamma \geq 1$. Integration by parts gives, for $u \in C^\infty(\overline{\Omega})$,

$$\int_0^1 |u(x + (1-t)e_i)|^\gamma \, dt$$
$$= |u(x)|^\gamma - \gamma \int_0^1 t |u(x + (1-t)e_i)|^{\gamma - 1} \frac{d}{dt} |u(x + (1-t)e_i)| \, dt. \tag{19}$$

Let $\hat{x}_i = (x_1, \ldots, x_{i-1}, x_{i+1}, \ldots, x_n)$ and set

$$F_i(\hat{x}_i) = \sup_{y \in w_i(x)} |u(y)|^{p/(n-p)}.$$

Then (19) gives

$$|F_i(\hat{x}_i)|^{n-1} \le \int_{w_i(x)} |u(x)|^\gamma \, dx_i + \gamma \int_{w_i(x)} |u(x)|^{\gamma-1} |D_i u(x)| \, dx_i. \tag{20}$$

Integration over $\Omega_i$, the projection of $\Omega$ onto the plane $x_i = 0$, now leads to

$$\int_{\Omega_i} |F_i(\hat{x}_i)|^{n-1} \, d\hat{x}_i \le \int_\Omega |u(x)|^\gamma \, dx + \gamma \int_\Omega |u(x)|^{\gamma-1} |D_i u(x)| \, dx.$$

If $p > 1$, then $\gamma > 1$ and an application of Hölder's inequality gives

$$\begin{aligned} \|F_i\|_{0,n-1,\Omega_i}^{n-1} &\le \gamma \left[ \int_\Omega (|u(x)| + |D_i u(x)|)^p \, dx \right]^{1/p} \left[ \int_\Omega |u(x)|^{(\gamma-1)p'} \, dx \right]^{1/p'} \\ &\le 2^{(p-1)/p} \gamma \, \|u\|_{1,p,\Omega} \|u\|_{0,q,\Omega}^{q/p'} \end{aligned}$$

since $(\gamma - 1)p' = q$.

We now apply Lemma 5.9 to the functions $F_i$, $1 \le i \le n$, noting that $k = n-1$ so that the exponent $\lambda$ of that lemma is itself $n-1$:

$$\begin{aligned} \|u\|_{0,q,\Omega}^q = \int_\Omega |u(x)|^{np/(n-p)} \, dx &\le \int_\Omega \prod_{i=1}^n F_i(\hat{x}_i) \, dx \le \prod_{i=1}^n \|F_i\|_{0,n-1,\Omega_i} \\ &\le (2^{(p-1)/p} \gamma \, \|u\|_{1,p,\Omega} \|u\|_{0,q,\Omega}^{q/p'})^{n/(n-1)}. \end{aligned}$$

Since $(n-1)q/n - q/p' = 1$, (18) follows by cancellation. The cancellation is justified, for since $u \in C^\infty(\overline{\Omega})$ and $\Omega$ is bounded, $\|u\|_{0,q,\Omega}$ is finite. Since $C^\infty(\overline{\Omega})$ is dense in $W^{1,p}(\Omega)$, (18) extends by continuity to all of $W^{1,p}(\Omega)$. ▮

**5.11 REMARK** Let $u \in C_0(\mathbb{R}^n)$ and let $q, r$ be as in the above proof. From the identity

$$\int_0^\infty \frac{d}{dt} |u(x + te_i)|^\gamma \, dt = -|u(x)|^\gamma,$$

we obtain

$$\sup_{y \in w_i(x)} |u(y)|^\gamma \le \gamma \int_{-\infty}^\infty |u(x)|^{\gamma-1} |D_i u(x)| \, dx_i,$$

where $w_i(x)$ is the line through $x$ parallel to the $x_i$-axis. Comparing this with (20), we see that the computations of the above proof can be reproduced to yield in this case

$$\|u\|_{0,q,\mathbb{R}^n} \le K |u|_{1,p,\mathbb{R}^n}, \tag{21}$$

where the seminorm $|\cdot|_{1,p}$ is defined in Section 4.11. Inequality (21) is known as *Sobolev's inequality*.

**5.12 LEMMA** Let $\Omega$ be a bounded domain in $\mathbb{R}^n$ having the cone property. If $mp < n$, then $W^{m,p}(\Omega) \to L^q(\Omega)$ for $p \le q \le np/(n-mp)$. The imbedding constant may be chosen to depend only on $m$, $p$, $n$, $q$, and the cone $C$ determining the cone property for $\Omega$.

PROOF Let $q_0 = np/(n-mp)$. We first prove by induction on $m$ that $W^{m,p}(\Omega) \to L^{q_0}(\Omega)$. Note that Lemma 5.10 establishes the case $m = 1$.

Assume, therefore, that $W^{m-1,p}(\Omega) \to L^r(\Omega)$ for $r = np/(n-mp+p)$ whenever $n > (m-1)p$. If $u \in W^{m,p}(\Omega)$, where $n > mp$, then $u$ and $D_j u$ $(1 \le j \le n)$ belong to $W^{m-1,p}(\Omega)$. It follows that $u \in W^{1,r}(\Omega)$ and

$$\|u\|_{1,r,\Omega} \le K_1 \|u\|_{m,p,\Omega}.$$

Since $mp < n$, we have $r < n$ and so by Lemma 5.10 we have $W^{1,r}(\Omega) \to L^{q_0}(\Omega)$ where $q_0 = nr/(n-r) = np/(n-mp)$ and

$$\|u\|_{0,q_0,\Omega} \le K_2 \|u\|_{1,r,\Omega} \le K_3 \|u\|_{m,p,\Omega}. \tag{22}$$

This completes the induction.

Now suppose $p \le q \le q_0$. We set

$$s = (q_0 - q)p/(q_0 - p) \qquad \text{and} \qquad t = p/s = (q_0 - p)/(q_0 - q)$$

and obtain by Hölder's inequality

$$\begin{aligned} \|u\|_{0,q,\Omega}^q &= \int_\Omega |u(x)|^s |u(x)|^{q-s}\,dx \\ &\le \left[\int_\Omega |u(x)|^{st}\,dx\right]^{1/t} \left[\int_\Omega |u(x)|^{(q-s)t'}\,dx\right]^{1/t'} \\ &= \|u\|_{0,p,\Omega}^{p/t} \|u\|_{0,q_0,\Omega}^{q_0/t'} \le K_3^{q_0/t'} \|u\|_{m,p,\Omega}^q \end{aligned} \tag{23}$$

by (22). ∎

**5.13 COROLLARY** If $mp = n$, then $W^{m,p}(\Omega) \to L^q(\Omega)$ for $p \le q < \infty$. The imbedding constant here may also depend on $\operatorname{vol}\Omega$.

PROOF If $q \ge p' = p/(p-1)$, then $q = ns/(n-ms)$, where $s = pq/(p+q)$ satisfies $1 \le s < p$. By Theorem 2.8, $W^{m,p}(\Omega) \to W^{m,s}(\Omega)$ with imbedding constant dependent on $\operatorname{vol}\Omega$. Since $ms < n$, $W^{m,s}(\Omega) \to L^q(\Omega)$ by Lemma 5.12. If $p \le q \le p'$ the desired imbedding follows by interpolation between $W^{m,p}(\Omega) \to L^p(\Omega)$ and $W^{m,p}(\Omega) \to L^{p'}(\Omega)$ as in (23). ∎

For $mp = n$ and $q \ge p$ the dependence of the imbedding constant on $\operatorname{vol}\Omega$ may be removed as we show in the following lemma which removes the restriction of boundedness of $\Omega$ from Lemma 5.12 and Corollary 5.13.

**5.14 LEMMA** Let $\Omega$ be an arbitrary domain in $\mathbb{R}^n$ having the cone property. If $mp < n$, then $W^{m,p}(\Omega) \to L^q(\Omega)$ for $p \le q \le np/(n-mp)$. If $mp = n$, then $W^{m,p}(\Omega) \to L^q(\Omega)$ for $p \le q < \infty$. If $p = 1$ and $m = n$, then $W^{m,p}(\Omega) \to C_B^0(\Omega)$. The constants for these imbeddings may depend on $m, p, n, q$, and the cone $C$ determining the cone property for $\Omega$.

PROOF We tesselate $\mathbb{R}^n$ by cubes of unit side. If $\lambda = (\lambda_1, \dots, \lambda_n)$ is an $n$-tuple of integers, let $H = \{x \in \mathbb{R}^n : \lambda_i \le x_i \le \lambda_i + 1;\ 1 \le i \le n\}$. Then $\mathbb{R}^n = \bigcup_\lambda H_\lambda$.

As remarked in the first paragraph of the proof of Theorem 4.8, even an unbounded domain $\Omega$ with the cone property can be expressed as a union of finitely many subdomains, say $\Omega = \bigcup_{j=1}^N \Omega_j$, such that $\Omega_j = \bigcup_{x \in A_j} (x + P_j)$, where $A_j \subset \Omega$ and $P_j$ is a parallelepiped with one vertex at the origin. The number $N$ and the dimensions of the parallelepipeds $P_j$ depend on $n$ and the cone $C$ determining the cone property for $\Omega$. For each $\lambda$ and for $1 \le j \le N$ let

$$\Omega_{\lambda,j} = \bigcup_{x \in A_j \cap H_\lambda} (x + P_j).$$

The domains $\Omega_{\lambda,j}$ evidently possess the following properties:

(i) $\Omega = \bigcup_{\lambda,j} \Omega_{\lambda,j}$;

(ii) $\Omega_{\lambda,j}$ is bounded;

(iii) there exists a finite cone $C'$ depending only on $P_1, \dots, P_N$ (and hence only on $n$ and $C$) such that each $\Omega_{\lambda,j}$ has the cone property determined by $C'$;

(iv) there exists a positive integer $R$ depending on $n$ and $C$ such that any $R+1$ of the domains $\Omega_{\lambda,j}$ have empty intersection;

(v) there exist constants $K'$ and $K''$ depending on $n$ and $C$ such that for each $\Omega_{\lambda,j}$,

$$K' \le \operatorname{vol} \Omega_{\lambda,j} \le K''.$$

Suppose $mp < n$ and let $u \in W^{m,p}(\Omega)$. If $p \le q \le np/(n-mp)$, then by (ii), (iii), and Lemma 5.12, we have

$$\|u\|_{0,q,\Omega_{\lambda,j}} \le K \|u\|_{m,p,\Omega_{\lambda,j}}, \tag{24}$$

where $K = K(m, p, n, q, C)$ is independent of $\lambda$ and $j$. Hence by (i) and (iv) and since $q \ge p$

$$\|u\|_{0,q,\Omega}^q \le \sum_{\lambda,j} \|u\|_{0,q,\Omega_{\lambda,j}}^q \le K^q \sum_{\lambda,j} [\|u\|_{m,p,\Omega_{\lambda,j}}^p]^{q/p}$$

$$\le K^q \left[ \sum_{\lambda,j} \|u\|_{m,p,\Omega_{\lambda,j}}^p \right]^{q/p} \le K^q R^{q/p} \|u\|_{m,p,\Omega}^q.$$

Thus $W^{m,p}(\Omega) \to L^q(\Omega)$ with imbedding constant $KR^{1/p}$.

If $mp = n$, (24) holds for any $q$ such that $p \le q < \infty$ by virtue of Corollary 5.13, and the constant $K$ can be chosen independent of $\lambda$ and $j$ thanks to (v). The rest of the above proof then carries over to this case.

Finally, if $p = 1$ and $m = n$, we have by Lemma 5.8 and a nonsingular linear transformation that $W^{n,1}(P) \to C^0(\bar{P})$ for any parallelepiped $P \subset \Omega$, the imbedding constant depending only on $n$ and the dimensions of $P$. Hence $W^{n,1}(\Omega) \to C_B{}^0(\Omega)$ by virtue of the decomposition $\Omega = \bigcup \Omega_{\lambda,j}$. ▮

We have now proved Part I, Cases A and B of Theorem 5.1 for the case $k = n$. Before completing these cases by considering the trace imbeddinge $(k < n)$, we establish the continuous function space imbeddings, Part I, Case C, and Part II.

**5.15 LEMMA** Let $\Omega$ be a domain in $\mathbb{R}^n$ having the cone property. If $mp > n$, then $W^{m,p}(\Omega) \to C_B{}^0(\Omega)$, the imbedding constant depending only on $m$, $p$, $n$, and the cone $C$ determining the cone property for $\Omega$.

PROOF Suppose that we can prove that for any $\phi \in C^\infty(\Omega)$,

$$\sup_{x \in \Omega} |\phi(x)| \le K \|\phi\|_{m,p,\Omega}, \tag{25}$$

where $K = K(m, p, n, C)$. If $u \in W^{m,p}(\Omega)$, then by Theorem 3.16 there exists a sequence $\{\phi_n\}$ in $C^\infty(\Omega)$ converging to $u$ in norm in $W^{m,p}(\Omega)$. Since $\{\phi_n\}$ is a Cauchy sequence in $W^{m,p}(\Omega)$, (25) implies that $\{\phi_n\}$ converges to a continuous function on $\Omega$. Thus $u$ must coincide a.e. with an element of $C_B{}^0(\Omega)$. It is therefore sufficient to establish (25).

First suppose $m = 1$ so that $p > n$. Let $x \in \Omega$ and let $C_x \subset \Omega$ be a finite cone congruent to $C$ and having vertex at $x$. Let $h$ be the height of $C$. Let $(r, \theta)$ denote spherical polar coordinates in $\mathbb{R}^n$ with origin at $x$ so that $C_x$ is specified by $0 < r < h$, $\theta \in A$. The volume element in this system is denoted by $r^{n-1}\omega(\theta)\, dr\, d\theta$. We have

$$\phi(x) = \phi(0, \theta) = \phi(r, \theta) - \int_0^r \frac{d}{dt} \phi(t, \theta)\, dt,$$

from which we conclude, for $0 < r < h$,

$$|\phi(x)| \le |\phi(r, \theta)| + \int_0^h |\operatorname{grad} \phi(t, \theta)|\, dt.$$

Multiplying this inequality by $r^{n-1}\omega(\theta)$ and integrating $r$ over $(0, h)$ and $\theta$ over $A$, we obtain

$$\begin{aligned}(\operatorname{vol} C_x)|\phi(x)| &\le \int_{C_x} |\phi(y)|\, dy + \frac{h^n}{n} \int_{C_x} \frac{|\operatorname{grad} \phi(y)|}{|x-y|^{n-1}}\, dy \\ &\le (\operatorname{vol} C_x)^{1/p'} \|\phi\|_{0,p,C_x} \\ &\quad + \frac{h^n}{n} \|\operatorname{grad} \phi\|_{0,p,C_x} \left| \int_{C_x} |x-y|^{-(n-1)p'}\, dy \right|^{1/p'},\end{aligned}$$

the last inequality following from two applications of Hölder's inequality. Since $p > n$ we have $(n-1)(1-p') > -1$ and so

$$\int_{C_x} |x-y|^{-(n-1)p'} \, dy = \int_A \omega(\theta) \, d\theta \int_0^h r^{(n-1)(1-p')} \, dr < \infty.$$

Hence

$$|\phi(x)| \le K\|\phi\|_{1,p,C_x} \le K\|\phi\|_{1,p,\Omega}$$

with $K = K(m,p,n,C_x) = K(m,p,n,C)$. Thus (25) is proved for $m = 1$.

If $m > 1$ but $p > n$, we still have

$$|\phi(x)| \le K\|\phi\|_{1,p,C_x} \le K\|\phi\|_{m,p,C_x} \le K\|\phi\|_{m,p,\Omega}.$$

If $p \le n < mp$, there exists an integer $j$ satisfying $1 \le j \le m-1$ such that $jp \le n < (j+1)p$. If $jp < n$, set $r = np/(n-jp)$; if $jp = n$, choose $r > \max(n,p)$. In either case we have by the result proved above and by Lemma 5.14 that

$$|\phi(x)| \le K_1\|\phi\|_{1,r,C_x} \le K_1\|\phi\|_{m-j,r,C_x} \le K\|\phi\|_{m,p,C_x} \le K\|\phi\|_{m,p,\Omega},$$

the constants depending only on $m$, $p$, $n$, and $C$. This completes the proof. ∎

**5.16 COROLLARY** If $mp > n$, then $W^{m,p}(\Omega) \to L^q(\Omega)$ for $p \le q \le \infty$. The imbedding constants depend only on $m$, $p$, $n$, $q$, and the cone $C$.

PROOF We have already established that

$$\|u\|_{0,\infty,\Omega} = \operatorname*{ess\,sup}_{x\in\Omega} |u(x)| \le K\|u\|_{m,p,\Omega}$$

for all $u \in W^{m,p}(\Omega)$. If $p \le q < \infty$, we have

$$\|u\|_{0,q,\Omega}^q = \int_\Omega |u(x)|^p |u(x)|^{q-p} \, dx$$

$$\le K^{q-p}\|u\|_{m,p,\Omega}^{q-p}\|u\|_{0,p,\Omega}^p \le K^{q-p}\|u\|_{m,p,\Omega}^q. \quad ∎$$

**5.17 LEMMA** Let $\Omega$ be a domain in $\mathbb{R}^n$ having the strong local Lipschitz property, and suppose that $mp > n \ge (m-1)p$. Then $W^{m,p}(\Omega) \to C^{0,\lambda}(\overline{\Omega})$ for:

(i) $0 < \lambda \le m - n/p$ if $n > (m-1)p$, or
(ii) $0 < \lambda < 1$ if $n = (m-1)p$, or
(iii) $0 < \lambda \le 1$ if $p = 1$, $n = m-1$.

In particular $W^{m,p}(\Omega) \to C^0(\overline{\Omega})$. The imbedding constants depend on $m$, $p$, $n$ and the parameters $\delta, M$ specified in the description of the strong local Lipschitz property for $\Omega$ (see Section 4.5).

PROOF Let $u \in W^{m,p}(\Omega)$. The strong local Lipschitz property implies the cone property so by Lemma 5.15 we may assume that $u$ is continuous on $\Omega$ and satisfies

$$\sup_{x \in \Omega} |u(x)| \le K_1 \|u\|_{m,p,\Omega}. \tag{26}$$

It is therefore sufficient to establish further that for suitable $\lambda$,

$$\sup_{\substack{x,y \in \Omega \\ x \neq y}} \frac{|u(x)-u(y)|}{|x-y|^{\lambda}} \le K_2 \|u\|_{m,p,\Omega}. \tag{27}$$

Since $mp > n \ge (m-1)p$ we have by Lemma 5.14 that $W^{m,p}(\Omega) \to W^{1,r}(\Omega)$ where:

(i) $r = np/(n-mp+p)$ and $1-(n/r) = m-(n/p)$ if $n > (m-1)p$, or
(ii) $r$ is arbitrary, $p < r < \infty$ and $0 < 1-(n/r) < 1$ if $n = (m-1)p$, or
(iii) $r = \infty$, $1-(n/r) = m-(n/p) = 1$ if $p = 1$ and $n = m-1$.

It is therefore sufficient to eastablish (27) for $m = 1$; that is, we wish to prove that if $n < p \le \infty$ and $0 < \lambda \le 1-(n/p)$, then

$$\sup_{\substack{x,y \in \Omega \\ x \neq y}} \frac{|u(x)-u(y)|}{|x-y|^{\lambda}} \le K_3 \|u\|_{1,p,\Omega}. \tag{28}$$

Suppose, for the moment, that $\Omega$ is a cube, which we may also assume without loss of generality to have unit edge. For $0 < t < 1$, $\Omega_t$ will denote a cube of edge $t$ with faces parallel to those of $\Omega$ and such that $\overline{\Omega}_t \subset \Omega$. Let $u \in C^{\infty}(\Omega)$.

Let $x, y \in \Omega$, $|x-y| = \sigma < 1$. Then there exists a fixed cube $\Omega_\sigma$ with $x, y \in \overline{\Omega}_\sigma \subset \Omega$. If $z \in \Omega_\sigma$, then

$$u(x) = u(z) - \int_0^1 \frac{d}{dt} u(x + t(z-x))\, dt,$$

so that

$$|u(x)-u(z)| \le \sqrt{n}\,\sigma \int_0^1 |\operatorname{grad} u(x + t(z-x))|\, dt.$$

Hence

$$\begin{aligned} \left| u(x) - \frac{1}{\sigma^n} \int_{\Omega_\sigma} u(z)\, dz \right| &\le \left| \frac{1}{\sigma^n} \int_{\Omega_\sigma} (u(x)-u(z))\, dz \right| \\ &\le \frac{\sqrt{n}}{\sigma^{n-1}} \int_{\Omega_\sigma} dz \int_0^1 |\operatorname{grad} u(x + t(z-x))|\, dt \end{aligned}$$

*equation continues*

$$= \frac{\sqrt{n}}{\sigma^{n-1}} \int_0^1 t^{-n}\, dt \int_{\Omega_{t\sigma}} |\operatorname{grad} u(z)|\, dz$$

$$\le \frac{\sqrt{n}}{\sigma^{n-1}} \|\operatorname{grad} u\|_{0,p,\Omega} \int_0^1 (\operatorname{vol} \Omega_{t\sigma})^{1/p'} t^{-n}\, dt \tag{29}$$

$$\le K_4 \sigma^{1-(n/p)} \|\operatorname{grad} u\|_{0,p,\Omega},$$

where $K_4 = K_4(n,p) = \sqrt{n} \int_0^1 t^{-n/p}\, dt < \infty$. A similar inequality holds with $y$ in place of $x$ and so

$$|u(x) - u(y)| \le 2K_4 |x-y|^{1-(n/p)} \|\operatorname{grad} u\|_{0,p,\Omega}.$$

It follows for $0 < \lambda \le 1-(n/p)$ that (28) holds for $\Omega$ a cube, and so, via a nonsingular linear transformation, for $\Omega$ a parallelepiped.

Now suppose that $\Omega$ has the strong local Lipschitz property. Let $\delta$, $M$, $\Omega_\delta$, $U_j$, and $\mathscr{V}_j$ be as specified in Section 4.5. There exists a parallelepiped $P$ of diameter $\delta$ whose dimensions depend only on $\delta$ and $M$ such that to each $j$ there corresponds a parallelepiped $P_j$ congruent to $P$ and having one vertex at the origin, such that for every $x \in \mathscr{V}_j \cap \Omega$ we have $x + P_j \subset \Omega$. Furthermore there exist constants $\delta_0$ and $\delta_1$ depending only on $\delta$ and $P$, with $\delta_0 \le \delta$, such that if $x, y \in \mathscr{V}_j \cap \Omega$ and $|x-y| < \delta_0$, then there exists $z \in (x+P_j) \cap (y+P_j)$ with $|x-z| + |y-z| \le \delta_1 |x-y|$. It follows from application of (28) to $x+P_j$ and $y+P_j$ that if $u \in C^\infty(\Omega)$, then

$$\begin{aligned} |u(x) - u(y)| &\le |u(x) - u(z)| + |u(y) - u(z)| \\ &\le K_5 |x-z|^\lambda \|u\|_{1,p,\Omega} + K_5 |y-z|^\lambda \|u\|_{1,p,\Omega} \\ &\le 2^{1-\lambda} K_5 \delta_1{}^\lambda |x-y|^\lambda \|u\|_{1,p,\Omega}. \end{aligned} \tag{30}$$

Now let $x, y \in \Omega$ be arbitrary. If $|x-y| < \delta_0 \le \delta$ and $x, y \in \Omega_\delta$, then $x, y \in \mathscr{V}_j$ for some $j$ and estimate (30) holds. If $|x-y| < \delta_0$, $x \in \Omega_\delta$, $y \in \Omega \sim \Omega_\delta$, then $x \in \mathscr{V}_j$ for some $j$ and (30) follows by application of (28) to $x+P_j$ and $y+P_j$ again. If $|x-y| < \delta_0$ and $x, y \in \Omega \sim \Omega_\delta$, then (30) follows from application of (28) to $x+P'$, $y+P'$, where $P'$ is any parallelepiped congruent to $P$ and having one vertex at the origin. Finally, if $|x-y| \ge \delta_0$, then we have

$$|u(x) - u(y)| \le |u(x)| + |u(y)| \le K_6 \|u\|_{1,p,\Omega} \le K_6 \delta_0^{-\lambda} |x-y|^\lambda \|u\|_{1,p,\Omega}.$$

This completes the proof of (28) for $u \in C^\infty(\Omega)$, and so by Theorem 3.16, for all continuous $u$. ∎

We have now completed the proof of all parts of the imbedding theorem 5.4 except the trace imbeddings of Cases A and B (corresponding to $k < n$). For the proof of these we will need the following interpolation result.

**5.18 LEMMA** Let $Q$ be a cube of edge length $l$, having edges parallel to the coordinate axes in $\mathbb{R}^n$. If $p > 1$, $q \geq 1$ and $mp - p < n < mp$, then there exists a constant $K = K(p, q, m, n, l)$ such that for every $u \in W^{m,p}(Q)$ we have (a.e. in $Q$)

$$|u(x)| \leq K \|u\|_{0,q,Q}^{s} \|u\|_{m,p,Q}^{1-s}, \tag{31}$$

where $s = (mp-n)q/[np + (mp-n)q]$.

PROOF It is sufficient to establish (31) for $u \in C^\infty(\bar{Q})$. Since each point of $\bar{Q}$ is a corner point of a cube contained in $\bar{Q}$, having edges parallel to those of $Q$, and having edge length $l/2$, we may assume without loss of generality that $x$ is itself a corner point of $Q$, say $Q = \{y \in \mathbb{R}^n : x_i < y_i < x_i + l;\ 1 \leq i \leq n\}$.

By Lemma 5.17 we have for $y \in Q$,

$$|u(x)| - |u(y)| \leq |u(x) - u(y)| \leq K_1 |x-y|^{m-(n/p)} \|u\|_{m,p,Q}. \tag{32}$$

Let $U = \|u\|_{m,p,Q}$, which we may assume to be positive; let $\rho = |x-y|$ and $\zeta = [|u(x)|/K_1 U]^{p/(mp-n)}$. Suppose for the moment that $\zeta \leq l$. We have for $\rho \leq \zeta$,

$$|u(y)| \geq |u(x)| - K_1 U \rho^{m-(n/p)} \geq 0.$$

Raising the above inequality to the power $q$ and integrating $y$ over $Q$, we obtain

$$\begin{aligned}\int_\Omega |u(y)|^q \, dy &\geq K_2 \int_0^\zeta (|u(x)| - K_1 U \rho^{m-(n/p)})^q \rho^{n-1} \, d\rho \\ &= K_2 \zeta^n |u(x)|^q \int_0^1 (1 - \sigma^{m-(n/p)}) \sigma^{n-1} \, d\sigma \\ &= K_3 |u(x)|^{q+(np/(mp-n))} U^{-np/(mp-n)},\end{aligned}$$

from which (31) follows at once.

If, on the other hand, $\zeta > l$, then from (32) we obtain

$$\begin{aligned}|u(y)| \geq |u(x)| - K_1 U \rho^{m-(n/p)} &\geq |u(x)| - |u(x)| (\rho/l)^{m-(n/p)} \\ &\geq 0 \qquad \text{if} \quad \rho \leq l.\end{aligned}$$

If $t > 0$, then

$$\int_Q |u(y)|^t \, dy \geq K_2 \int_0^l |u(x)|^t (1 - (\rho/l)^{m-(n/p)})^t \rho^{n-1} \, d\rho = K_4 |u(x)|^t.$$

Set $t = [(mp-n)q + np]/mp$. Then

$$\begin{aligned}|u(x)|^{[(mp-n)q+np]/mp} &\leq (1/K_4) \int_\Omega [|u(y)|^q]^{(mp-n)/mp} [|u(y)|^p]^{n/mp} \, dy \\ &\leq (1/K_4) \|u\|_{0,q,\Omega}^{q(mp-n)/mp} \|u\|_{0,p,\Omega}^{n/m},\end{aligned}$$

by an application of Hölder's inequality. Since $\|u\|_{0,p,\Omega} \le \|u\|_{m,p,\Omega}$, (31) follows at once. ▌

We remark that the above lemma also holds for the case $p = 1$, $m = n$. In this case we have from Lemma 5.14 that $W^{n,1}(\Omega) \to L^\infty(\Omega)$ so that $|u(x)| \le K\|u\|_{n,1,Q}$ a.e. in $Q$, which is (31) in this case.

**5.19 LEMMA** Let $\Omega$ be a domain in $\mathbb{R}^n$ having the cone property, and let $\Omega^k$ denote the intersection of $\Omega$ with some $k$-dimensional plane, where $1 \le k \le n$ ($\Omega^n \equiv \Omega$). If $n \ge mp$ and $n - mp < k \le n$, then

$$W^{m,p}(\Omega) \to L^q(\Omega^k) \tag{33}$$

for $p \le q \le kp/(n-mp)$ if $n > mp$, or for $p \le q < \infty$ if $n = mp$. If $p = 1$, $n > m$ and $n - m \le k \le n$, then (33) holds for $1 \le q \le k/(n-m)$.

The imbedding constants depend only on $m$, $p$, $k$, $n$, $q$, and the cone $C$ determining the cone property for $\Omega$.

PROOF It is sufficient to establish the above conclusions for $\Omega$ bounded, $n > mp$, and $q = kp/(n-mp)$, as extension to the other cases can be carried out in the same manner as was described for the case $k = n$ in Corollary 5.13 and Lemma 5.14. We may also assume, as in Lemma 5.10, that $\Omega$ is a union of coordinate cubes of edge 2 units.

Let $\mathbb{R}_0^{\ k}$ be a $k$-dimensional coordinate subspace of $\mathbb{R}^n$ on which $\Omega^k$ has a one-to-one projection $\Omega_0^{\ k}$. Suppose, for the moment, that $p > 1$. Let $\nu$ be the largest integer less than $mp$. Then $mp - p < \nu < mp$ and since $n - mp < k$ we have $n - \nu \le k$. (Note that if $p = 1$, the same conclusion holds with $k = n - m$, $\nu = m$.) Let $\mu = \binom{k}{n-\nu}$ and let $E_i$ $(1 \le i \le \mu)$ denote the various coordinate subspaces of $\mathbb{R}_0^{\ k}$ having dimension $n - \nu$. Let $\Omega_i$ denote the projection of $\Omega_0^{\ k}$ (and hence of $\Omega^k$) onto $E_i$. Also, for each $x \in \Omega_i$ let $\Omega_{i,x}$ denote the intersection of $\Omega$ with the $\nu$-dimensional plane through $x$ perpendicular to $E_i$. Then $\Omega_{i,x}$ contains a $\nu$-dimensional coordinate cube of unit edge with one vertex at $x$. By Lemma 5.18, with $q = q_0 = np/(n-mp)$, we have for $u \in C^\infty(\Omega)$

$$\sup_{y \in \Omega_{i,x}} |u(y)|^{(n-\nu)p/(n-mp)} \le K_1 \|u\|_{0,q_0,\Omega_{i,x}}^{(mp-\nu)q_0/mp} \|u\|_{m,p,\Omega_{i,x}}^{\nu/m}. \tag{34}$$

Let $dx^i$ and $dx_*^{\ i}$ denote the volume elements in $E_i$ and the orthogonal complement of $E_i$, respectively. Integration of (34) over $\Omega_i$ leads to

$$\int_{\Omega_i} \sup_{y \in \Omega_{i,x}} |u(y)|^{(n-\nu)p/(n-mp)}\, dx^i$$
$$\le K_1 \int_{\Omega_i} \left[ \int_{\Omega_{i,x}} |u(x)|^{q_0}\, dx_*^{\ i} \right]^{(mp-\nu)/mp}$$

*equation continues*

$$\times \left[ \int_{\Omega_{i,x}} \sum_{|\alpha| \le m} |D^\alpha u(x)|^p \, dx_*{}^i \right]^{\nu/mp} dx^i$$

$$\le K_1 \left[ \int_\Omega |u(x)|^{q_0} \, dx \right]^{(mp-\nu)/mp} \left[ \int_\Omega \sum_{|\alpha| \le m} |D^\alpha u(x)|^p \, dx \right]^{\nu/mp}$$

$$= K_1 \|u\|_{0,q_0,\Omega}^{q_0(mp-\nu)/mp} \|u\|_{m,p,\Omega}^{\nu/m} \tag{35}$$

by Hölder's inequality.

Finally, we apply Lemma 5.9 to the subspaces $E_i$ of $\mathbb{R}_0{}^k$. Note that the constant $\lambda$ of that lemma is here equal to $\binom{k-1}{n-\nu-1}$. Letting $dx^{(k)}$ denote the volume element in $\mathbb{R}_0{}^k$ and setting $q = kp/(n-mp)$, we obtain

$$\|u\|_{0,q,\Omega^k}^q \le K_2 \int_{\Omega_0{}^k} \prod_{i=1}^{\mu} \sup_{y \in \Omega_{i,x}} |u(y)|^{q/\mu} \, dx^{(k)}$$

$$\le K_2 \prod_{i=1}^{\mu} \left[ \int_{\Omega_i} \sup_{y \in \Omega_{i,x}} |u(y)|^{q\lambda/\mu} \, dx^i \right]^{1/\lambda}. \tag{36}$$

Since $q\lambda/\mu = (n-\nu)p/(n-mp)$, it follows from (35) and (36) and from Lemma 5.14 that

$$\|u\|_{0,q,\Omega^k} \le K_3 \prod_{i=1}^{\mu} \|u\|_{0,q_0,\Omega}^{q_0(mp-\nu)/mp\lambda q} \|u\|_{m,p,\Omega}^{\nu/m\lambda q}$$

$$\le K_4 \left[ \|u\|_{m,p,\Omega}^{q_0(mp-\nu)/mp} \|u\|_{m,p,\Omega}^{\nu/m} \right]^{\mu/\lambda q} = K_4 \|u\|_{m,p,\Omega}.$$

This establishes the desired imbedding. ∎

We have now completed the proof of Theorem 5.4.

## Traces of Functions in $W^{m,p}(\Omega)$ on the Boundary of $\Omega$

**5.20** Of importance in the study of boundary value problems for differential operators defined on a domain $\Omega$ is the determination of spaces of functions defined on the boundary of $\Omega$ containing the traces $u|_{\text{bdry}\,\Omega}$ of functions $u$ in $W^{m,p}(\Omega)$. For example, if $W^{m,p}(\Omega) \to C(\overline{\Omega})$, then clearly $u|_{\text{bdry}\,\Omega}$ belongs to $C(\text{bdry}\,\Omega)$. We outline below an $L^q$-imbedding theorem for such traces which can be obtained as a corollary of Theorem 5.4.

The problem of characterizing the image of $W^{m,p}(\Omega)$ under the operator $u \to u|_{\text{bdry}\,\Omega}$ has been extensively studied by many authors. The solution, which involves Sobolev spaces of fractional order $m$ will be given in Chapter VII (see in particular Theorem 7.53). The approach used in that chapter is due to Lions [37, 38].

**5.21** Let $\Omega$ be a domain in $\mathbb{R}^n$ having the uniform $C^m$-regularity property. Thus there exists a locally finite open cover $\{U_j\}$ of bdry $\Omega$, and corresponding $m$-smooth transformations $\Psi_j$ mapping $B = \{y \in \mathbb{R}^n : |y| < 1\}$ onto $U_j$ such that $U_j \cap \text{bdry}\,\Omega = \Psi_j(B_0)$; $B_0 = \{y \in B : y_n = 0\}$. If $f$ is a function having support in $U_j$, we may define the integral of $f$ over bdry $\Omega$ via

$$\int_{\text{bdry}\,\Omega} f(x)\,d\sigma = \int_{U_j \cap \text{bdry}\,\Omega} f(x)\,d\sigma = \int_{B_0} f \circ \Psi_j(y', 0) J_j(y')\,dy',$$

where $y' = (y_1, \ldots, y_{n-1})$ and if $x = \Psi_j(y)$, then

$$J_j(y') = \left\{ \sum_{k=1}^{n} \left( \frac{\partial(x_1, \ldots, \hat{x}_k, \ldots, x_n)}{\partial(y_1, \ldots, y_{n-1})} \right)^2 \right\}^{1/2} \Bigg|_{y_n = 0}.$$

If $f$ is an arbitrary function defined on bdry $\Omega$, we may set

$$\int_{\text{bdry}\,\Omega} f(x)\,d\sigma = \sum_j \int_{\text{bdry}\,\Omega} f(x) v_j(x)\,d\sigma,$$

where $\{v_j\}$ is a partition of unity for bdry $\Omega$ subordinate to $\{U_j\}$.

**5.22 THEOREM** Let $\Omega$ be a domain in $\mathbb{R}^n$ having the uniform $C^m$-regularity property, and suppose there exists a simple $(m, p)$-extension operator $E$ for $\Omega$. If $mp < n$ and $p \le q \le (n-1)p/(n-mp)$, then

$$W^{m,p}(\Omega) \to L^q(\text{bdry}\,\Omega). \tag{37}$$

If $mp = n$, then (37) holds for $p \le q < \infty$.

PROOF Imbedding (37) should be interpreted in the following sense: If $u \in W^{m,p}(\Omega)$, then $Eu$ has a trace on bdry $\Omega$ in the sense described in the final paragraph of Section 5.2, and $\|Eu\|_{0,q,\text{bdry}\,\Omega} \le K \|u\|_{m,p,\Omega}$ with $K$ independent of $u$. [Note that since $C_0(\mathbb{R}^n)$ is dense in $W^{m,p}(\Omega)$, $\|Eu\|_{0,q,\text{bdry}\,\Omega}$ is independent of the particular extension operator $E$ used.]

It is sufficient to prove the theorem for $mp < n$ and $q = (n-1)p/(n-mp)$. There exists a constant $K_1$ such that for every $u \in W^{m,p}(\Omega)$

$$\|Eu\|_{m,p,\mathbb{R}^n} \le K_1 \|u\|_{m,p,\Omega}.$$

By the conditions of the uniform $C^m$-regularity property (Section 4.6), there exists a constant $K_2$ such that for each $j$ and every $y \in B$, $x = \Psi_j(y) \in U_j$

$$|J_j(y')| \le K_2 \qquad \text{and} \qquad \left| \frac{\partial(y_1, \ldots, y_n)}{\partial(x_1, \ldots, x_n)} \right| \le K_2.$$

Noting that $0 \le v_j(x) \le 1$ on $\mathbb{R}^n$, and using imbedding (2) of Theorem 5.4

applied over $B$, we have for $u \in W^{m,p}(\Omega)$,

$$\begin{aligned}\int_{bdry\,\Omega} |Eu(x)|^q \, d\sigma &\le \sum_j \int_{U_j \cap bdry\,\Omega} |Eu(x)|^q \, d\sigma \\ &\le K_2 \sum_j \|Eu \circ \Psi_j\|_{0,q,B_0}^q \\ &\le K_3 \sum_j \left( \|Eu \circ \Psi_j\|_{m,p,B}^p \right)^{q/p} \\ &\le K_4 \left( \sum_j \|Eu\|_{m,p,U_j}^p \right)^{q/p} \\ &\le K_4 R \|Eu\|_{m,p,\mathbb{R}^n}^q \\ &\le K_5 \|u\|_{m,p,\Omega}^q .\end{aligned}$$

The second last inequality above makes use of the finite intersection property possessed by the cover $\{U_j\}$. The constant $K_4$ is independent of $j$ since $|D^\alpha \Psi_{j,i}(y)| \le \text{const}$ for all $i, j$, where $\Psi_j = (\Psi_{j,1}, \ldots, \Psi_{j,n})$. This completes the proof. ∎

## $W^{m,p}(\Omega)$ as a Banach Algebra

Given functions $u$ and $v$ in $W^{m,p}(\Omega)$, where $\Omega$ is a domain in $\mathbb{R}^n$, one cannot in general expect their pointwise product $uv$ to belong to $W^{m,p}(\Omega)$. [$(uv)(x) = u(x)v(x)$ a.e. in $\Omega$.] It is, however, a straightforward application of the Sobolev imbedding theorem to show that this is the case provided $mp > n$ and $\Omega$ has the cone property.

**5.23 THEOREM** Let $\Omega$ be a domain in $\mathbb{R}^n$ having the cone property. If $mp > n$, then there exists a constant $K^*$ depending on $m$, $p$, $n$, and the finite cone $C$ determining the cone property for $\Omega$, such that for all $u, v \in W^{m,p}(\Omega)$ the product $uv$, defined pointwise a.e. in $\Omega$, belongs to $W^{m,p}(\Omega)$ and satisfies

$$\|uv\|_{m,p,\Omega} \le K^* \|u\|_{m,p,\Omega} \|v\|_{m,p,\Omega} . \tag{38}$$

In particular, $W^{m,p}(\Omega)$ is a commutative Banach algebra with respect to pointwise multiplication and the equivalent norm

$$\|u\|_{m,p,\Omega}^* = K^* \|u\|_{m,p,\Omega} .$$

PROOF In order to establish (38) it is sufficient to show that if $|\alpha| \le m$, then

$$\int_\Omega |D^\alpha[u(x)v(x)]|^p \, dx \le K_\alpha \|u\|_{m,p,\Omega}^p \|v\|_{m,p,\Omega}^p ,$$

where $K = K(m, p, n, C)$. Let us assume for the moment that $u \in C^\infty(\Omega)$. By Leibniz's rule for distributional derivatives, that is,

$$D^\alpha[uv] = \sum_{\beta \le \alpha} \binom{\alpha}{\beta} D^\beta u D^{\alpha-\beta} v,$$

it is sufficient to show that for any $\beta \le \alpha$, $|\alpha| \le m$, we have

$$\int_\Omega |D^\beta u(x)\, D^{\alpha-\beta} v(x)|^p\, dx \le K_{\alpha,\beta} \|u\|_{m,p,\Omega}^p \|v\|_{m,p,\Omega}^p,$$

where $K_{\alpha,\beta} = K_{\alpha,\beta}(m, p, n, C)$. By the imbedding theorem there exists, for each $\beta$ with $|\beta| \le m$, a constant $K(\beta) = K(\beta, m, p, n, C)$ such that for any $w \in W^{m,p}(\Omega)$

$$\int_\Omega |D^\beta w(x)|^r\, dx \le K(\beta) \|w\|_{m,p,\Omega}^r \tag{39}$$

provided $(m - |\beta|)p \le n$ and $p \le r \le np/(n - (m - |\beta|)p)$ [or $p \le r < \infty$ if $(m - |\beta|)p = n$], or

$$|D^\beta w(x)| \le K(\beta) \|w\|_{m,p,\Omega} \qquad \text{a.e. in } \Omega$$

provided $(m - |\beta|)p > n$.

Let $k$ be the largest integer such that $(m-k)p > n$. Since $mp > n$ we have that $k \ge 0$. If $|\beta| \le k$, then $(m - |\beta|)p > n$, so

$$\begin{aligned}\int_\Omega |D^\beta u(x)\, D^{\alpha-\beta} v(x)|^p\, dx &\le K(\beta)^p \|u\|_{m,p,\Omega}^p \|D^{\alpha-\beta} v\|_{0,p,\Omega}^p \\ &\le K(\beta)^p \|u\|_{m,p,\Omega}^p \|v\|_{m,p,\Omega}^p.\end{aligned}$$

Similarly, if $|\alpha - \beta| \le k$, then

$$\int_\Omega |D^\beta u(x)\, D^{\alpha-\beta} v(x)|^p\, dx \le K(\alpha-\beta)^p \|u\|_{m,p,\Omega}^p \|v\|_{m,p,\Omega}^p.$$

Now if $|\beta| > k$ and $|\alpha - \beta| > k$, then, in fact, $|\beta| \ge k+1$ and $|\alpha - \beta| \ge k+1$ so that $n \ge (m - |\beta|)p$ and $n \ge (m - |\alpha - \beta|)p$. Moreover,

$$\frac{n - (m - |\beta|)p}{n} + \frac{n - (m - |\alpha-\beta|)p}{n} = 2 - \frac{(2m - |\alpha|)p}{n} < 2 - \frac{mp}{n} < 1.$$

Hence there exist positive numbers $r, r'$ with $(1/r) + (1/r') = 1$ such that

$$p \le rp < \frac{np}{n - (m - |\beta|)p}, \qquad p \le r'p < \frac{np}{n - (m - |\alpha - \beta|)p}.$$

Thus by Hölder's inequality and (39) we have

$$\int_\Omega |D^\beta u(x)\, D^{\alpha-\beta} v(x)|^p\, dx \le \left[\int_\Omega |D^\beta u(x)|^{rp}\, dx\right]^{1/r} \left[\int_\Omega |D^{\alpha-\beta} v(x)|^{r'p}\, dx\right]^{1/r'}$$
$$\le [K(\beta)]^{1/r} [K(\alpha-\beta)]^{1/r'} \|u\|_{m,p,\Omega}^p \|v\|_{m,p,\Omega}^p.$$

This completes the proof of (38) for $u \in C^\infty(\Omega)$, $v \in W^{m,p}(\Omega)$.

If $u \in W^{m,p}(\Omega)$, then by Theorem 3.16 there exists a sequence $\{u_n\}$ of $C^\infty(\Omega)$ functions converging to $u$ in $W^{m,p}(\Omega)$. Then by the above argument $\{u_n v\}$ is a Cauchy sequence in $W^{m,p}(\Omega)$ so converges to an element $w$ of that space. Since $mp > n$, $u$ and $v$ may be assumed continuous and bounded on $\Omega$. Thus

$$\|w - uv\|_{0,p,\Omega} \le \|w - u_n v\|_{0,p,\Omega} + \|(u_n - u)v\|_{0,p,\Omega}$$
$$\le \|w - u_n v\|_{0,p,\Omega} + \|v\|_{0,\infty,\Omega} \|u_n - u\|_{0,p,\Omega}$$
$$\to 0 \quad \text{as} \quad n \to \infty.$$

Hence $w = uv$ in $L^p(\Omega)$ and so $w = uv$ in the sense of distributions. Therefore $w = uv$ in $W^{m,p}(\Omega)$ and

$$\|uv\|_{m,p,\Omega} = \|w\|_{m,p,\Omega} \le \limsup_{n\to\infty} \|u_n v\|_{m,p,\Omega} = \|u\|_{m,p,\Omega} \|v\|_{m,p,\Omega}.$$

This completes the proof of the theorem. ∎

We remark that Banach algebra $W^{m,p}(\Omega)$ has an identity if and only if $\Omega$ is bounded, that is, the function $e(x) \equiv 1$ belongs to $W^{m,p}(\Omega)$ if and only if $\operatorname{vol} \Omega < \infty$, but there are no unbounded domains with finite volume having the cone property.

## Counterexamples and Nonimbedding Theorems

**5.24** Consideration of the statement of the Sobolev imbedding Theorem 5.4 may lead the reader to speculate on several directions of possible generalization. Before exploring the possibility of proving imbedding theorems for domains not satisfying the conditions of Theorem 5.4, we first construct examples showing that in certain respects that theorem gives "best possible" imbedding results for the domains considered, and indeed for any domain.

Let $\Omega$ be an arbitrary domain in $\mathbb{R}^n$ and assume, without loss of generality, that the origin belongs to $\Omega$. Let $R > 0$ be such that the closed ball $\overline{B_{2R}}$ is contained in $\Omega$. (Here $B_R = \{x \in \mathbb{R}^n : |x| < R\}$.) In each of the following examples we construct a function $u \in C^\infty(B_{3R} \sim \{0\})$ depending only on $\rho = |x|$. If $f \in C^\infty(0, \infty)$ satisfies $f(t) = 1$ if $t \le R$ while $f(t) = 0$ if $t \ge 2R$, then the function $w$ defined by

$$w(x) = \begin{cases} 0 & \text{if} \quad \rho = |x| \geq 3R \\ f(\rho)u(\rho) & \text{if} \quad 0 < \rho < 3R \end{cases}$$

has compact support in $\Omega$ and belongs to $W^{m,p}(\Omega)$ if and only if $u \in W^{m,p}(B_R)$.

**5.25 EXAMPLE** Let $k$ be an integer such that $1 \leq k \leq n$ and suppose that $mp < n$ and $q > kp/(n-mp)$. We construct $u$ so that $u \in W^{m,p}(B_R)$ but $u \notin L^q(B_R{}^k)$, where $B_R{}^k = \{x \in B_R : x_{k+1} = \cdots = x_n = 0\}$. Hence no imbedding of the type $W^{m,p}(\Omega) \to L^q(\Omega^k)$ is possible if $q > kp/(n-mp)$.

Let $u(x) = \rho^\lambda$, where $\rho = |x|$ and the exponent $\lambda$ will be specified below. It is readily checked by induction on $|\alpha|$ that

$$D^\alpha u(x) = P_\alpha(x)\rho^{\lambda - 2|\alpha|} \tag{40}$$

where $P_\alpha$ is a polynomial homogeneous of degree $|\alpha|$ in the components of $x$. Hence $|D^\alpha u(x)| \leq K_\alpha \rho^{\lambda - |\alpha|}$ and

$$\int_{B_R} |D^\alpha u(x)|^p \, dx \leq \text{const} \int_0^R \rho^{(\lambda - |\alpha|)p + n - 1} \, dx.$$

Therefore $u$ belongs to $W^{m,p}(B_R)$ provided

$$(\lambda - m)p + n > 0. \tag{41}$$

On the other hand, if $\sigma = (x_1{}^2 + \cdots + x_k{}^2)^{1/2}$, then

$$\int_{B_R{}^k} |u(x)|^q \, dx_1 \cdots dx_k = \text{const} \int_0^R \sigma^{\lambda q + k - 1} \, d\sigma$$

so that $u \notin L^q(B_R{}^k)$ if

$$\lambda q + k < 0. \tag{42}$$

Since $q > kp/(n-mp)$, it is possible to select $\lambda$ to satisfy both (41) and (42) as required. ∎

Since the function $u$ constructed above is unbounded near 0, no imbedding of the form $W^{m,p}(\Omega) \to C_B{}^0(\Omega)$ is possible if $mp < n$.

**5.26 EXAMPLE** Suppose $p > 1$ and $mp = n$. We construct $u \in W^{m,p}(B_R)$ so that $u \notin L^\infty(B_R)$. Hence the imbeddings $W^{m,p}(\Omega) \to L^q(\Omega)$, valid for $p \leq q < \infty$ if $mp = n$ and $\Omega$ has the cone property, cannot be extended to yield $W^{m,p}(\Omega) \to L^\infty(\Omega)$ or $W^{m,p}(\Omega) \to C_B{}^0(\Omega)$ unless $p = 1$ and $n = m$ (see, however, Theorem 8.25).

Let $u(x) = \log(\log 4R/\rho)$, where $\rho = |x|$. Clearly $u \notin L^\infty(B_R)$. Again it is easily checked by induction that

$$D^\alpha u(x) = \sum_{j=1}^{|\alpha|} P_{\alpha,j}(x)\rho^{-2|\alpha|}[\log(4R/\rho)]^{-j}, \tag{43}$$

where $P_{\alpha,j}$ is a polynomial homogeneous of degree $|\alpha|$ in the components of $x$. Since $p = n/m$ we have

$$|D^\alpha u(x)|^p \le \sum_{j=1}^{|\alpha|} K_{\alpha,j}\rho^{-|\alpha|n/m}[\log(4R/\rho)]^{-jp},$$

so that

$$\int_{B_R} |D^\alpha u(x)|^p\,dx \le \text{const} \sum_{j=1}^{|\alpha|} \int_0^R [\log(4R/\rho)]^{-jp}\rho^{-|\alpha|n/m+n-1}\,d\rho.$$

The right side of the above inequality is certainly finite if $|\alpha| < m$. If $|\alpha| = m$, we have, setting $\sigma = \log(4R/\rho)$,

$$\int_{B_R} |D^\alpha u(x)|^p\,dx \le \text{const} \sum_{j=1}^{|\alpha|} \int_{\log 4}^{\infty} \sigma^{-jp}\,d\sigma$$

which is finite since $p > 1$. Thus $u \in W^{m,p}(B_R)$. ∎

It is interesting that the function $u$ above is independent of the choice of $m$ and $p$ with $mp = n$.

**5.27 EXAMPLE** Suppose $mp > n > (m-1)p$, and let $\lambda > m-(n/p)$. We construct $u \in W^{m,p}(B_R)$ such that $u \notin C^{0,\lambda}(\overline{B_R})$. Hence no imbedding of the form $W^{m,p}(\Omega) \to C^{0,\lambda}(\overline{\Omega})$ is possible if $mp > n > (m-1)p$ and $\lambda > m-(n/p)$.

As in Example 5.25 we take $u(x) = \rho^\mu$, $\rho = |x|$. From (41) we have $u \in W^{m,p}(B_R)$ provided $\mu > m-(n/p)$. Now $|u(x)-u(0)|/|x-0|^\lambda = \rho^{\mu-\lambda}$ so that $u \notin C^{0,\lambda}(\overline{B_R})$ when $\mu < \lambda$. Thus $u$ has the required properties if we choose $\mu$ to satisfy $m-(n/p) < \mu < \lambda$. ∎

**5.28 EXAMPLE** Suppose $(m-1)p = n$ and $p > 1$. We construct $u \in W^{m,p}(B_R)$ such that $u \notin C^{0,1}(\overline{B_R})$. Hence the imbedding $W^{m,p}(\Omega) \to C^{0,\lambda}(\overline{\Omega})$, valid for $0 < \lambda < 1$ whenever $\Omega$ has the strong local Lipschitz property, cannot be extended to $\lambda = 1$ unless $p = 1$, $m-1 = n$.

Let $u(x) = \rho\log(\log 4R/\rho)$ where $\rho = |x|$. Since

$$|u(x)-u(0)|/|x-0| = \log(\log 4R/\rho) \to \infty \qquad \text{as} \quad x \to 0$$

it is clear that $u \notin C^{0,1}(\overline{B_R})$. Following (40) and (43) we have

$$D^\alpha u(x) = \sum_{j=1}^{|\alpha|} P_{\alpha,j}(x)\rho^{1-2|\alpha|}[\log(4R/\rho)]^{-j},$$

where $P_{\alpha,j}$ is a polynomial homogeneous of degree $|\alpha|$. Hence

$$|D^\alpha u(x)|^p \le \sum_{j=1}^{|\alpha|} K_{\alpha,j}\,\rho^{p(1-|\alpha|)}[\log(4R/\rho)]^{-jp}.$$

It then follows as in Example 5.26 that $u \in W^{m,p}(B_R)$. ∎

**5.29** The above examples show that even for very regular domains there can exist no imbeddings of the types considered in Theorem 5.4, except those explicitly stated there. It remains to be seen whether any imbeddings of these types can exist for irregular domains not having the cone property. We shall show that Theorem 5.4 can be extended, with weakened conclusions, to certain types of such irregular domains, but we first show that no extension is possible if the domain is "too irregular."

An unbounded domain $\Omega$ in $\mathbb{R}^n$ may have a smooth boundary and still fail to have the cone property if it becomes narrow at infinity, that is, if

$$\lim_{\substack{|x|\to\infty \\ x\in\Omega}} \operatorname{dist}(x, \operatorname{bdry}\Omega) = 0.$$

The following theorem shows that Parts I and II of Theorem 5.4 fail completely for any such unbounded $\Omega$ which has finite volume.

**5.30 THEOREM** Let $\Omega$ be an unbounded domain in $\mathbb{R}^n$ having finite volume, and let $q > p$. Then $W^{m,p}(\Omega)$ is not imbedded in $L^q(\Omega)$.

PROOF We construct a function $u(x)$, depending only on the distance $\rho = |x|$ of $x$ from the origin, whose growth as $\rho$ increases is rapid enough to preclude membership of $u$ in $L^q(\Omega)$ but not so rapid as to prevent $u \in W^{m,p}(\Omega)$.

Without loss of generality we assume $\operatorname{vol}\Omega = 1$. Let $A(\rho)$ denote the surface area [$(n-1)$-measure] of the intersection of $\Omega$ with the spherical surface of radius $r$ centered at the origin. Then

$$\int_0^\infty A(\rho)\,d\rho = 1.$$

Let $r_0 = 0$ and define $r_n$ for $n = 1, 2, \ldots$ by

$$\int_{r_n}^\infty A(\rho)\,d\rho = 1/2^n = \int_{r_{n-1}}^{r_n} A(\rho)\,d\rho.$$

Clearly $r_n$ increases to infinity with $n$. Let $\Delta r_n = r_{n+1} - r_n$ and fix $\varepsilon$ such that $0 < \varepsilon < [1/(mp)] - [1/(mq)]$. There must exist an increasing sequence $\{n_j\}_{j=1}^\infty$ such that $\Delta r_{n_j} \ge 2^{-\varepsilon n_j}$, for otherwise $\Delta r_n < 2^{-\varepsilon n}$ for all but possibly finitely many values of $n$ whence we would have $\sum_{n=0}^\infty \Delta r_n < \infty$, a contradiction. For convenience we assume $n_1 \ge 1$ so that $n_j \ge j$ for all $j$. Let $a_0 = 0$ $a_j = r_{n_j+1}$, and $b_j = r_{n_j}$. Note that $a_{j-1} \le b_j < a_j$ and $a_j - b_j = \Delta r_{n_j} \ge 2^{-\varepsilon n_j}$.

Let $f$ be a nonnegative, infinitely differentiable function on $\mathbb{R}$ having the properties:

(i) $0 \le f(t) \le 1$ for all $t$,
(ii) $f(t) = 0$ if $t \le 0$, $f(t) = 1$ if $t \ge 1$,
(iii) $|(d/dt)^k f(t)| \le M$ for all $t$ if $1 \le k \le m$.

If $x \in \Omega$ and $\rho = |x|$, set

$$u(x) = \begin{cases} 2^{n_{j-1}/q} & \text{for} \quad a_{j-1} \le \rho \le b_j \\ 2^{n_{j-1}/q} + (2^{n_j/q} - 2^{n_{j-1}/q})\, f\left(\dfrac{\rho - b_j}{a_j - b_j}\right) & \text{for} \quad b_j \le \rho \le a_j. \end{cases}$$

Clearly $u \in C^\infty(\Omega)$. Denoting $\Omega_j = \{x \in \Omega : a_{j-1} \le \rho \le a_j\}$, we have

$$\begin{aligned} \int_{\Omega_j} |u(x)|^p \, dx &= \left\{ \int_{a_{j-1}}^{b_j} + \int_{b_j}^{a_j} \right\} [u(x)]^p A(\rho)\, d\rho \\ &\le 2^{n_{j-1} p/q} \int_{a_{j-1}}^{\infty} A(\rho)\, d\rho + 2^{n_j p/q} \int_{b_j}^{a_j} A(\rho)\, d\rho \\ &= \tfrac{1}{2}[2^{-n_{j-1}(1-p/q)} + 2^{-n_j(1-p/q)}] \le 2^{-(j-1)(1-p/q)}. \end{aligned}$$

Since $p < q$ the above inequality forces

$$\int_\Omega |u(x)|^p \, dx = \sum_{j=1}^{\infty} \int_{\Omega_j} |u(x)|^p \, dt < \infty.$$

Also, if $1 \le k \le m$, we have

$$\begin{aligned} \int_{\Omega_j} \left| \frac{d^k u}{d\rho^k} \right|^p dx &= \int_{b_j}^{a_j} \left| \frac{d^k u}{d\rho^k} \right|^p A(\rho)\, d\rho \\ &\le M^p 2^{n_j p/q} [a_j - b_j]^{-kp} \int_{b_j}^{a_j} A(\rho)\, d\rho \\ &= \tfrac{1}{2} M^p 2^{-n_j(1 - p/q - \varepsilon kp)} \le \tfrac{1}{2} M^p 2^{-Cj}, \end{aligned}$$

where $C = 1 - p/q - \varepsilon kp > 0$ since $\varepsilon < [1/(mp)] - [1/(mq)]$. Hence $D^\alpha u \in L^p(\Omega)$ for $|\alpha| \le m$, that is, $u \in W^{m,p}(\Omega)$. However, $u \notin L^q(\Omega)$, for we have for each $j$,

$$\begin{aligned} \int_{\Omega_j} |u(x)|^q \, dx &\ge 2^{n_{j-1}} \int_{a_{j-1}}^{a_j} A(\rho)\, d\rho \\ &= 2^{n_{j-1}} [2^{-n_{j-1}-1} - 2^{-n_j - 1}] \ge \tfrac{1}{4}. \end{aligned}$$

Therefore $W^{m,p}(\Omega)$ cannot be imbedded in $L^q(\Omega)$. ∎

The conclusions of the above theorem can be extended (see Section 6.35) to unbounded domains $\Omega$ having infinite volume but satisfying

$$\limsup_{N\to\infty} \operatorname{vol}\{x\in\Omega : N \le |x| \le N+1\} = 0.$$

**5.31** Parts I and II of Theorem 5.4 also fail completely for domains with sufficiently sharp boundary cusps. If $\Omega$ is a domain in $\mathbb{R}^n$ and $x_0$ is a point on the boundary of $\Omega$, let $B_r = B_r(x_0)$ denote the open ball of radius $r$ about $x_0$, let $\Omega_r = B_r \cap \Omega$, let $S_r = (\operatorname{bdry} B_r) \cap \Omega$, and let $A(r,\Omega)$ be the surface area [$(n-1)$-measure] of $S_r$. We shall say that $\Omega$ has an *exponential cusp* at $x_0 \in \operatorname{bdry}\Omega$ if for every real number $k$, we have

$$\lim_{r\to 0+} \frac{A(r,\Omega)}{r^k} = 0. \tag{44}$$

**5.32 THEOREM** If $\Omega$ is a domain in $\mathbb{R}^n$ having an exponential cusp at the point $x_0$ on $\operatorname{bdry}\Omega$, and if $q > p$, then $W^{m,p}(\Omega)$ is not imbedded in $L^q(\Omega)$.

PROOF We construct $u \in W^{m,p}(\Omega)$ which fails to belong to $L^q(\Omega)$ because it becomes unbounded too rapidly near $x_0$. Without loss of generality we may assume $x_0 = 0$ so that $r = |x|$. Let $\Omega^* = \{y = x/|x|^2 : x \in \Omega, |x| < 1\}$. It is easily seen that $\Omega^*$ is unbounded and has finite volume, and that

$$A(r,\Omega^*) = r^{2(n-1)} A(1/r,\Omega).$$

Let $t$ satisfy $p < t < q$. By Theorem 5.30 there exists a function $\tilde{v} \in C^m(0,\infty)$ such that

(i) $\tilde{v}(r) = 0$ if $0 < r \le 1$,

(ii) $\displaystyle\int_1^\infty |v^{(j)}(r)|^t A(r,\Omega^*)\, dr < \infty$ if $0 \le j \le m$,

(iii) $\displaystyle\int_1^\infty |\tilde{v}(r)|^q A(r,\Omega^*)\, dr = \infty.$

[If $r = |y|$, then $v(y) = \tilde{v}(r)$ defines $v \in W^{m,p}(\Omega^*)$ but $v \notin L^q(\Omega^*)$.] Let $x = y/|y|^2$ so that $\rho = |x| = 1/|y| = 1/r$. Set $\lambda = 2n/q$ and define $u(x) = \tilde{u}(\rho) = r^\lambda \tilde{v}(r) = |y|^\lambda v(y)$. It follows for $|\alpha| = j \le m$ that

$$|D^\alpha u(x)| \le |\tilde{u}^{(j)}(\rho)| \le \sum_{i=1}^{j} c_{ij} r^{\lambda+j+i} \tilde{v}^{(i)}(r),$$

where the coefficients $c_{ij}$ depend only on $\lambda$. Now $u(x)$ vanishes for $|x| \ge 1$ and so

$$\int_\Omega |u(x)|^q\, dx = \int_0^1 |\tilde{u}(\rho)|^q A(\rho,\Omega)\, d\rho = \int_1^\infty |\tilde{v}(r)|^q A(r,\Omega^*)\, dr = \infty.$$

On the other hand, if $0 \le |\alpha| = j \le m$, we have

$$\int_\Omega |D^\alpha u(x)|^p\,dx \le \int_0^1 |\tilde{u}^{(j)}(\rho)|^p A(\rho,\Omega)\,d\rho$$

$$\le \text{const} \sum_{i=0}^{j} \int_1^\infty |\tilde{v}^{(i)}(r)|^p r^{(\lambda+j+i)p-2n} A(r,\Omega^*)\,dr. \tag{45}$$

If it happens that $(\lambda+2m)p \le 2n$, then, since $p < t$ and $\operatorname{vol}\Omega^* < \infty$, all the integrals in (45) are finite by Hölder's inequality and so $u \in W^{m,p}(\Omega)$. Otherwise let

$$k = [(\lambda+2m)p - 2n][t/(t-p)] + 2n.$$

By (44) there exists $a \le 1$ such that if $\rho \le a$, then $A(\rho,\Omega) \le \rho^k$. It follows that if $r \ge 1/a$, then

$$r^{k-2n} A(r,\Omega^*) \le r^{k-2} \rho^k = r^{-2}.$$

Thus

$$\int_1^\infty |\tilde{v}^{(i)}(r)|^p r^{(\lambda+j+i)p-2n} A(r,\Omega^*)\,dr$$

$$= \int_1^\infty |\tilde{v}^{(i)}(r)|^p r^{(k-2n)(t-p)/t} A(r,\Omega^*)\,dr$$

$$\le \left\{\int_1^\infty |\tilde{v}^{(i)}(r)|^t A(r,\Omega^*)\,dr\right\}^{p/t} \left\{\int_1^\infty r^{k-2n} A(r,\Omega^*)\,dr\right\}^{(t-p)/t}$$

which is finite. Hence $u \in W^{m,p}(\Omega)$ and the proof is complete. ∎

## Imbedding Theorems for Domains with Cusps

**5.33** Having proved that Theorem 5.4 fails completely for domains that are sufficiently irregular we now show that certain imbeddings of the types considered in that theorem do hold for less irregular domains that however fail to have the cone property. Questions of this sort have been considered by several writers—see, for instance, the work of Globenko [26, 27] and Maz'ja [44, 45]. The treatment given below follows that given in one of the author's papers [1].

We consider domains $\Omega \subset \mathbb{R}^n$ whose boundaries consist only of $(n-1)$-dimensional surfaces, and it is assumed that $\Omega$ lies on only one side of its boundary. $\Omega$ is said to have a *cusp* at the point $x_0 \in \operatorname{bdry}\Omega$ if no finite open cone of positive volume contained in $\Omega$ can have vertex at $x_0$. The failure of a domain $\Omega$ to have any cusps does not, of course, imply that $\Omega$ has the cone property. We consider, for the moment, a family of special domains that we call *standard cusps* and that have cusps of power sharpness (less sharp than exponential cusps).

**5.34** If $1 \le k \le n-1$ and $\lambda > 1$, let $Q_{k,\lambda}$ denote the standard cusp in $E_n$ specified by the inequalities

$$\begin{gathered} x_1{}^2 + \cdots + x_k{}^2 < x_{k+1}^{2\lambda}, \quad x_{k+1} > 0, \dots, x_n > 0, \\ (x_1{}^2 + \cdots + x_k{}^2)^{1/\lambda} + x_{k+1}^2 + \cdots + x_n{}^2 < a^2, \end{gathered} \tag{46}$$

where $a$ is the radius of the ball of unit volume in $\mathbb{R}^n$. We note that $a < 1$, $Q_{k,\lambda}$ has axial plane spanned by the $x_{k+1}, \dots, x_n$ axes, and verticial plane (cusp plane) spanned by $x_{k+2}, \dots, x_n$. If $k = n-1$, the origin is the only vertex point of $Q_{k,\lambda}$. The outer boundary surface of $Q_{k,\lambda}$ is taken to be of the form (46) in order to simplify calculations. A sphere, or other suitable surface bounded away from the origin, could be used instead.

Corresponding to the standard cusp $Q_{k,\lambda}$ we consider the associated *standard cone* $Q_k = Q_{k,1}$ specified in terms of Cartesian coordinates $y_1, \dots, y_n$ by

$$\begin{gathered} y_1{}^2 + \cdots + y_k{}^2 < y_{k+1}^2, \quad y_{k+1} > 0, \dots, y_n > 0, \\ y_1{}^2 + \cdots + y_n{}^2 < a^2. \end{gathered}$$

Figure 4 illustrates standard cusps and their associated standard cones in $\mathbb{R}^2$ and $\mathbb{R}^3$. In $\mathbb{R}^3$ the cusp $Q_{2,2}$ has a single cusp point at the origin, while $Q_{1,2}$ has a cusp line along the $x_3$-axis.

It is convenient to adopt generalized "cylindrical" coordinates $(r_k, \phi_1, \dots, \phi_{k-1}, y_{k+1}, \dots, y_n)$ in $E_n$ so that $r_k \ge 0$, $-\pi \le \phi_1 \le \pi$, $0 \le \phi_2, \dots, \phi_{k-1} \le \pi$, and

$$\begin{aligned} y_1 &= r_k \sin\phi_1 \sin\phi_2 \cdots \sin\phi_{k-1}, \\ y_2 &= r_k \cos\phi_1 \sin\phi_2 \cdots \sin\phi_{k-1}, \\ y_3 &= \qquad\quad r_k \cos\phi_2 \cdots \sin\phi_{k-1}, \\ &\;\vdots \\ y_k &= \qquad\qquad\qquad\quad r_k \cos\phi_{k-1}. \end{aligned} \tag{47}$$

In terms of these coordinates $Q_k$ is represented by

$$\begin{gathered} 0 \le r_k < y_{k+1}, \quad y_{k+1} > 0, \dots, y_n > 0, \\ r_k{}^2 + y_{k+1}^2 + \cdots + y_n{}^2 < a^2. \end{gathered}$$

The standard cusp $Q_{k,\lambda}$ may be transformed into the associated cone $Q_k$ by

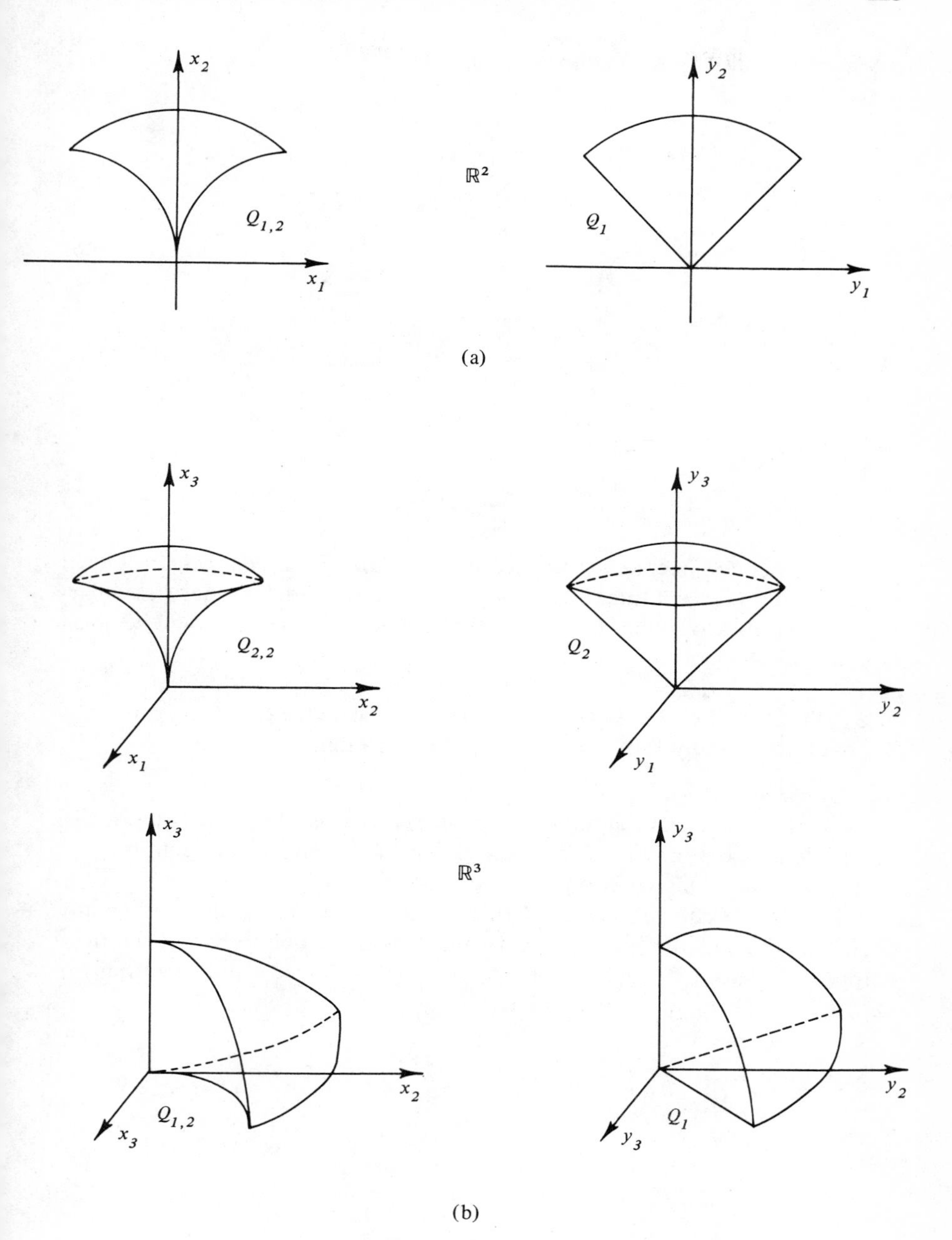


FIG. 4 Standard cusps and cones in (a) $\mathbb{R}^2$ and (b) $\mathbb{R}^3$.

means of the one-to-one transformation

$$\begin{aligned}
x_1 &= r_k^{\lambda} \sin\phi_1 \sin\phi_2 \cdots \sin\phi_{k-1}, \\
x_2 &= r_k^{\lambda} \cos\phi_1 \sin\phi_2 \cdots \sin\phi_{k-1}, \\
x_3 &= \qquad\quad r_k^{\lambda} \cos\phi_2 \cdots \sin\phi_{k-1}, \\
&\vdots \\
x_k &= \qquad\qquad\qquad\quad r_k^{\lambda} \cos\phi_{k-1}, \\
x_{k+1} &= y_{k+1}, \\
&\vdots \\
x_n &= y_n,
\end{aligned} \tag{48}$$

which has Jacobian determinant

$$\left|\frac{\partial(x_1, \dots, x_n)}{\partial(y_1, \dots, y_n)}\right| = \lambda r_k^{(\lambda-1)k}. \tag{49}$$

We now state three theorems extending imbeddings of the types considered in Theorem 5.4 (except trace imbeddings) to domains with boundary irregularities comparable to standard cusps. The proof of these theorems will be given later in this chapter.

**5.35 THEOREM** Let $\Omega$ be a domain in $\mathbb{R}^n$ having the following property: There exists a family $\Gamma$ of open subsets of $\Omega$ such that

(i) $\Omega = \bigcup_{G\in\Gamma} G$;

(ii) $\Gamma$ has the finite intersection property, that is, there exists a positive integer $N$ such that any $N+1$ distinct sets in $\Gamma$ have empty intersection;

(iii) at most one set $G \in \Gamma$ has the cone property;

(iv) there exist positive constants $\nu > mp-n$ and $A$ such that for any $G \in \Gamma$ not having the cone property there exists a one to one function $\psi$ mapping $G$ onto a standard cusp $Q_{k,\lambda}$, where $(\lambda-1)k \le \nu$ and such that for all $i, j$ $(1 \le i, j \le n)$, all $x \in G$, and all $y \in Q_{k,\lambda}$,

$$\left|\frac{\partial \psi_j}{\partial x_i}\right| \le A \qquad \text{and} \qquad \left|\frac{\partial (\psi^{-1})_j}{\partial y_i}\right| \le A.$$

Then

$$W^{m,p}(\Omega) \to L^q(\Omega), \qquad p \le q \le \frac{(\nu+n)p}{\nu+n-mp}. \tag{50}$$

[If $\nu = mp-n$, (50) holds for $p \le q < \infty$ (and $q = \infty$ if $p = 1$). If $\nu < mp-n$, (50) holds for $p \le q \le \infty$.]

**5.36 THEOREM** Let $\Omega$ be a domain in $\mathbb{R}^n$ having the following property: There exist positive constants $\nu < mp-n$ and $A$ such that for each $x \in \Omega$ there exists an open set $G$ with $x \in G \subset \Omega$ and a one-to-one mapping $\psi$ of $G$ onto a standard cusp $Q_{k,\lambda}$ with $(\lambda-1)k \le \nu$ and such that for all $i,j$ $(1 \le i,j \le n)$, all $x \in G$, and all $y \in Q_{k,\lambda}$,

$$\left|\frac{\partial \psi_j}{\partial x_i}\right| \le A \qquad \text{and} \qquad \left|\frac{\partial (\psi^{-1})_j}{\partial y_i}\right| \le A.$$

Then

$$W^{m,p}(\Omega) \to C_B{}^0(\Omega). \tag{51}$$

More generally, if $\nu < (m-j)p-n$, where $0 \le j \le m-1$, then

$$W^{m,p}(\Omega) \to C_B{}^j(\Omega).$$

**5.37 THEOREM** Let $\Omega$ be a domain in $\mathbb{R}^n$ with the following property: There exist positive constants $\nu$, $\delta$, and $A$ such that for each pair of points $x, y \in \Omega$ with $|x-y| \le \delta$ there exists an open set $G$ with $x, y \in G \subset \Omega$, and a one to one mapping $\psi$ of $G$ onto some standard cusp $Q_{k,\lambda}$ with $(\lambda-1)k \le \nu$, and such that for all $i,j$ $(1 \le i,j \le n)$, all $x \in G$, and all $y \in Q_{k,\lambda}$

$$\left|\frac{\partial \psi_j}{\partial x_i}\right| \le A \qquad \text{and} \qquad \left|\frac{\partial (\psi^{-1})_j}{\partial y_i}\right| \le A.$$

Suppose that for some $j$ with $0 \le j \le m-1$ we have $(m-j-1)p < \nu+n < (m-j)p$. Then

$$W^{m,p}(\Omega) \to C^{j,\mu}(\overline{\Omega}), \qquad 0 < \mu \le m-j-[(n+\nu)/p]. \tag{52}$$

If $(m-j-1)p = \nu+n$, then (52) holds with $0 < \mu < 1$. In either event we have $W^{m,p}(\Omega) \to C^j(\overline{\Omega})$.

**5.38 REMARKS** (1) The reader may wish to construct examples similar to those of Sections 5.25–5.28 to show that the three theorems above give the best possible imbeddings for the domains considered.

(2) The following example may help to illustrate Theorem 5.35: Let $\Omega = \{(x_1, x_2, x_3) \in \mathbb{R}^3 : x_2 > 0,\ x_2{}^2 < x_1 < 3x_2{}^2\}$. Setting $a = (3/4\pi)^{1/3}$, the radius of the ball of unit volume in $\mathbb{R}^3$, we may readily verify that the transformation

$$y_1 = x_1 + 2x_2{}^2, \qquad y_2 = x_2, \qquad y_3 = x_3 - (k/a), \qquad k = 0, \pm 1, \pm 2, \ldots,$$

transforms a subdomain $G_k$ of $\Omega$ onto the standard cusp $\Omega_{1,2}$ in the manner required of the functions $\psi$ in the statement of Theorem 5.35. Moreover, $\{G_k\}_{k=-\infty}^{\infty}$ has the finite intersection property and covers $\Omega$ up to a set with the cone property. Hence $W^{m,p}(\Omega) \to L^q(\Omega)$ for $p \le q \le 4p/(4-mp)$ if $mp < 4$, or for $p \le q < \infty$ if $mp = 4$, or $p \le q \le \infty$ if $mp > 4$.

## Imbedding Inequalities Involving Weighted Norms

**5.39** The technique of mapping a standard cusp onto its associated standard cone via (47) and (48) is central to the proof of Theorem 5.35. Such a transformation introduces into any integrals involved a weight factor in the form of the Jacobian determinant (49). Accordingly, we must obtain imbedding inequalities for such standard cones, corresponding to $L^p$-norms weighted by powers of the distance from the axial plane of the cone. Such inequalities are also useful in extending the imbedding theorem 5.4 to more general Sobolev spaces involving weighted norms.

We begin with some one-dimensional inequalities for functions continuously differentiable on a fixed open interval $(0, T)$ in $\mathbb{R}$.

**5.40 LEMMA** If $v > 0$ and $u \in C^1(0, T)$, and if $\int_0^T |u'(t)|\, t^v\, dt < \infty$, then $\lim_{t\to 0+} |u(t)|\, t^v = 0$.

PROOF Let $\varepsilon > 0$ be given and fix $s$, $0 < s < T/2$, small enough so that for any $t$, $0 < t < s$, we have

$$\int_t^s |u'(\tau)|\, \tau^v\, d\tau < \varepsilon/3.$$

Now there exists $\delta$, $0 < \delta < s$, such that

$$\delta^v |u'(T/2)| < \varepsilon/3 \qquad \text{and} \qquad (\delta/s)^v \int_s^{T/2} |u'(\tau)|\, \tau^v\, d\tau < \varepsilon/3.$$

If $0 < t \le \delta$, we have

$$|u(t)| \le |u(T/2)| + \int_t^{T/2} |u'(\tau)|\, d\tau$$

so that

$$t^v |u(t)| \le \delta^v |u(T/2)| + \int_t^s |u'(\tau)|\, \tau^v\, d\tau + (\delta/s)^v \int_s^{T/2} |u'(\tau)|\, \tau^v\, d\tau < \varepsilon.$$

Hence $\lim_{t\to 0+} t^v |u(t)| = 0$. ∎

**5.41 LEMMA** If $v > 0$, $p \ge 1$, and $u \in C^1(0, T)$, then

$$\int_0^T |u(t)|^p t^{v-1}\, dt \le \frac{v+1}{vT} \int_0^T |u(t)|^p t^v\, dt + \frac{p}{v} \int_0^T |u(t)|^{p-1} |u'(t)|\, t^v\, dt. \quad (53)$$

PROOF We may assume without loss of generality that the right side of (53) is finite, and that $p = 1$. Integration by parts gives

$$\int_0^T |u(t)|\left[\nu t^{\nu-1} - \frac{\nu+1}{T} t^\nu\right] dt = -\int_0^T \left[t^\nu - \frac{1}{T} t^{\nu+1}\right] \frac{d}{dt}|u(t)|\, dt;$$

Lemma 5.40 assures the vanishing of the integrated term at zero. Transposition and estimation of the term on the right now yields

$$\nu \int_0^T |u(t)|\, t^{\nu-1}\, dt \le \frac{\nu+1}{T} \int_0^T |u(t)|\, t^\nu\, dt + \int_0^T |u'(t)|\, t^\nu\, dt,$$

which is (53) for $p = 1$. ∎

**5.42 LEMMA** If $\nu > 0$, $p \ge 1$, and $u \in C^1(0, T)$, we have the following pair of inequalities:

$$\sup_{0<t<T} |u(t)|^p \le \frac{2}{T} \int_0^T |u(t)|^p\, dt + p \int_0^T |u(t)|^{p-1}\, |u'(t)|\, dt, \tag{54}$$

$$\sup_{0<t<T} |u(t)|^p t^\nu \le \frac{\nu+3}{T} \int_0^T |u(t)|^p t^\nu\, dt + 2p \int_0^T |u(t)|^{p-1}\, |u'(t)|\, t^\nu\, dt. \tag{55}$$

PROOF Again the inequalities need only be proved for $p = 1$. If $0 < t \le T/2$, we obtain by integration by parts

$$\int_0^{T/2} \left|u\left(t + \frac{T}{2} - \tau\right)\right| d\tau = \frac{T}{2}|u(t)| - \int_0^{T/2} \tau \frac{d}{d\tau}\left|u\left(t + \frac{T}{2} - \tau\right)\right| d\tau$$

whence

$$|u(t)| \le \frac{2}{T} \int_0^T |u(\sigma)|\, d\sigma + \int_0^T |u'(\sigma)|\, d\sigma.$$

For $T/2 \le t < T$ the same inequality results from partial integration of $\int_0^{T/2} |u(t+\tau-T/2)|\, d\tau$. This proves (54) for $p = 1$. Replacing $u(t)$ by $u(t)t^\nu$ in this inequality, we obtain

$$\begin{aligned}\sup_{0<t<T} |u(t)|\, t^\nu &\le \frac{2}{T} \int_0^T |u(t)|\, t^\nu\, dt + \int_0^T [|u'(t)|\, t^\nu + \nu\, |u(t)|\, t^{\nu-1}]\, dt \\ &\le \frac{2}{T} \int_0^T |u(t)|\, t^\nu\, dt + \int_0^T |u'(t)|\, t^\nu\, dt \\ &\quad + \nu\left\{\frac{\nu+1}{\nu T} \int_0^T |u(t)|\, t^\nu\, dt + \frac{1}{\nu} \int_0^T |u'(t)|\, t^\nu\, dt\right\},\end{aligned}$$

where (53) has been used to obtain the last inequality. This is the desired result (55) for $p = 1$. ∎

**5.43** Now we turn to $\mathbb{R}^n$, $n \geq 2$. If $x \in \mathbb{R}^n$, we shall make use of the spherical polar coordinate representation

$$x = (\rho, \phi) = (\rho, \phi_1, \ldots, \phi_{n-1}),$$

where $\rho \geq 0$, $-\pi \leq \phi_1 \leq \pi$, $0 \leq \phi_2, \ldots, \phi_{n-1} \leq \pi$, and

$$\begin{aligned} x_1 &= \rho \sin\phi_1 \sin\phi_2 \cdots \sin\phi_{n-1}, \\ x_2 &= \rho \cos\phi_1 \sin\phi_2 \cdots \sin\phi_{n-1}, \\ x_3 &= \rho \cos\phi_2 \cdots \sin\phi_{n-1}, \\ &\vdots \\ x_n &= \rho \cos\phi_{n-1}. \end{aligned}$$

The volume element is

$$dx = dx_1\, dx_2 \cdots dx_n = \rho^{n-1} \prod_{j=1}^{n-1} \sin^{j-1}\phi_j \, d\rho \, d\phi,$$

where $d\phi = d\phi_1 \cdots d\phi_{n-1}$.

We define functions $r_k = r_k(x)$ for $1 \leq k \leq n$ as follows:

$$\begin{aligned} r_1(x) &= \rho |\sin\phi_1| \prod_{j=2}^{n-1} \sin\phi_j, \\ r_k(x) &= \rho \prod_{j=k}^{n-1} \sin\phi_j, \qquad k = 2, 3, \ldots, n-1, \\ r_n(x) &= \rho. \end{aligned}$$

For $1 \leq k \leq n-1$, $r_k(x)$ is the distance of $x$ from the coordinate plane spanned by the axes $x_{k+1}, \ldots, x_n$; $r_n(x)$ being just the distance of $x$ from the origin. In connection with the use of product symbols of the form $P = \prod_{j=k}^{m} P_j$, be it agreed hereafter that $P = 1$ if $m < k$.

Let $Q$ be an open, conical domain in $\mathbb{R}^n$ specified by the inequalities

$$0 < \rho < a, \qquad -\beta_1 < \phi_1 < \beta_1, \qquad 0 \leq \phi_j < \beta_j, \qquad j = 2, 3, \ldots, n-1, \tag{56}$$

where $0 < \beta_i \leq \pi$. [Inequalities "$<$" in (56), corresponding to any $\beta_i = \pi$, are replaced by "$\leq$." If all $\beta_i = \pi$, the first inequality is replaced by $0 \leq \rho < a$.] It should be noted that any standard cone $Q_k$ (introduced in Section 5.34) is of the form (56) for some choice of the parameters $\beta_i$, $1 \leq i \leq n-1$.

The following lemma generalizes Lemma 5.41 in a manner suitable for our purposes.

**5.44 LEMMA** Let $Q$ be specified by (56) and let $p \geq 1$. Suppose that either $m = k = 1$, or $2 \leq m \leq n$ and $1 \leq k \leq n$. Suppose also that $1-k < v_1 \leq v \leq$

$\nu_2 < \infty$. Then there exists a constant $K = K(m, k, n, p, \nu_1, \nu_2, \beta_1, \ldots, \beta_{n-1})$ independent of $\nu$ and $a$, such that for every function $u \in C^1(Q)$ we have

$$\int_Q |u(x)|^p [r_k(x)]^\nu [r_m(x)]^{-1} \, dx$$
$$\leq K \int_Q |u(x)|^{p-1} [(1/a)\,|u(x)| + |\operatorname{grad} u(x)|] [r_k(x)]^\nu \, dx. \tag{57}$$

Proof Once again it is sufficient to establish (57) for $p = 1$. Let $Q_+ = \{x = (\rho, \phi) \in Q : \phi_1 \geq 0\}$, $Q_- = \{x \in Q : \phi_1 \leq 0\}$. Then $Q = Q_+ \cup Q_-$. We shall prove (57) only for $Q_+$ (which, however, we continue to call $Q$); a similar proof holds for $Q_-$ so that (57) holds for the given $Q$. Accordingly, assume $Q = Q_+$.

For $k \leq m$ we may write (57) in the form (taking $p = 1$)

$$\int_Q |u| \prod_{j=2}^{k-1} \sin^{j-1} \phi_j \prod_{j=k}^{m-1} \sin^{\nu+j-1} \phi_j \prod_{j=m}^{n-1} \sin^{\nu+j-2} \phi_j \, \rho^{\nu+n-2} \, d\rho \, d\phi$$
$$\leq K \int_Q [(1/a)\,|u| + |\operatorname{grad} u|] \prod_{j=2}^{k-1} \sin^{j-1} \phi_j \prod_{j=k}^{n-1} \sin^{\nu+j-1} \phi_j \, \rho^{\nu+n-1} \, d\rho \, d\phi. \tag{58}$$

For $k > m \geq 2$ we may write (57) in the form

$$\int_Q |u| \prod_{j=2}^{m-1} \sin^{j-1} \phi_j \prod_{j=m}^{k-1} \sin^{j-2} \phi_j \prod_{j=k}^{n-1} \sin^{\nu+j-2} \phi_j \, \rho^{\nu+n-2} \, d\rho \, d\phi$$
$$\leq K \int_Q [(1/a)\,|u| + |\operatorname{grad} u|] \prod_{j=2}^{k-1} \sin^{j-1} \phi_j \prod_{j=k}^{n-1} \sin^{\nu+j-1} \phi_j \, \rho^{\nu+n-1} \, d\rho \, d\phi. \tag{59}$$

By virtue of the restrictions placed on $\nu$, $m$, and $k$ in the statement of the lemma, (58) and (59) are both special cases of

$$\int_Q |u| \prod_{j=1}^{i-1} \sin^{\mu_j} \phi_j \prod_{j=i}^{n-1} \sin^{\mu_j - 1} \phi_j \, \rho^{\nu+n-2} \, d\rho \, d\phi$$
$$\leq K \int_Q [(1/a)\,|u| + |\operatorname{grad} u|] \prod_{j=1}^{n-1} \sin^{\mu_j} \phi_j \, \rho^{\nu+n-1} \, d\rho \, d\phi, \tag{60}$$

where $1 \leq i \leq n$, $\mu_j \geq 0$, and $0 < \mu_j^* \leq \mu_j$ if $j \geq i$. We prove (60) by backwards induction on $i$. For $i = n$, (60) is obtained by applying Lemma 5.41 to $u$ considered as a function of $\rho$ on $(0, a)$ and then integrating the remaining variables with the appropriate weights. Assume, therefore, that (60) has been proved for $i = k+1$ where $1 \leq k \leq n-1$. We prove it now holds for $i = k$.

If $\beta_k < \pi$, we have

$$\sin \phi_k \leq \phi_k \leq K_1 \sin \phi_k, \qquad 0 \leq \phi_k \leq \beta_k, \tag{61}$$

where $K_1 = K_1(\beta_k)$. By Lemma 5.41, and since

$$|\partial u/\partial \phi_k| \le \rho\,|\mathrm{grad}\, u| \prod_{j=k+1}^{n-1} \sin\phi_j,$$

we have

$$\begin{aligned}
&\int_0^{\beta_k} |u(\rho,\phi)| \sin^{\mu_k - 1}\phi_k \, d\phi_k \\
&\qquad \le \int_0^{\beta_k} |u|\, \phi_k^{\mu_k - 1}\, d\phi_k \\
&\qquad \le K_2 \int_0^{\beta_k} \left[ |u| + |\mathrm{grad}\, u|\, \rho \prod_{j=k+1}^{n-1} \sin\phi_j \right] \phi_k^{\mu_k}\, d\phi_k \\
&\qquad \le K_3 \int_0^{\beta_k} \left[ |u| + |\mathrm{grad}\, u|\, \rho \prod_{j=k+1}^{n-1} \sin\phi_j \right] \sin^{\mu_k}\phi_k\, d\phi_k . \qquad (62)
\end{aligned}$$

Note that $K_2$, and hence $K_3$, depend on $\beta_k$ but may be chosen independent of $\mu_k$, and hence of $\nu$, under the conditions of the lemma. If $\beta_k = \pi$, we obtain (62) by writing $\int_0^\pi = \int_0^{\pi/2} + \int_{\pi/2}^\pi$ and using, in place of (61) ,the inequalities

$$\begin{aligned}
&\sin\phi_k \le \phi_k \le (\pi/2)\sin\phi_k && \text{if} \quad 0 \le \phi_k \le \pi/2 \\
&\sin\phi_k \le \pi - \phi_k \le (\pi/2)\sin\phi_k && \text{if} \quad \pi/2 \le \phi_k \le \pi.
\end{aligned} \qquad (63)$$

We now have, using (62) and the induction hypothesis,

$$\begin{aligned}
&\int_Q |u| \prod_{j=1}^{k-1} \sin^{\mu_j}\phi_j \prod_{j=k}^{n-1} \sin^{\mu_j - 1}\phi_j\, \rho^{\nu+n-2}\, d\rho\, d\phi \\
&\qquad \le \int_0^a \rho^{\nu+n-2}\, d\rho \prod_{j=1}^{k-1} \int_0^{\beta_j} \sin^{\mu_j}\phi_j\, d\phi_j \\
&\qquad\qquad \times \prod_{j=k+1}^{n-1} \int_0^{\beta_j} \sin^{\mu_j - 1}\phi_j\, d\phi_j \int_0^{\beta_k} |u| \sin^{\mu_k - 1}\phi_k\, d\phi_k \\
&\qquad \le K_3 \int_Q |\mathrm{grad}\, u| \prod_{j=1}^{n-1} \sin^{\mu_j}\phi_j\, \rho^{\nu+n-1}\, d\rho\, d\phi \\
&\qquad\qquad + K_3 \int_Q |u| \prod_{j=1}^{k} \sin^{\mu_j}\phi_j \prod_{j=k+1}^{n-1} \sin^{\mu_j - 1}\phi_j\, \rho^{\nu+n-2}\, d\rho\, d\phi \\
&\qquad \le K \int_Q [(1/a)\,|u| + |\mathrm{grad}\, u|] \prod_{j=1}^{n-1} \sin^{\mu_j}\phi_j\, \rho^{\nu+n-1}\, d\rho\, d\phi.
\end{aligned}$$

This completes the induction establishing (60) and hence the lemma. ∎

In the following lemma we obtain an imbedding inequality similar to that of Lemma 5.10 for the domain $Q$ and appropriately weighted $L^p$-norms.

**5.45 LEMMA** Let $Q$ be as specified by (56) and let $p \geq 1$ and $1 \leq k \leq n$. Suppose that $\max(1-k, p-n) < v_1 < v_2 < \infty$. Then there exists a constant $K = K(k, n, p, v_1, v_2, \beta_1, \dots, \beta_{n-1})$, independent of $a$, such that for every $v$ satisfying $v_1 \leq v \leq v_2$, and every function $u \in C^1(Q) \cap C(\overline{Q})$ we have

$$\left\{\int_Q |u(x)|^q [r_k(x)]^v \, dx\right\}^{1/q} \leq K\left\{\int_Q [(1/a^p)|u(x)|^p + |\operatorname{grad} u(x)|^p][r_k(x)]^v \, dx\right\}^{1/p}, \tag{64}$$

where $q = (v+n)p/(v+n-p)$.

PROOF Let $\delta = (v+n-1)p/(v+n-p)$, $s = (v+n-1)/v$, $s' = (v+n-1)/(n-1)$. We have by Hölder's inequality and Lemma 5.44 (the case $m = k$)

$$\begin{aligned}\int_Q |u(x)|^q [r_k(x)]^v \, dx &\leq \left\{\int_Q |u|^\delta r_k^{v-1} \, dx\right\}^{1/s} \left\{\int_Q |u|^{n\delta/(n-1)} r_k^{nv/(n-1)} \, dx\right\}^{1/s'} \\ &\leq K_1 \left\{\int_Q |u|^{\delta-1} [(1/a)|u| + |\operatorname{grad} u|] r_k{}^v \, dx\right\}^{1/s} \\ &\quad \times \left\{\int_Q |u|^{n\delta/(n-1)} r_k^{nv/(n-1)} \, dx\right\}^{1/s'}. \end{aligned} \tag{65}$$

In order to estimate the last integral above we adopt the notation

$$\rho^* = (\phi_1, \dots, \phi_{n-1}), \qquad \phi_j{}^* = (\rho, \phi_1, \dots, \hat{\phi}_j, \dots, \phi_{n-1}), \qquad j = 1, 2, \dots, n-1,$$

where the caret denotes omission of a component. Let

$$\begin{aligned} Q_0{}^* &= \{\rho^* : (\rho, \rho^*) \in Q \text{ for } 0 < \rho < a\} \\ Q_j{}^* &= \{\phi_j{}^* : (\rho, \phi) \in Q \text{ for } 0 < \phi_j < \beta_j\}. \end{aligned}$$

$Q_0{}^*$ and $Q_j{}^*$ $(1 \leq j \leq n-1)$ are domains in $\mathbb{R}^{n-1}$. We define functions $F_0$ and $F_j$ on $Q_0{}^*$ and $Q_j{}^*$, respectively, as follows:

$$\begin{aligned} [F_0(\rho^*)]^{n-1} &= [F_0(\phi_1, \dots, \phi_{n-1})]^{n-1} \\ &= \sup_{0<\rho<a} [|u|^\delta \rho^{v+n-1}] \prod_{i=k}^{n-1} \sin^v \phi_i \prod_{i=2}^{n-1} \sin^{i-1} \phi_i, \\ [F_j(\phi_j{}^*)]^{n-1} &= [F_j(\rho, \phi_1, \dots, \hat{\phi}_j, \dots, \phi_{n-1})]^{n-1} \\ &= \sup_{0<\phi_j<\beta_j} [|u|^\delta \sin^{v+j-1} \phi_j] \rho^{v+n-2} \\ &\quad \times \prod_{i=k}^{n-1} \sin^v \phi_i \prod_{i=2}^{j-1} \sin^{i-1} \phi_i \prod_{i=j+1}^{n-1} \sin^{i-2} \phi_i. \end{aligned}$$

Then we have

$$|u|^{n\delta/(n-1)} r_k^{n\nu/(n-1)} \rho^{n-1} \prod_{i=2}^{n-1} \sin^{i-1} \phi_i \le F_0(\rho^*) \prod_{j=1}^{n-1} F_j(\phi_j^*).$$

Applying the combinatorial Lemma 5.9, we obtain

$$\begin{aligned}
\int_Q |u|^{n\delta/(n-1)} r_k^{n\nu/(n-1)}\, dx
&\le \int_Q F_0(\rho^*) \prod_{j=1}^{n-1} F_j(\phi_j^*)\, d\rho\, d\phi \\
&\le \left\{ \int_{Q_0^*} [F_0(\rho^*)]^{n-1}\, d\phi \prod_{j=1}^{n-1} \int_{Q_j^*} [F_j(\phi_j^*)]^{n-1}\, d\rho\, d\hat{\phi}_j \right\}^{1/(n-1)}
\end{aligned} \tag{66}$$

Now by Lemma 5.42, and since $|\partial u/\partial\rho| \le |\operatorname{grad} u|$,

$$\sup_{0<\rho<a} |u|^\delta \rho^{\nu+n-1} \le K_2 \int_0^a |u|^{\delta-1} [(1/a)|u| + |\operatorname{grad} u|] \rho^{\nu+n-1}\, d\rho,$$

where $K_2$ is independent of $\nu$ for $1-n < \nu_1 \le \nu \le \nu_2 < \infty$. It follows that

$$\int_{Q_0^*} [F_0(\rho^*)]^{n-1}\, d\phi \le K_2 \int_Q |u|^{\delta-1} [(1/a)|u| + |\operatorname{grad} u|] r_k^\nu\, dx. \tag{67}$$

Similarly, by making use of inequality (61) or (63) as in Lemma 5.44, we obtain from Lemma 5.42

$$\begin{aligned}
\sup_{0<\phi_j<\beta_j} |u|^\delta \sin^{\nu+j-1} \phi_j
&\le K_{2,j} \int_0^{\beta_j} |u|^{\delta-1} \left[ |u| + \left| \frac{\partial u}{\partial \phi_j} \right| \right] \sin^{\nu+j-1} \phi_j\, d\phi_j \\
&\le K_{2,j} \int_0^{\beta_j} |u|^{\delta-1} \left[ |u| + |\operatorname{grad} u|\, \rho \prod_{i=j+1}^{n-1} \sin \phi_i \right] \sin^{\nu+j-1} \phi_j\, d\phi_j
\end{aligned}$$

since $|\partial u/\partial\phi_j| \le \rho \prod_{i=j+1}^{n-1} \sin\phi_i$. Hence

$$\begin{aligned}
\int_{Q_j^*} [F_j(\phi_j^*)]^{n-1}\, d\rho\, d\hat{\phi}_j
&\le K_{2,j} \int_Q |\operatorname{grad} u| |u|^{\delta-1} r_k^\nu\, dx + K_{2,j} \int_Q |u|^\delta r_k^\nu r_{j+1}^{-1}\, dx \\
&\le K_{3,j} \int_Q |u|^{\delta-1} [(1/a)|u| + |\operatorname{grad} u|] r_k^\nu\, dx
\end{aligned} \tag{68}$$

where we have used Lemma 5.44 again to obtain the last inequality. Note that the constants $K_{2,j}$ and $K_{3,j}$ can be chosen independent of $v$ for the values of $\nu$ allowed. Substitution of (67) and (68) into (66) and then into (65) leads to

$$\int_Q |u|^q r_k^{\nu}\,dx$$

$$\le K_4 \left\{ \int_Q |u|^{\delta-1} [(1/a)\,|u| + |\operatorname{grad} u|]\, r_k^{\nu}\,dx \right\}^{1/s+n/(n-1)s'}$$

$$\le K_4 \left\{ \left\{ \int_Q |u|^q r_k^{\nu}\,dx \right\}^{(p-1)/p} \right.$$

$$\left. \times \left\{ 2^{p-1} \int_Q [(1/a^p)\,|u|^p + |\operatorname{grad} u|^p]\, r_k^{\nu}\,dx \right\}^{1/p} \right\}^{(\nu+n)/(\nu+n-1)}.$$

Since $(\nu+n-1)/(\nu+n)-(p-1)/p = 1/q$, inequality (64) follows by cancellation for, since $u$ is bounded on $Q$ and $\nu > 1-n$, $\int_Q |u|^q r_k^{\nu}\,dx$ is finite. ∎

**5.46 REMARKS** (1) The assumption that $u \in C(\bar{Q})$ was made only to ensure that the above cancellation was justified. In fact the lemma holds for any $u \in C^1(Q)$.

(2) If $1-k < \nu_1 < \nu_2 < \infty$ and $\nu_1 \le \nu \le \nu_2$, where $\nu \le p-n$, then (64) holds for any $q$ satisfying $1 \le q < \infty$. It is sufficient to prove this for large $q$. If $q \ge (\nu+n)/(\nu+n-1)$, then $q = (\nu+n)s/(\nu+n-s)$ for some $s$ satisfying $1 \le s < p$. Thus

$$\left\{ \int_Q |u|^q r_k^{\nu}\,dx \right\}^{s/q}$$

$$\le K \int_Q [(1/a^s)\,|u|^s + |\operatorname{grad} u|^s]\, r_k^{\nu}\,dx$$

$$\le K \left\{ 2^{(p-s)/s} \int_Q [(1/a^p)\,|u|^p + |\operatorname{grad} u|^p]\, r_k^{\nu}\,dx \right\}^{s/p} \left\{ \int_Q r_k^{\nu}\,dx \right\}^{(p-s)/p},$$

which yields (64) since the last factor on the right is finite.

(3) If $\nu = m$, a positive integer, then (64) can be obtained very simply as follows. Let $y = (x, z) = (x_1, \ldots, x_n, z_1, \ldots, z_m)$ denote a point in $\mathbb{R}^{n+m}$ and define $u^*(y) = u(x)$ for $x \in Q$. If

$$Q^* = \{y \in \mathbb{R}^{n+m} : y = (x, z),\ x \in Q,\ 0 < z_j < r_k(x),\ 1 \le j \le m\},$$

then $Q^*$ has the cone property in $\mathbb{R}^{n+m}$, whence by Theorem 5.4 we have, putting $q = (n+m)p/(n+m-p)$,

$$\left\{\int_Q |u|^q r_k^{\ m}\, dx\right\}^{1/q} = \left\{\int_{Q^*} |u^*(y)|^q\, dy\right\}^{1/q}$$
$$\leq K\left\{\int_{Q^*} [(1/a^p)\,|u^*(y)|^p + |\operatorname{grad} u^*(y)|^p]\, dy\right\}^{1/p}$$
$$= K\left\{\int_Q [(1/a^p)\,|u|^p + |\operatorname{grad} u|^p]\, r_k^{\ m}\, dx\right\}^{1/p}$$

since $|\operatorname{grad} u^*(y)| = |\operatorname{grad} u(x)|$, $u^*$ being independent of $z$.

(4) Suppose that $u \in C_0^{\infty}(\mathbb{R}^n)$, or, more generally, that

$$\int_{\mathbb{R}^n} |u(x)|^p [r_k(x)]^{\nu}\, dx < \infty$$

with $\nu$ as in the above lemma. If we take $\beta_i = \pi$, $1 \leq i \leq n-1$, and let $a \to \infty$ in (64), we obtain

$$\left\{\int_{\mathbb{R}^n} |u(x)|^q [r_k(x)]^{\nu}\, dx\right\}^{1/q} \leq K\left\{\int_{\mathbb{R}^n} |\operatorname{grad} u(x)|^p [r_k(x)]^{\nu}\, dx\right\}^{1/p}.$$

This generalizes Sobolev's inequality as given in Section 5.11.

We now generalize Lemma 5.15 to allow for weighted norms. It is convenient to deal here with arbitrary domains having the cone property, rather than the special case $Q$ considered above. The following elementary result will be needed.

**5.47 LEMMA** Let $z \in \mathbb{R}^k$ and let $\Omega$ be a domain of finite volume in $\mathbb{R}^k$. If $0 \leq \nu < k$, then

$$\int_\Omega |x-z|^{-\nu}\, dx \leq K(\operatorname{vol}\Omega)^{1-\nu/k}, \tag{69}$$

where the constant $K$ depends on $\nu$ and $k$ but not on $z$ or $\Omega$.

PROOF Let $B$ be the ball in $\mathbb{R}^k$ having center $z$ and the same volume as $\Omega$. It is readily verified that the left side of (69) does not exceed $\int_B |x-z|^{-\nu}\, dx$. But (69) clearly holds for $\Omega = B$. ∎

**5.48 LEMMA** Let $\Omega$ be a domain with the cone property in $\mathbb{R}^n$. Let $1 \leq k \leq n$ and let $P$ be an $(n-k)$-dimensional plane in $\mathbb{R}^n$. Denote by $r(x)$ the distance from $x$ to $P$. If $0 \leq \nu < p-n$, then for all $u \in C^1(\Omega)$ we have

$$\sup_{x\in\Omega} |u(x)| \leq K\left\{\int_\Omega [|u(x)|^p + |\operatorname{grad} u(x)|^p][r(x)]^{\nu}\, dx\right\}^{1/p}, \tag{70}$$

where the constant $K$ may depend on $\nu, n, p, k$ and the cone $C$ determining the cone property for $\Omega$, but not on $u$.

PROOF Throughout this proof $A_i$ and $K_i$ will denote various constants depending on one or more of the parameters on which $K$ is allowed to depend in (70). It is sufficient to prove that if $C$ is a finite cone contained in $\Omega$ having vertex at, say, the origin, then

$$|u(0)| \le K\left\{\int_C [|u(x)|^p + |\operatorname{grad} u(x)|^p][r(x)]^\nu\, dx\right\}^{1/p}. \tag{71}$$

For $0 \le j \le n$ let $A_j$ denote the supremum of the Lebesgue $j$-dimensional measure of the projection of $C$ onto $\mathbb{R}^j$, taken over all $j$-dimensional subspaces $\mathbb{R}^j$ of $\mathbb{R}^n$. Writing $x = (x', x'')$ where $x' = (x_1, \ldots, x_{n-k})$ and $x'' = (x_{n-k+1}, \ldots, x_n)$, we may assume, without loss of generality, that $P$ is orthogonal to the coordinate axes corresponding to the components of $x''$. Define

$$S = \{x' \in \mathbb{R}^{n-k} : (x', x'') \in C \text{ for some } x'' \in \mathbb{R}^k\},$$
$$R(x') = \{x'' \in \mathbb{R}^k : (x', x'') \in C\} \text{ for each } x' \in S.$$

For $0 \le t \le 1$ we denote by $C_t$ the cone $\{tx : x \in C\}$ so that $C_t \subset C$ and $C_t = C$ if $t = 1$. For $C_t$ we define the quantities $A_{t,j}$, $S_t$, and $R_t(x')$ analogously to the similar quantities defined above for $C$. Clearly $A_{t,j} = t^j A_j$. If $x \in C$, we have

$$u(x) = u(0) + \int_0^1 \frac{d}{dt} u(tx)\, dt,$$

so that

$$|u(0)| \le |u(x)| + |x| \int_0^1 |\operatorname{grad} u(tx)|\, dt.$$

Setting $V = \operatorname{vol} C$ and $a = \sup_{x \in C} |x|$, and integrating the above inequality over $C$, we obtain

$$\begin{aligned} V|u(0)| &\le \int_C |u(x)|\, dx + a \int_C \int_0^1 |\operatorname{grad} u(tx)|\, dt \\ &= \int_C |u(x)|\, dx + a \int_0^1 t^{-n}\, dt \int_{C_t} |\operatorname{grad} u(x)|\, dx. \end{aligned} \tag{72}$$

Let $z$ denote the orthogonal projection of $x$ onto $P$. Then $r(x) = |x'' - z''|$. Since $0 \le \nu < p - n$ we have $p > 1$, and so by Lemma 5.47

$$\begin{aligned} \int_{C_t} [r(x)]^{-\nu/(p-1)}\, dx &= \int_{S_t} dx' \int_{R_t(x')} |x'' - z''|^{-\nu/(p-1)}\, dx'' \\ &\le K_1 \int_{S_t} [A_{t,k}]^{1-\nu/k(p-1)}\, dx' \\ &\le K_1 [A_{t,k}]^{1-\nu/k(p-1)} A_{t,n-k} = K_2 t^{n-\nu/(p-1)}. \end{aligned}$$

It follows that

$$\int_{C_t} |\operatorname{grad} u(x)|\, dx \le \left\{ \int_{C_t} |\operatorname{grad} u(x)|^p [r(x)]^\nu\, dx \right\}^{1/p} \times \left\{ \int_{C_t} [r(x)]^{-\nu/(p-1)}\, dx \right\}^{1/p'}$$
$$\le K_3 t^{n-(\nu+n)/p} \left\{ \int_C |\operatorname{grad} u(x)|^p [r(x)]^\nu\, dx \right\}^{1/p}. \qquad (73)$$

Hence, since $\nu < p-n$,

$$\int_0^1 t^{-n}\, dt \int_{C_t} |\operatorname{grad} u(x)|\, dx \le K_4 \left\{ \int_C |\operatorname{grad} u(x)|^p [r(x)]^\nu\, dx \right\}^{1/p}. \qquad (74)$$

Similarly,

$$\int_C |u(x)|\, dx \le \left\{ \int_C |u(x)|^p [r(x)]^\nu\, dx \right\}^{1/p} \left\{ \int_C [r(x)]^{-\nu/(p-1)}\, dx \right\}^{1/p'}$$
$$\le K_5 \left\{ \int_C |u(x)|^p [r(x)]^\nu\, dx \right\}^{1/p}. \qquad (75)$$

Inequality (71) now follows from (72), (74), and (75). ∎

**5.49 LEMMA** Suppose all the conditions of Lemma 5.48 are satisfied and, in addition, $\Omega$ has the strong local Lipschitz property. Then for all $u \in C^1(\Omega)$ we have

$$\sup_{\substack{x, y \in \Omega \\ x \ne y}} \frac{|u(x)-u(y)|}{|x-y|^\mu} \le K \left\{ \int_\Omega [|u(x)|^p + |\operatorname{grad} u(x)|^p]\, [r(x)]^\nu\, dx \right\}^{1/p}, \qquad (76)$$

where $\mu = 1-(\nu+n)/p$ satisfies $0 < \mu < 1$ and $K$ is independent of $u$.

PROOF The proof is the same as that given for inequality (28) in Lemma 5.17, except that the inequality

$$\int_{\Omega_{t\sigma}} |\operatorname{grad} u(z)|\, dz \le K_1 t^{n-(\nu+n)/p} \left\{ \int_\Omega |\operatorname{grad} u(z)|^p [r(z)]^\nu\, dz \right\}^{1/p} \qquad (77)$$

is used in (29) in place of the special case $\nu = 0$ actually used there. Inequality (77) is obtained in the same way as (73) above. ∎

## Proofs of Theorems 5.35–5.37

**5.50 LEMMA** Let $\bar{\nu} \ge 0$. If $\bar{\nu} > p-n$, let $1 \le q \le (\bar{\nu}+n)p/(\bar{\nu}+n-p)$; otherwise let $1 \le q < \infty$. There exists a constant $K = K(n, p, \bar{\nu})$ such that for every standard cusp domain $Q_{k,\lambda}$ (see Section 5.34) for which $(\lambda-1)k \equiv \nu \le \bar{\nu}$, and every $u \in C^1(Q_{k,\lambda})$, we have

$$\|u\|_{0,q,Q_{k,\lambda}} \le K\|u\|_{1,p,Q_{k,\lambda}}. \tag{78}$$

PROOF Since each $Q_{k,\lambda}$ has the segment property, it suffices to prove (78) for $u \in C^1(\overline{Q_{k,\lambda}})$. We first do so for given $k$ and $\lambda$ and then show that $K$ may be chosen so as to be independent of these parameters.

First suppose $\bar{\nu} > p-n$. It is sufficient to prove (78) for

$$q = (\bar{\nu}+n)/(\bar{\nu}+n-p).$$

For $u \in C^1(\overline{Q_{k,\lambda}})$ define $\tilde{u}(y) = u(x)$, where $y$ is related to $x$ by (47) and (48). Thus $\tilde{u} \in C^1(Q_k) \cap C(Q_k)$, where $Q_k$ is the standard cone associated with $Q_{k,\lambda}$. By Lemma 5.45, and since $q \le (\nu+n)p/(\nu+n-p)$ we have

$$\begin{aligned}\|u\|_{0,q,Q_{k,\lambda}} &= \left\{\lambda \int_{Q_k} |\tilde{u}(y)|^q [r_k(y)]^\nu \, dy\right\}^{1/q} \\ &\le K_1 \left\{\int_{Q_k} [|\tilde{u}(y)|^p + |\operatorname{grad} \tilde{u}(y)|^p][r_k(y)]^\nu \, dy\right\}^{1/q}.\end{aligned} \tag{79}$$

Now $x_j = r_k^{\lambda-1} y_j$ if $1 \le j \le k$; $x_j = y_j$ if $k+1 \le j \le n$. Since $r_k^2 = y_1^2 + \cdots + y_k^2$ we have

$$\frac{\partial x_j}{\partial y_i} = \begin{cases} \delta_{ij} r_k^{\lambda-1} + (\lambda-1) r_k^{\lambda-3} y_i y_j & \text{if } 1 \le i,\ j \le k \\ \delta_{ij} & \text{otherwise.} \end{cases}$$

Since $r_k(y) \le 1$ on $Q_k$ it follows that

$$|\operatorname{grad} \tilde{u}(y)| \le K_2 |\operatorname{grad} u(x)|.$$

Hence (78) follows from (79) in this case. For $\bar{\nu} \le p-n$ and arbitrary $q$ the proof is similar, being based on Remark 5.46(2).

In order to show that the constant $K$ in (78) can be chosen independent of $k$ and $\lambda$ provided $\nu = (\lambda-1)k \le \bar{\nu}$, we note that it is sufficient to prove that there is a constant $\tilde{K}$ such that for any such $k, \lambda$ and all $v \in C^1(Q_k) \cap C(\bar{Q}_k)$ we have

$$\begin{aligned}&\left\{\int_{Q_k} |v(y)|^q [r_k(y)]^\nu \, dy\right\}^{1/q} \\ &\qquad \le \tilde{K} \left\{\int_{Q_k} [|v(y)|^p + |\operatorname{grad} v(y)|^p][r_k(y)]^\nu \, dy\right\}^{1/p}.\end{aligned} \tag{80}$$

In fact, it is sufficient to establish (80) with $\tilde{K}$ depending on $k$ as we can then use the maximum of $\tilde{K}(k)$ over the finitely many values of $k$ allowed. We distinguish three cases.

CASE I $\bar{\nu} < p-n$, $1 \le q < \infty$. By Lemma 5.48 we have for $0 \le \nu \le \bar{\nu}$,

$$\sup_{x \in Q_k} |v(x)| \le K(\nu)\left\{\int_{Q_k} [|v(y)|^p + |\operatorname{grad} v(y)|^p][r_k(y)]^\nu \, dy\right\}^{1/p}. \tag{81}$$

Since the integral on the right decreases as $\nu$ increases we have $K(\nu) \le K(\bar{\nu})$ and (80) now follows from (81) and the boundedness of $Q_k$.

CASE II $\bar{\nu} > p-n$ Again it is enough to deal with $q = (\bar{\nu}+n)p/(\bar{\nu}+n-p)$. From Lemma 5.45 we obtain

$$\left\{\int_{Q_k} |v|^s r_k{}^\nu \, dy\right\}^{1/s} \le K_1\left\{\int_{Q_k} [|v|^p + |\operatorname{grad} v|^p] r_k{}^\nu \, dy\right\}^{1/p}, \tag{82}$$

where $s = (\nu+n)p/(\nu+n-p) \ge q$ and $K_1$ is independent of $\nu$ for $p-n < \nu_0 \le \nu \le \bar{\nu}$. By Hölder's inequality, and since $r_k(y) \le 1$ on $Q_k$, we have

$$\left\{\int_{Q_k} |v|^q r_k{}^\nu \, dy\right\}^{1/q} \le \left\{\int_{Q_k} |v|^s r_k{}^\nu \, dy\right\}^{1/s} [\operatorname{vol} Q_k]^{(s-q)/sq}$$

so that if $\nu_0 \le \nu \le \bar{\nu}$, then (80) follows from (82).

If $p-n < 0$, we can take $\nu_0 = 0$ and be done. Otherwise $p \ge n \ge 2$. Fixing $\nu_0 = (\bar{\nu}-n+p)/2$, we can find $\nu_1$ such that $0 \le \nu_1 \le p-n$ (or $\nu_1 = 0$ if $p = n$) such that for $\nu_1 \le \nu \le \nu_0$ we have

$$1 \le t = \frac{(\nu+n)(\bar{\nu}+n)p}{(\nu+n)(\bar{\nu}+n) + (\bar{\nu}-\nu)p} \le \frac{p}{1+\varepsilon_0},$$

where $\varepsilon_0 > 0$ and depends only on $\bar{\nu}$, $n$, and $p$. Because of the latter inequality we may also assume $t-n < \nu_1$. Since $(\nu+n)t/(\nu+n-t) = q$ we have, again by Lemma 5.45 and Hölder's inequality,

$$\begin{aligned}
&\left\{\int_{Q_k} |v|^q r_k{}^\nu \, dy\right\}^{1/q} \\
&\quad \le K_2\left\{\int_{Q_k} [|v|^t + |\operatorname{grad} v|^t] r_k{}^\nu \, dy\right\}^{1/t} \\
&\quad \le 2^{(p-t)/pt} K_2 \left\{\int_{Q_k} [|v|^p + |\operatorname{grad} v|^p] r_k{}^\nu \, dy\right\}^{1/p} [\operatorname{vol} Q_k]^{(p-t)/pt}, \qquad (83)
\end{aligned}$$

where $K_2$ is independent of $\nu$ for $\nu_1 \le \nu \le \nu_0$.

In the case $\nu_1 > 0$ we can obtain a similar (uniform) estimate for $0 \le \nu \le \nu_1$ by the method of Case I. Combining this with (82) and (83), we prove (80) for this case.

CASE III $\bar{\nu} = p-n$, $1 \le q < \infty$ Fix $s \ge \max(q, n/(n-1))$ and let $t = (\nu+n)s/(\nu+n+s)$ so that $s = (\nu+n)t/(\nu+n-t)$. Then $1 \le t \le ps/(p+s) < p$

for $0 \le \nu \le \bar{\nu}$. Hence we can select $\nu_1 \ge 0$ such that $t-n < \nu_1 < p-n$. The rest of the proof is similar to Case II. This completes the proof. ∎

**5.51** PROOF OF THEOREM 5.35 By the same argument used in the proof of Lemma 5.12 it is sufficient to consider here only the special case $m = 1$. Let $q$ satisfy $p \le q \le (\nu+n)p/(\nu+n-p)$ if $\nu+n > p$, or $p \le q < \infty$ otherwise. Clearly $q < np/(n-p)$ if $n > p$ so in either case we have by Theorem 5.4

$$\|u\|_{0,q,G} \le K_1 \|u\|_{1,p,G}$$

for every $u \in C^1(\Omega)$ and that element $G$ of $\Gamma$ which has the cone property (if such $G$ exists.) If $G \in \Gamma$ does not have the cone property, and if $\psi: G \to Q_{k,\lambda}$, where $(\lambda-1)k \le \nu$, is the 1-smooth mapping specified in the statement of the theorem, then by Theorem 3.35 and Lemma 5.50

$$\|u\|_{0,q,G} \le K_2 \|u \circ \psi^{-1}\|_{0,q,Q_{k,\lambda}} \le K_3 \|u \circ \psi^{-1}\|_{1,p,Q_{k,\lambda}} \le K_4 \|u\|_{1,p,G},$$

where $K_4$ is independent of $G$. We have, therefore, noting that $q/p \ge 1$,

$$\begin{aligned}\|u\|_{0,q,\Omega}^q &\le \sum_{G\in\Gamma} \|u\|_{0,q,G}^q \le K_5 \sum_{G\in\Gamma} \left(\|u\|_{1,p,G}^p\right)^{q/p} \\ &\le K_5 \left(\sum_{G\in\Gamma} \|u\|_{1,p,G}^p\right)^{q/p} \le K_5 N^{q/p} \|u\|_{1,p,\Omega}^q,\end{aligned}$$

where we have used the finite intersection property of $\Gamma$ to obtain the final inequality. Imbedding (50) now follows by completion. [If $\nu < mp-n$, we require that (50) hold for $q = \infty$. This is a consequence of Theorem 5.36 proved below.] ∎

**5.52 LEMMA** Let $0 \le \bar{\nu} < mp-n$. There exists a constant $K = K(m,p,n,\bar{\nu})$ such that if $Q_{k,\lambda}$ is any standard cusp domain for which $(\lambda-1)k = \nu \le \bar{\nu}$ and if $u \in C^m(Q_{k,\lambda})$, then

$$\sup_{x\in Q_{k,\lambda}} |u(x)| \le K\|u\|_{m,p,Q_{k,\lambda}}. \tag{84}$$

PROOF It is sufficient to prove the lemma for the case $m = 1$; the proof for general $m$ then follows by the same type of argument used in the last paragraph of the proof of Lemma 5.15.

If $u \in C^1(Q_{k,\lambda})$, $(\lambda-1)k = \nu \le \bar{\nu}$, we have by Lemma 5.48 and via the method of the second paragraph of the proof of Lemma 5.50,

$$\begin{aligned}\sup_{x\in Q_{k,\lambda}} |u(x)| &= \sup_{y\in Q_k} |\tilde{u}(y)| \\ &\le K_1 \left\{\int_{Q_k} [|\tilde{u}(y)|^p + |\operatorname{grad} \tilde{u}(y)|^p][r_k(y)]^\nu\, dy\right\}^{1/p} \\ &\le K_2 \left\{\int_{Q_{k,\lambda}} [|u(x)|^p + |\operatorname{grad} u(x)|^p]\, dx\right\}^{1/p}.\end{aligned} \tag{85}$$

Since $r_k(y) \le 1$ for $y \in Q_k$ it is evident that $K_1$, and hence $K_2$, can be chosen independent of $k, \lambda$ provided $0 \le \nu = (\lambda-1)k \le \bar{\nu}$. ▮

**5.53** PROOF OF THEOREM 5.36 It is sufficient to prove (51). Let $u \in C^m(\Omega)$. If $x \in \Omega$, then $x \in G \subset \Omega$ for some domain $G$ for which there exists a 1-smooth transformation $\psi: G \to Q_{k,\lambda}$, $(\lambda-1)k \le \nu$, as specified in the statement of the theorem. Thus

$$\begin{aligned} |u(x)| &\le \sup_{x \in G} |u(x)| = \sup_{y \in Q_{k,\lambda}} |u \circ \psi^{-1}(y)| \\ &\le K_1 \|u \circ \psi^{-1}\|_{m,p,Q_{k,\lambda}} \le K_2 \|u\|_{m,p,G} \\ &\le K_2 \|u\|_{m,p,\Omega}, \end{aligned} \tag{86}$$

where $K_1$ and $K_2$ are independent of $G$. The rest of the proof is similar to the first paragraph of the proof of Lemma 5.15. ▮

**5.54** PROOF OF THEOREM 5.37 As in Lemma 5.17 it is sufficient to prove that (52) holds when $j = 0$ and $m = 1$, that is, that

$$\sup_{\substack{x, y \in \Omega \\ x \ne y}} \frac{|u(x) - u(y)|}{|x-y|^\mu} \le K \|u\|_{1,p,\Omega} \tag{87}$$

holds when $\nu + n < p$ and $0 < \mu \le 1 - (\nu+n)/p$. For $x, y \in \Omega$, $|x-y| > \delta$, (87) holds by virtue of (86). If $|x-y| < \delta$, there exists $G$ with $x, y \in G \subset \Omega$, and a 1-smooth transformation $\psi$ from $G$ onto a standard cusp $Q_{k,\lambda}$ with $(\lambda-1)k \le \nu$, satisfying the conditions of the theorem. Inequality (87) can then be derived from Lemma 5.49 by the same method used in the proof of Lemma 5.52. The details are left to the reader. ▮

VI

# Compact Imbeddings of $W^{m,p}(\Omega)$

## The Rellich–Kondrachov Theorem

**6.1** Let $\Omega$ be a domain in $\mathbb{R}^n$ and let $\Omega_0$ be a subdomain of $\Omega$. Let $X(\Omega)$ denote any of the possible target spaces for imbeddings of $W^{m,p}(\Omega)$, that is, $X(\Omega)$ is a space of the form $C_B{}^j(\Omega)$, $C^{j,\lambda}(\overline{\Omega})$, $L^q(\Omega^k)$, or $W^{j,q}(\Omega^k)$, where $\Omega^k$, $1 \le k \le n$, is the intersection of $\Omega$ with a $k$-dimensional plane in $\mathbb{R}^n$. Since the linear restriction operator $i_{\Omega_0}\colon u \to u|_{\Omega_0}$ is bounded from $X(\Omega)$ into $X(\Omega_0)$ [in fact $\|i_{\Omega_0}u; X(\Omega_0)\| \le \|u; X(\Omega)\|$] any imbedding of the form

$$W^{m,p}(\Omega) \to X(\Omega) \tag{1}$$

can be composed with this restriction to yield the imbedding

$$W^{m,p}(\Omega) \to X(\Omega_0) \tag{2}$$

and (2) has imbedding constant no larger than (1).

If $\Omega$ satisfies the hypotheses of the Sobolov imbedding Theorem 5.4 and if $\Omega_0$ is bounded, then, with the exception of certain extreme cases, all imbeddings (2) (corresponding to imbeddings asserted in Theorem 5.4) are compact. The most important of these compact imbedding results originated in a lemma of Rellich [57] and was proved specifically for Sobolev spaces by Kondrachov [33]. Such compact imbeddings have many important applications in analysis, especially to showing the discreteness of the spectra of linear elliptic partial differential operators defined over bounded domains.

We summarize the various compact imbeddings of $W^{m,p}(\Omega)$ in the following theorem.

**6.2 THEOREM** (*The Rellich–Kondrachov theorem*) Let $\Omega$ be a domain in $\mathbb{R}^n$, $\Omega_0$ a bounded subdomain of $\Omega$, and $\Omega_0{}^k$ the intersection of $\Omega_0$ with a $k$-dimensional plane in $\mathbb{R}^n$. Let $j, m$ be integers, $j \geq 0$, $m \geq 1$, and let $1 \leq p < \infty$.

**PART I** If $\Omega$ has the cone property and $mp \leq n$, then the following imbeddings are compact:

$$W^{j+m,p}(\Omega) \to W^{j,q}(\Omega_0{}^k) \quad \text{if} \quad 0 < n - mp < k \leq n \quad \text{and} \quad 1 \leq q < kp/(n-mp), \tag{3}$$

$$W^{j+m,p}(\Omega) \to W^{j,q}(\Omega_0{}^k) \quad \text{if} \quad n = mp, \quad 1 \leq k \leq n \quad \text{and} \quad 1 \leq q < \infty. \tag{4}$$

**PART II** If $\Omega$ has the cone property and $mp > n$, then the following imbeddings are compact:

$$W^{j+m,p}(\Omega) \to C_B{}^j(\Omega), \tag{5}$$

$$W^{j+m,p}(\Omega) \to W^{j,q}(\Omega_0{}^k) \quad \text{if} \quad 1 \leq q \leq \infty. \tag{6}$$

**PART III** If $\Omega$ has the strong local Lipschitz property, then the following imbeddings are compact:

$$W^{j+m,p}(\Omega) \to C^j(\overline{\Omega}_0) \quad \text{if} \quad mp > n, \tag{7}$$

$$W^{j+m,p}(\Omega) \to C^{j,\lambda}(\overline{\Omega}_0) \quad \text{if} \quad mp > n \geq (m-1)p \quad \text{and} \quad 0 < \lambda < m-(n/p). \tag{8}$$

**PART IV** If $\Omega$ is an arbitrary domain in $\mathbb{R}^n$, all imbeddings (3)–(8) are compact provided $W^{j+m,p}(\Omega)$ is replaced by $W_0^{j+m,p}(\Omega)$.

**6.3 REMARKS** (1) If $X$, $Y$, and $Z$ are spaces for which we have the imbeddings $X \to Y$ and $Y \to Z$ and if one of these imbeddings is compact, then the composite imbedding $X \to Z$ is compact. Thus, for example, if $Y \to Z$ is compact, then any sequence $\{u_i\}$ bounded in $X$ will be bounded in $Y$ and therefore have a subsequence $\{u_i'\}$ convergent in $Z$.

Since the extension operator $u \to \tilde{u}$ where $\tilde{u}(x) = u(x)$ if $x \in \Omega$, $\tilde{u}(x) = 0$ if $x \notin \Omega$ defines an imbedding $W_0^{j+m,p}(\Omega) \to W^{j+m,p}(\mathbb{R}^n)$ by Lemma 3.22, Part IV of Theorem 6.2 follows from the application of Parts I–III to $\mathbb{R}^n$.

(2) In proving the compactness of any of the imbeddings (3)–(8) it is sufficient to consider only the case $j = 0$. Suppose, for example, that (3) has been proven compact if $j = 0$. For $j \geq 1$ and $\{u_i\}$ a bounded sequence in $W^{j+m,p}(\Omega)$ it is clear that $\{D^\alpha u_i\}$ is bounded in $W^{m,p}(\Omega)$ for each $\alpha$ such that $|\alpha| \leq j$. Hence $\{D^\alpha u_i\}$ is precompact in $L^q(\Omega_0{}^k)$ with $q$ specified as in (3). It is possible, therefore, to select (by finite induction) a subsequence $\{u_i'\}$ of $\{u_i\}$ for which $\{D^\alpha u_i'\}$ converges in $L^q(\Omega_0{}^k)$ for each $\alpha$ such that $|\alpha| \leq j$. Thus $\{u_i'\}$ converges in $W^{j,q}(\Omega_0{}^k)$ and (3) is compact.

(3) Since $\Omega_0$ is bounded, $C_B{}^0(\Omega_0{}^k) \to L^q(\Omega_0{}^k)$ for $1 \leq q \leq \infty$; in fact, $\|u\|_{0,q,\Omega_0{}^k} \leq \|u; C_B{}^0(\Omega_0{}^k)\| [\mathrm{vol}_k \Omega_0{}^k]^{1/q}$. Thus the compactness of (6) (for $j = 0$) follows from that of (5).

(4) For the purpose of proving Theorem 6.2 the bounded subdomain $\Omega_0$ of $\Omega$ may always be assumed to have the cone property if $\Omega$ does. If $C$ is a finite cone determining the cone property for $\Omega$, let $\tilde{\Omega}$ be the union of all finite cones congruent to $C$, contained in $\Omega$ and having nonempty intersection with $\Omega_0$. Then $\Omega_0 \subset \tilde{\Omega} \subset \Omega$ and $\tilde{\Omega}$ is bounded and has the cone property. If $W^{m,p}(\Omega) \to X(\tilde{\Omega})$ is compact, then so is $W^{m,p}(\Omega) \to X(\Omega_0)$ by restriction.

Note that if $\Omega$ is bounded, we may have $\Omega_0 = \Omega$ in the statement of the theorem.

**6.4** Proof of Theorem 6.2, Part III If $mp > n \geq (m-1)p$ and $0 < \lambda < (m-n)/p$, then there exists $\mu$ such that $\lambda < \mu < m - (n/p)$. Since $\Omega_0$ is bounded, the imbedding $C^{0,\mu}(\overline{\Omega}_0) \to C^{0,\lambda}(\overline{\Omega}_0)$ is compact by Theorem 1.31. Since $W^{m,p}(\Omega) \to C^{0,\mu}(\overline{\Omega}) \to C^{0,\mu}(\overline{\Omega}_0)$ by Theorem 5.4 and restriction, imbedding (8) is compact for $j = 0$ by Remark 6.3(1).

If $mp > n$, let $j^*$ be the nonnegative integer satisfying $(m - j^*)p > n \geq (m - j^* - 1)p$. Then we have the imbedding chain

$$W^{m,p}(\Omega) \to W^{m-j^*,p}(\Omega) \to C^{0,\mu}(\overline{\Omega}_0) \to C(\overline{\Omega}_0) \tag{9}$$

where $0 < \mu < m - j^* - (n/p)$. The last imbedding in (9) is compact by Theorem 1.31. Thus (7) is compact for $j = 0$. ∎

**6.5** Proof of Theorem 6.2, Part II As noted in Remark 6.3(4), $\Omega_0$ may be assumed to have the cone property. Since $\Omega_0$ is also bounded it can, by Theorem 4.8, be written as a finite union, $\Omega_0 = \bigcup_{k=1}^M \Omega_k$, where each $\Omega_k$ has the strong local Lipschitz property. If $mp > n$, then $W^{m,p}(\Omega) \to W^{m,p}(\Omega_k) \to C(\overline{\Omega}_k)$, the latter imbedding being compact as proved above. If $\{u_i\}$ is a sequence bounded in $W^{m,p}(\Omega)$, we may select (by finite induction on $k$) a subsequence $\{u_i'\}$ whose restriction to $\Omega_k$ converges in $C(\overline{\Omega}_k)$ for each $k$, $1 \leq k \leq M$. But then $\{u_i'\}$ converges in $C_B{}^0(\Omega_0)$ proving that (5) is compact for $j = 0$. Therefore (6) is also compact by Remark 6.3(3). ∎

**6.6 LEMMA** Let $\Omega$ be a domain in $\mathbb{R}^n$, $\Omega_0$ a subdomain of $\Omega$, and $\Omega_0{}^k$ the intersection of $\Omega_0$ with a $k$-dimensional plane in $\mathbb{R}^n$ $(1 \le k \le n)$. Let $1 \le q_1 < q_0$ and suppose

$$W^{m,p}(\Omega) \to L^{q_0}(\Omega_0{}^k), \tag{10}$$

$$W^{m,p}(\Omega) \to L^{q_1}(\Omega_0{}^k). \tag{11}$$

Suppose also that (11) is compact. If $q_1 \le q < q_0$, then the imbedding

$$W^{m,p}(\Omega) \to L^{q}(\Omega_0{}^k) \tag{12}$$

(exists and) is compact.

PROOF Let $\lambda = q_1(q_0 - q)/q(q_0 - q_1)$ and $\mu = q_0(q - q_1)/q(q_0 - q_1)$. Clearly $\lambda > 0$ and $\mu \ge 0$. By Hölder's inequality and (10) there exists a constant $K$ such that for all $u \in W^{m,p}(\Omega)$,

$$\begin{aligned} \|u\|_{0,q,\Omega_0{}^k} &\le \|u\|^{\lambda}_{0,q_1,\Omega_0{}^k} \|u\|^{\mu}_{0,q_0,\Omega_0{}^k} \\ &\le K \|u\|^{\lambda}_{0,q_1,\Omega_0{}^k} \|u\|^{\mu}_{m,p,\Omega}. \end{aligned} \tag{13}$$

Let $\{u_i\}$ be a sequence bounded in $W^{m,p}(\Omega)$. Since (11) is compact there exists a subsequence $\{u_i'\}$ that converges, and is therefore a Cauchy sequence in $L^{q_1}(\Omega_0{}^k)$. By (13), $\{u_i'\}$ is a Cauchy sequence in $L^q(\Omega_0{}^k)$ as well. Hence (12) is compact. ▮

**6.7** PROOF OF THEOREM 6.2, PART I First we deal with the case $j = 0$ of imbeddings (3). Assume, for the moment, that $k = n$ and let $q_0 = np/(n - mp)$. In order to prove that the imbeddings

$$W^{m,p}(\Omega) \to L^q(\Omega_0), \qquad 1 \le q < q_0, \tag{14}$$

are compact it suffices, by Lemma 6.6, to do so only for $q = 1$. For $j = 1, 2, 3, \ldots$ let

$$\Omega_j = \{x \in \Omega_0 : \operatorname{dist}(x, \operatorname{bdry} \Omega) > 2/j\}.$$

Let $S$ be a set of functions bounded in $W^{m,p}(\Omega)$. We show that $S$ (when restricted to $\Omega_0$) is precompact in $L^1(\Omega_0)$ by showing that $S$ satisfies the conditions of Theorem 2.21. Accordingly, let $\varepsilon > 0$ be given and for each $u \in W^{m,p}(\Omega)$ set

$$\tilde{u}(x) = \begin{cases} u(x) & \text{if} \quad x \in \Omega_0 \\ 0 & \text{otherwise.} \end{cases}$$

By Hölder's inequality and since $W^{m,p}(\Omega) \to L^{q_0}(\Omega_0)$, we have

$$\begin{aligned} \int_{\Omega_0 \sim \Omega_j} |u(x)|\, dx &\le \left\{ \int_{\Omega_0 \sim \Omega_j} |u(x)|^{q_0}\, dx \right\}^{1/q_0} \left\{ \int_{\Omega_0 \sim \Omega_j} 1\, dx \right\}^{1 - 1/q_0} \\ &\le K_1 \|u\|_{m,p,\Omega} [\operatorname{vol}(\Omega_0 \sim \Omega_j)]^{1 - 1/q_0}, \end{aligned}$$

with $K_1$ independent of $u$. Since $\Omega_0$ has finite volume, $j$ may be selected large enough to ensure that for every $u \in S$,

$$\int_{\Omega_0 \sim \Omega_j} |u(x)|\, dx < \varepsilon$$

and also, for every $h \in \mathbb{R}^n$,

$$\int_{\Omega_0 \sim \Omega_j} |\tilde{u}(x+h) - \tilde{u}(x)|\, dx < \varepsilon/2. \tag{15}$$

Now if $|h| < 1/j$, then $x + th \in \Omega_{2j}$ provided $x \in \Omega_j$ and $0 \le t \le 1$. If $u \in C^\infty(\Omega)$, it follows that

$$\begin{aligned} \int_{\Omega_j} |u(x+h) - u(x)|\, dx &\le \int_{\Omega_j} dx \int_0^1 \left| \frac{d}{dt} u(x+th) \right| dt \\ &\le |h| \int_0^1 dt \int_{\Omega_{2j}} |\operatorname{grad} u(y)|\, dy \\ &\le |h|\, \|u\|_{1,1,\Omega_0} \le K_2\, |h|\, \|u\|_{m,p,\Omega}, \end{aligned} \tag{16}$$

where $K_2$ is independent of $u$. Since $C^\infty(\Omega)$ is dense in $W^{m,p}(\Omega)$, (16) holds for any $u \in W^{m,p}(\Omega)$. Hence if $|h|$ is sufficiently small, we have from (15) and (16) that

$$\int_{\Omega_0} |\tilde{u}(x+h) - \tilde{u}(x)| < \varepsilon.$$

Hence $S$ is precompact in $L^1(\Omega_0)$ by Theorem 2.21, and imbeddings (14) are compact.

Next suppose $k < n$ but $p > 1$. Let $r$ be chosen so that $1 < r < p$ and $n - mr < k$. Let $v$ be the largest integer less than $mr$; let $s = kr/(n - mr)$, and let $q = nr/(n - mr)$. Assuming, as we may, that $\Omega_0$ has the cone property we obtain from inequalities (35) and (36) in the proof of Lemma 5.19,

$$\begin{aligned} \|u\|_{0,1,\Omega_0{}^k} &\le K_3\, \|u\|_{0,s,\Omega_0{}^k} \\ &\le K_4\, \|u\|_{0,q,\Omega_0}^{\lambda}\, \|u\|_{m,r,\Omega_0}^{1-\lambda} \\ &\le K_5\, \|u\|_{0,q,\Omega_0}^{\lambda}\, \|u\|_{m,p,\Omega}^{1-\lambda}, \end{aligned} \tag{17}$$

where $\lambda = n(mr - v)/mr(n - v)$ satisfies $0 < \lambda < 1$, and where $K_3$, $K_4$, and $K_5$ are independent of $u$. Note that $1 < q < q_0$. If $\{u_i\}$ is a sequence bounded in $W^{m,p}(\Omega)$, we have shown that it must have a subsequence $\{u_i'\}$ which converges in $L^q(\Omega_0)$. From (17), $\{u_i'\}$ must be a Cauchy sequence in $L^1(\Omega_0{}^k)$ so that $W^{m,p}(\Omega) \to L^1(\Omega_0{}^k)$ is compact. By Lemma 6.6 so are the imbeddings $W^{m,p}(\Omega) \to L^q(\Omega_0{}^k)$ for $1 \le q < kp/(n - mp)$.

Finally, suppose $p = 1$ and $0 \le n-m < k < n$. Then clearly $n-m+1 \le k < n$ so that $2 \le m \le n$. By Theorem 5.4, $W^{m,1}(\Omega) \to W^{m-1,r}(\Omega)$ where $r = n/(n-1) > 1$. Also, $k \ge n-(m-1) > n-(m-1)r$ so the imbedding $W^{m-1,r}(\Omega) \to L^1(\Omega_0{}^k)$ is compact as proved above. This is sufficient to complete the proof of the compactness of (3).

To show that (4) is compact we proceed as follows. If $n = mp$, $p > 1$, and $1 \le q < \infty$, we may select $r$ such that $1 \le r < p$, $k > n-mr > 0$, and $kr/(n-mr) > q$. Assuming again that $\Omega_0$ has the cone property, we have

$$W^{m,p}(\Omega) \to W^{m,r}(\Omega_0) \to L^q(\Omega_0{}^k). \tag{18}$$

The latter imbedding in (18) is compact by (3). If $p = 1$ and $n = m \ge 2$, then, setting $r = n/(n-1) > 1$ so that $n = (n-1)r$, we have for $1 \le q < \infty$,

$$W^{n,1}(\Omega) \to W^{n-1,r}(\Omega) \to L^q(\Omega_0{}^k),$$

the latter imbedding being compact as proved in (18). Finally, if $n = m = p = 1$, then of necessity $k = 1$. Letting $q_0 > 1$ be arbitrary chosen, we prove the compactness of $W^{1,1}(\Omega) \to L^1(\Omega_0)$ exactly as in the case $k = n$ of (3) considered above. Since $W^{1,1}(\Omega) \to L^q(\Omega_0)$ for $1 \le q < \infty$ all these imbeddings are compact by Lemma 6.6. ∎

**6.8** The reader may find it instructive to carry out the obvious generalization of Theorem 6.2 to the imbeddings supplied by Theorems 5.35–5.37.

## Two Counterexamples

**6.9** Two obvious questions arise from consideration of the statement of the Rellich–Kondrachov Theorem 6.2. First, can that theorem be extended to cover unbounded $\Omega_0$? Second, can the "extreme cases"

$$W^{j+m,p}(\Omega) \to W^{j,q}(\Omega_0{}^k), \qquad 0 < n - mp < k \le n, \quad q = kp/(n-mp) \tag{19}$$

and

$$W^{j+m,p}(\Omega) \to C^{j,\lambda}(\overline{\Omega}_0), \qquad mp > n > (m-1)p, \quad \lambda = m-(n/p) \tag{20}$$

ever be compact?

The first of these questions will be investigated later in this chapter. For the moment we show that, at least for $k = n$, the answer is certainly "no" unless $\Omega_0$ is *quasibounded*, that is, unless

$$\lim_{\substack{x \in \Omega_0 \\ |x| \to \infty}} \operatorname{dist}(x, \operatorname{bdry} \Omega_0) = 0.$$

**6.10 EXAMPLE** Let $\Omega$ be an unbounded domain in $\mathbb{R}^n$ which is not quasibounded. Then there exists a sequence $\{B_i\}$ of mutually disjoint open balls contained in $\Omega$ and all having the same positive radius. Let $\phi_1 \in C_0^\infty(B_1)$ and suppose $\|\phi_1\|_{k,p,B_1} = A_{k,p} > 0$ for each $k = 0, 1, 2, \ldots$ and $p \geq 1$. Let $\phi_i$ be a translate of $\phi_1$ having support in $B_i$. Then clearly $\{\phi_i\}$ is a bounded sequence in $W_0^{j+m,p}(\Omega)$ for any fixed $j, m, p$. But for any $q$,

$$\|\phi_i - \phi_k\|_{j,q,\Omega} = [\|\phi_i\|_{j,q,B_i}^q + \|\phi_k\|_{j,q,B_k}^q]^{1/q} = 2^{1/q} A_{j,q} > 0$$

so that $\{\phi_i\}$ cannot have a subsequence converging in $W^{j,q}(\Omega)$. Thus no imbedding of the form $W_0^{j+m,p}(\Omega) \to W^{j,q}(\Omega)$ can be compact. The noncompactness of the other imbeddings of Theorem 6.2 is proved similarly. ∎

We now show that the second question raised in Section 6.9 always has a negative answer.

**6.11 EXAMPLE** Let $\Omega$ be any domain in $\mathbb{R}^n$ and $\Omega_0$ any bounded subdomain of $\Omega$. Let $\Omega_0{}^k$ be the intersection of $\Omega_0$ with a $k$-dimensional plane in $\mathbb{R}^n$, say (without loss of generality) the plane spanned by the $x_1, \ldots, x_k$ coordinate axes. Let $\{a_1, a_2, \ldots\}$ be a sequence of distinct points in $\Omega_0{}^k$, and $\{r_1, r_2, \ldots\}$ a sequence of numbers such that $0 < r_i \leq 1$, such that $B_{r_i}(a_i) = \{x \in \mathbb{R}^n : |x - a_i| < r_i\} \subset \Omega_0$, and such that all the balls $B_{r_i}(a_i)$ are mutually disjoint.

Let $\phi \in C_0^\infty(B_1(0))$ satisfy the following conditions:

(i) For each nonnegative integer $h$, each real $q \geq 1$, and each $k$, $1 \leq k \leq n$, we have

$$|\phi|_{h,q,\mathbb{R}^k} = |\phi|_{h,q,\mathbb{R}^k \cap B_1(0)}$$

$$= \left\{ \sum_{\substack{|\alpha| = h \\ \alpha_{k+1} = \cdots = \alpha_n = 0}} \|D^\alpha \phi\|_{0,q,\mathbb{R}^k \cap B_1(0)}^q \right\}^{1/q} = A_{h,q,k} > 0.$$

(ii) There exists $a \in B_1(0)$, $a \neq 0$, such that for each nonnegative integer $h$,

$$|D_1{}^h \phi(a)| = B_h > 0. \tag{21}$$

Fix $p \geq 1$ and integers $j \geq 0$ and $m \geq 1$. For each $i$ let

$$\phi_i(x) = r_i^{j+m-n/p} \phi((x - a_i)/r_i).$$

Then clearly $\phi_i \in C_0^\infty(B_{r_i}(a_i))$ and a simple computation shows that

$$|\phi_i|_{h,q,\mathbb{R}^k} = r_i^{j+m-n/p-h+k/q} A_{h,q,k}. \tag{22}$$

If $h \leq j+m$, it follows from (22) and $r_i \leq 1$ that

$$|\phi_i|_{h,p,\mathbb{R}^n} \leq A_{h,p,n}$$

so $\{\phi_i\}$ is a bounded sequence in $W^{j+m,p}(\Omega)$.

Suppose $mp < n$ and $n-mp < k \le n$. Taking $q = kp/(n-mp)$, we obtain from (22)

$$\|\phi_i\|_{j,q,\Omega_0{}^k} \ge |\phi_i|_{j,q,\mathbb{R}^k} = A_{j,q,k}.$$

Since the functions $\phi_i$ have disjoint supports, we have

$$\|\phi_i - \phi_h\|_{j,q,\Omega_0{}^k} \ge 2^{1/q} A_{j,q,k} > 0$$

and so no subsequence of $\{\phi_i\}$ can converge in $W^{j,q}(\Omega_0{}^k)$. Thus imbedding (19) cannot be compact.

On the other hand, suppose $mp > n > (m-1)p$ and let $\lambda = m-(n/p)$. Letting $b_i = a_i + r_i a$, we obtain from (21)

$$|D_1{}^j \phi_i(b_i)| = r^{m-(n/p)} |D_1{}^j \phi(a)| = r^\lambda B_j > 0.$$

Let $c_i = a_i + ar_i/|a|$ so that $c_i \in \operatorname{bdry} B_{r_i}(a_i)$ and $|b_i - c_i| = (1-|a|)r_i$. Again since $\phi_i$ have disjoint supports,

$$\begin{aligned}\|\phi_i - \phi_h; C^{j,\lambda}(\overline{\Omega}_0)\| &\ge \max_{|\alpha|=j} \sup_{\substack{x,y\in\Omega_0\\ x\ne y}} \frac{|D^\alpha(\phi_i(x)-\phi_h(x)) - D^\alpha(\phi_i(y)-\phi_h(y))|}{|x-y|^\lambda} \\ &\ge \frac{|D_1{}^j\phi_i(b_i) - D_1{}^j\phi_h(b_i) - D_1{}^j\phi_i(c_i) + D_1{}^j\phi_h(c_i)|}{|b_i - c_i|^\lambda} \\ &= \frac{B_j}{(1-|a|)^\lambda} > 0.\end{aligned}$$

Thus no subsequence of $\{\phi_i\}$ can converge in $C^{j,\lambda}(\overline{\Omega}_0)$ and imbedding (20) cannot be compact. ∎

## Unbounded Domains—Compact Imbeddings of $W_0^{m,p}(\Omega)$

**6.12** Let $\Omega$ be an unbounded domain in $\mathbb{R}^n$. We shall be concerned below with determining whether the imbedding

$$W_0^{m,p}(\Omega) \to L^p(\Omega) \tag{23}$$

is compact. If (23) is compact, it will follow as in Remark 6.3(2) and the second paragraph of Section 6.7 that the imbeddings

$$W_0^{j+m,p}(\Omega) \to W^{j,q}(\Omega^k), \qquad 0 < n-mp < k \le n, \quad p \le q < kp/(n-mp),$$

and

$$W_0^{j+m,p}(\Omega) \to W^{j,q}(\Omega^k), \qquad n = mp, \quad 1 \le k \le n, \quad p \le q < \infty$$

are also compact.

As was shown in Example 6.10, imbedding (23) cannot be compact unless $\Omega$ is quasibounded. In Theorem 6.13 we give a geometric condition on $\Omega$ that is sufficient to guarantee the compactness of (23), and in Theorem 6.16 we give an analytic condition that is necessary and sufficient for the compactness of (23). Both theorems are from the work of Adams [2].

Let $\Omega_r$ denote the set $\{x \in \Omega : |x| \geq r\}$. Be it agreed that in the following discussion any cube $H$ referred to will have its faces parallel to the coordinate planes.

**6.13 THEOREM** Let $v$ be an integer such that $1 \leq v \leq n$ and $mp > v$ (or $p = m = v = 1$). Suppose that for every $\varepsilon > 0$ there exist numbers $h$ and $r$ with $0 < h \leq 1$ and $r \geq 0$ such that for every cube $H \subset \mathbb{R}^n$ having edge length $h$ and having nonempty intersection with $\Omega_r$ we have

$$\mu_{n-v}(H,\Omega)/h^{\,n-v} \geq h^p/\varepsilon,$$

where $\mu_{n-v}(H,\Omega)$ is the maximum, taken over all projections $P$ onto $(n-v)$-dimensional faces of $H$, of the area [that is, $(n-v)$-measure] of $P(H \sim \Omega)$. Then imbedding (23) is compact.

**6.14** The above theorem shows that for given quasibounded $\Omega$ the compactness of (23) may depend in an essential way on the dimension of bdry $\Omega$. Let us consider the two extreme cases $v = 1$ and $v = n$. For $v = n$ the condition of the theorem places on $\Omega$ only the minimal restriction of quasiboundedness. Thus if $mp > n$, then (23) is compact for any quasibounded $\Omega$. It can also be shown that if $p > 1$ and $\Omega$ is quasibounded and has boundary consisting entirely of isolated points with no finite accumulation point, then (23) cannot be compact unless $mp > n$.

If $v = 1$, the conditions of Theorem 6.13 make no requirement of $m$ and $p$ but do require that bdry $\Omega$ be "essentially $(n-1)$ dimensional." Any quasibounded domain whose boundary consists of reasonably regular $(n-1)$-dimensional surfaces will satisfy these conditions. An example of such a domain is the "spiny urchin" (Fig. 5), a domain in $\mathbb{R}^2$ obtained by deleting from the plane the union of all the sets $S_k$ $(k = 1, 2, \ldots)$ specified in polar coordinates by

$$S_k = \{(r,\theta) : r \geq k,\ \theta = n\pi/2^k,\ n = 1, 2, \ldots, 2^{k+1}\}.$$

Note that this domain, though quasibounded, is simply connected and has empty exterior.

More generally, if $v$ is the largest integer less than $mp$ the conditions of Theorem 6.13 require that in a certain sense the part of the boundary of $\Omega$ having dimension at least $n-v$ should bound a quasibounded domain.

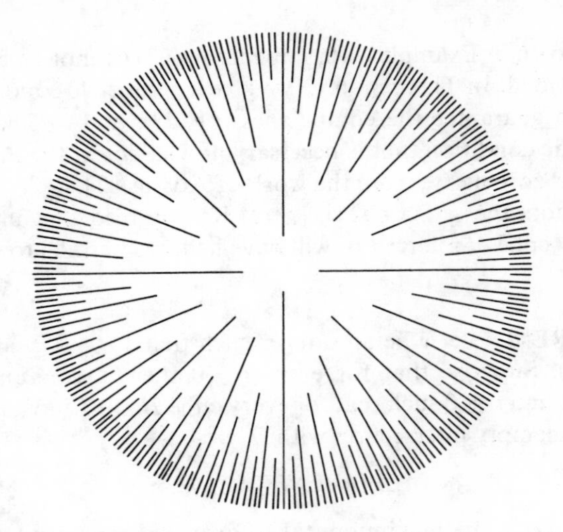

FIG. 5 A "spiny urchin."

**6.15** Let $H$ be a cube of side $h$ in $\mathbb{R}^n$ and $E$ a closed subset of $H$. Given $m$ and $p$ we define a functional $I_H^{m,p}$ on $C^\infty(H)$ by

$$I_H^{m,p}(u) = \sum_{1 \le j \le m} h^{jp} |u|_{j,p,H}^p = \sum_{1 \le |\alpha| \le m} h^{|\alpha| p} \int_H |D^\alpha u(x)|^p \, dx.$$

We denote by $C^{m,p}(H, E)$ the "$(m,p)$-capacity" of $E$ in $H$ defined by

$$C^{m,p}(H, E) = \inf_{u \in C^\infty(H, E)} \frac{I_H^{m,p}(u)}{\|u\|_{0,p,H}^p},$$

where $C^\infty(H, E)$ is the set of all functions $u \in C^\infty(H)$ that vanish identically in a neighborhood of $E$. Clearly $C^{m,p}(H, E) \le C^{m+1,p}(H, E)$ and for $E \subset F \subset H$, $C^{m,p}(H, E) \le C^{m,p}(H, F)$. Theorem 6.13 will be deduced from the following theorem characterizing in terms of the above capacity those domains for which (23) is compact.

**6.16 THEOREM** Imbedding (23) is compact if and only if $\Omega$ satisfies the following condition: For every $\varepsilon > 0$ there exists $h \le 1$ and $r \ge 0$ such that the inequality

$$C^{m,p}(H, H \sim \Omega) \ge h^p/\varepsilon$$

holds for every $n$-cube $H$ having edge length $h$ and having nonempty intersection with $\Omega_r$. (This condition clearly implies that $\Omega$ is quasibounded.)

**6.17 LEMMA** There exists a constant $K = K(n, p)$ such that for any $n$-cube $H$ of edge length $h$, any measurable subset $A$ of $H$ with positive volume, and any $u \in C^1(H)$, we have

$$\|u\|_{0,p,H}^p \leq \frac{2^{p-1}h^n}{\operatorname{vol} A} \|u\|_{0,p,A}^p + K\frac{h^{n+p}}{\operatorname{vol} A} \|\operatorname{grad} u\|_{0,p,H}^p. \tag{24}$$

PROOF Let $y \in A$ and $x = (\rho, \phi) \in H$, where $(\rho, \phi)$ are spherical polar coordinates centered at $y$, in terms of which the volume element is given by $dx = \omega(\phi)\rho^{n-1}\,d\rho\,d\phi$. Let bdry $H$ be specified by $\rho = f(\phi)$, $\phi \in \Sigma$. Clearly $f(\phi) \leq \sqrt{n}\,h$. Since

$$u(x) = u(y) + \int_0^\rho \frac{d}{dr} u(r, \phi)\,dr,$$

we have by Hölder's inequality,

$$\begin{aligned}
&\int_H |u(x)|^p\,dx \\
&\quad \leq 2^{p-1}|u(y)|^p h^n + 2^{p-1}\int_H \left|\int_0^\rho \frac{d}{dr}u(r,\phi)\,dr\right|^p dx \\
&\quad \leq 2^{p-1}h^n|u(y)|^p + 2^{p-1}\int_\Sigma \omega(\phi)\,d\phi \int_0^{f(\phi)} \rho^{n+p-2}\,d\rho \int_0^\rho |\operatorname{grad} u(r,\phi)|^p\,dr \\
&\quad \leq 2^{p-1}h^n|u(y)|^p + \frac{2^{p-1}}{n+p-1}(\sqrt{n}\,h)^{n+p-1}\int_H \frac{|\operatorname{grad} u(z)|^p}{|z-y|^{n-1}}\,dz.
\end{aligned}$$

Integration of $y$ over $A$ using Lemma 5.47 leads to

$$(\operatorname{vol} A)\|u\|_{0,p,H}^p \leq 2^{p-1}h^n\|u\|_{0,p,A}^p + Kh^{n+p}\int_H |\operatorname{grad} u(x)|^p\,dx$$

from which (24) follows at once. ∎

**6.18** PROOF OF THEOREM 6.16 (*Necessity*) Suppose $\Omega$ does not satisfy the condition stated in the theorem. Then there exists a finite constant $K_1 = 1/\varepsilon$ such that for every $h$ with $0 < h \leq 1$ there exists a sequence $\{H_j\}$ of mutually disjoint cubes of edge length $h$ which intersect $\Omega$ and for which

$$C^{m,p}(H_j, H_j \sim \Omega) < K_1 h^p.$$

By the definition of capacity, for each such cube $H_j$ there exists a function $u_j \in C^\infty(H_j, H_j \sim \Omega)$ such that $\|u_j\|_{0,p,H_j}^p = h^n$, $\|\operatorname{grad} u_j\|_{0,p,H_j}^p \leq K_1 h^n$, and $\|u_j\|_{m,p,H_j}^p \leq K_2(h)$. Let $A_j = \{x \in H_j : |u_j(x)| < \frac{1}{2}\}$. By Lemma 6.17 we have

$$h^n \leq \frac{2^{p-1}h^n}{\operatorname{vol} A_j} \cdot \frac{\operatorname{vol} A_j}{2^p} + \frac{KK_1}{\operatorname{vol} A_j} h^{2n+p}$$

from which it follows that $\operatorname{vol} A_j \leq K_3 h^{n+p}$. Let us choose $h$ so small that $K_3 h^p \leq \frac{1}{3}$, whence $\operatorname{vol} A_j \leq \frac{1}{3} \operatorname{vol} H_j$. Choose functions $w_j \in C_0^\infty(H_j)$ such that $w_j(x) = 1$ on a subset $S_j$ of $H_j$ having volume no less than $\frac{2}{3} \operatorname{vol} H_j$, and such that

$$\sup_j \max_{|\alpha| \leq m} \sup_{x \in H_j} |D^\alpha w_j(x)| = K_4 < \infty.$$

Then $v_j = u_j w_j \in C_0^\infty(H_j \cap \Omega) \subset C_0^\infty(\Omega)$ and $|v_j(x)| \geq \frac{1}{2}$ on $S_j \cap (H_j \sim A_j)$, a set of volume not less than $h^n/3$. Hence $\|v_j\|_{0,p,H_j}^p \geq h^n/3 \cdot 2^p$. On the other hand,

$$\int_{H_j} |D^\alpha u_j(x)|^p \cdot |D^\beta w_j(x)|^p \, dx \leq K_4{}^p K_2(h)$$

provided $|\alpha|, |\beta| \leq m$. Hence $\{v_j\}$ is a bounded sequence in $W_0^{m,p}(\Omega)$. Since the functions $v_j$ have disjoint supports, $\|v_j - v_k\|_{0,p,\Omega}^p \geq 2h^n/3 \cdot 2^p$ so imbedding (23) cannot be compact.

(*Sufficiency*) Now suppose $\Omega$ satisfies the condition in the statement of the theorem. Let $\varepsilon > 0$ be given and choose $r \geq 0$ and $h \leq 1$ such that for every cube $H$ of edge $h$ meeting $\Omega_r$ we have $C^{m,p}(H, H \sim \Omega) \geq h^p/\varepsilon^p$. Then for every $u \in C_0^\infty(\Omega)$ we obtain

$$\|u\|_{0,p,H}^p \leq (\varepsilon^p/h^p) \, I_H^{m,p}(u) \leq \varepsilon^p \, \|u\|_{m,p,H}^p.$$

Since a neighborhood of $\Omega_r$ can be tesselated by such cubes $H$ we have by summation

$$\|u\|_{0,p,\Omega_r} \leq \varepsilon \, \|u\|_{m,p,\Omega}.$$

That any bounded $S$ in $W_0^{m,p}(\Omega)$ is precompact in $L^p(\Omega)$ now follows at once from Theorems 2.22 and 6.2. ∎

**6.19 LEMMA** There is a constant $K$ independent of $h$ such that for any cube $H$ in $\mathbb{R}^n$ having edge length $h$, for every $q$ satisfying $p \leq q \leq np/(n-mp)$ (or $p \leq q < \infty$ if $n = mp$, or $p \leq q \leq \infty$ if $n < mp$), and for every $u \in C^\infty(H)$ we have

$$\|u\|_{0,q,H} \leq K \left\{ \sum_{|\alpha| \leq m} h^{|\alpha|p - n + np/q} \, \|D^\alpha u\|_{0,p,H}^p \right\}^{1/p}. \tag{25}$$

PROOF We may suppose $H$ to be centered at the origin and let $\tilde{H}$ be the cube of unit edge concentric with $H$. Inequality (25) holds for $\tilde{H}$ by the Sobolev imbedding theorem. For given $u \in C^\infty(H)$ we define corresponding $\tilde{u} \in C^\infty(\tilde{H})$ by $\tilde{u}(y) = u(x)$ where $x = hy$. Since

$$\left\{ \int_{\tilde{H}} |\tilde{u}(y)|^q \, dy \right\}^{1/q} \leq K \left\{ \sum_{|\alpha| \leq m} \int_{\tilde{H}} |D_y{}^\alpha u(y)|^p \, dy \right\}^{1/p}$$

it follows by change of variable that

$$h^{-n/q}\left\{\int_H |u(x)|^q\,dx\right\}^{1/q} \le K\left\{\sum_{|\alpha|\le m} h^{|\alpha|p-n}\int_H |D_x{}^\alpha u(x)|^p\,dx\right\}^{1/p},$$

whence (25) follows. ∎

**6.20 LEMMA** If $mp > n$ (or $m = p = n = 1$), there exists a constant $K = K(m,p,n)$ such that for every cube $H$ of edge length $h$ in $\mathbb{R}^n$ and every $u \in C^\infty(H)$ vanishing in a neighborhood of some point $y \in H$, we have

$$\|u\|_{0,p,H}^p \le KI_H^{m,p}(u).$$

PROOF The proof is somewhat similar to that of Lemma 5.15. First suppose $p \le n < mp$. Let $(\rho, \phi)$ denote polar coordinates centered at $y$. Then

$$u(\rho,\phi) = \int_0^\rho \frac{d}{dt}u(t,\phi)\,dt.$$

If $n > (m-1)p$, let $q = np/(n-mp+p)$ so that $q > n$. Otherwise let $q > n$ be arbitrary. If $(\rho,\phi) \in H$, then by Hölder's inequality

$$\begin{aligned}|u(\rho,\phi)|^q\rho^{n-1} &\le (\sqrt{n}\,h)^{n-1}\int_0^\rho \left|\frac{d}{dt}u(t,\phi)\right|^q t^{n-1}\,dt\left\{\int_0^{\sqrt{n}h} t^{-(n-1)/(q-1)}\,dt\right\}^{q-1}\\ &\le K_1 h^{q-1}\int_0^\rho \left|\frac{d}{dt}u(t,\phi)\right|^q t^{n-1}\,dt.\end{aligned}$$

It follows, using Lemma 6.19 with $m-1$ replacing $m$, that

$$\begin{aligned}\|u\|_{0,q,H}^q &\le K_2 h^q\int_H |\operatorname{grad} u(x)|^q\,dx\\ &\le K_2 h^q \sum_{|\alpha|=1}\|D^\alpha u\|_{0,q,H}^q\\ &\le K_3 h^q \sum_{|\alpha|=1}\left\{\sum_{|\beta|\le m-1} h^{|\beta|p-n+np/q}\|D^{\alpha+\beta}u\|_{0,p,H}^p\right\}^{q/p}.\end{aligned} \tag{26}$$

A further application of Hölder's inequality yields

$$\begin{aligned}\|u\|_{0,p,H}^p &\le \|u\|_{0,q,H}^p(\operatorname{vol} H)^{(q-p)/q}\\ &\le K_3^{p/q}\sum_{1\le|\gamma|\le m} h^{|\gamma|p}\|D^\gamma u\|_{0,p,H}^p = KI_H^{m,p}(u).\end{aligned}$$

If $p > n$ (or $p = n = 1$), the result follows directly from (26) with $q = p$,

$$\|u\|_{0,p,H}^p \le KI_H^{1,p}(u) \le KI_H^{m,p}(u).$$

**6.21** PROOF OF THEOREM 6.13 Let $H$ be a cube for which, for $mp > \nu$ (or $m = p = \nu = 1$) and $\mu_{n-\nu}(H, \Omega)/h^{n-\nu} \geq h^p/\varepsilon$. Let $P$ be the maximal projection referred to in the statement of the theorem, and let $E = P(H \sim \Omega)$. Without loss of generality we assume that the $(n-\nu)$ dimensional face $F$ of $H$ containing $E$ is parallel to the $x_{\nu+1}, \ldots, x_n$ coordinate plane. For each point $x = (x', x'')$ in $E$, where $x' = (x_1, \ldots, x_\nu)$ and $x'' = (x_{\nu+1}, \ldots, x_n)$, let $H_{x''}$ be the $\nu$-dimensional cube of edge $h$ in which $H$ intersects the $\nu$-plane through $x$ normal to $F$. By definition of $P$ there exists $y \in H_{x''} \sim \Omega$. If $u \in C^\infty(H, H \sim \Omega)$, then $u(\cdot, x'') \in C^\infty(H_{x''}, y)$. Applying Lemma 6.20 to $u(\cdot, x'')$, we obtain

$$\int_{H_{x''}} |u(x', x'')|^p \, dx' \leq K_1 \sum_{1 \leq |\alpha| \leq m} h^{|\alpha| p} \int_{H_{x''}} |D^\alpha u(x', x'')|^p \, dx',$$

where $K_1$ is independent of $h$, $x''$, and $u$. Integrating this inequality over $E$ and denoting $H' = \{x' : x = (x', x'') \in H \text{ for some } x''\}$, we obtain

$$\|u\|_{0,p,H' \times E}^p \leq K_1 I_{H' \times E}^{m,p}(u) \leq K_1 I_H^{m,p}(u).$$

Now we apply Lemma 6.17 with $A = H' \times E$ so that vol $A = h^\nu \mu_{n-\nu}(H, \Omega)$. This yields

$$\|u\|_{0,p,H}^p \leq K_2 \frac{h^{n-\nu}}{\mu_{n-\nu}(H, \Omega)} I_H^{m,p}(u),$$

where $K_2$ is independent of $h$. It follows that

$$C^{m,p}(H, H \sim \Omega) \geq \frac{\mu_{n-\nu}(H, \Omega)}{K_2 h^{n-\nu}} \geq \frac{h^p}{\varepsilon K_2}.$$

Hence $\Omega$ satisfies the hypothesis of Theorem 6.16 if it satisfies that of Theorem 6.13. ∎

The following two interpolation lemmas enable us to extend Theorem 6.13 to cover imbeddings of $W_0^{m,p}(\Omega)$ into continuous function spaces.

**6.22 LEMMA** Let $1 \leq p < \infty$ and $0 < \mu \leq 1$. There exists a constant $K = K(n, p, \mu)$ such that for every $u \in C_0^\infty(\mathbb{R}^n)$ we have

$$\sup_{x \in \mathbb{R}^n} |u(x)| \leq K \|u\|_{0,p,\mathbb{R}^n}^\lambda \left\{ \sup_{\substack{x, y \in \mathbb{R}^n \\ x \neq y}} \frac{|u(x) - u(y)|}{|x - y|^\mu} \right\}^{1-\lambda}, \tag{27}$$

where $\lambda = p\mu/(n + p\mu)$.

PROOF We may assume

$$\sup_{x \in \mathbb{R}^n} |u(x)| = N > 0 \qquad \text{and} \qquad \sup_{x, y \in \mathbb{R}^n} \frac{|u(x) - u(y)|}{|x - y|^\mu} = M < \infty.$$

Let $\varepsilon$ satisfy $0 < \varepsilon \le N/2$. There exists $x_0 \in \mathbb{R}^n$ such that $|u(x_0)| \ge N - \varepsilon \ge N/2$. Now $|u(x_0) - u(x)|/|x_0 - x|^\mu \le M$ for all $x$, so

$$|u(x)| \ge |u(x_0)| - M|x_0 - x|^\mu \ge \tfrac{1}{2}|u(x_0)|$$

provided $|x - x_0| \le (N/4M)^{1/\mu} = r$. Hence

$$\int_{\mathbb{R}^n} |u(x)|^p\, dx \ge \int_{B_r(x_0)} \left(\frac{|u(x_0)|}{2}\right)^p dx \ge K_1 \left(\frac{N-\varepsilon}{2}\right)^p \left(\frac{N}{4M}\right)^{n/\mu}.$$

Since this holds for arbitrarily small $\varepsilon$ we have

$$\|u\|_{0,p,\mathbb{R}^n} \ge (K_1^{1/p}/2 \cdot 4^{n/\mu p}) N^{1+n/p\mu} M^{-n/p\mu}$$

from which (27) follows at once. ∎

**6.23 LEMMA** Let $\Omega$ be an arbitrary domain in $\mathbb{R}^n$, and let $0 < \lambda < \mu \le 1$. For every function $u \in C^{0,\mu}(\overline{\Omega})$ we have

$$\|u; C^{0,\lambda}(\overline{\Omega})\| \le 3^{1-\lambda/\mu} \|u; C(\overline{\Omega})\|^{1-\lambda/\mu} \|u; C^{0,\mu}(\overline{\Omega})\|^{\lambda/\mu}. \tag{28}$$

PROOF Let $p = \mu/\lambda$, $p' = p/(p-1)$, and let

$$A_1 = \|u; C(\overline{\Omega})\|^{1/p}, \qquad B_1 = \sup_{\substack{x,y \in \Omega \\ x \ne y}} \left\{\frac{|u(x) - u(y)|}{|x-y|^\mu}\right\}^{1/p},$$

$$A_2 = \|u; C(\overline{\Omega})\|^{1/p'}, \qquad B_2 = \sup_{\substack{x,y \in \Omega \\ x \ne y}} |u(x) - u(y)|^{1/p'}.$$

Clearly $A_1{}^p + B_1{}^p = \|u; C^{0,\mu}(\overline{\Omega})\|$ and $B_2^{p'} \le 2\|u; C(\overline{\Omega})\|$. We have

$$\begin{aligned}
\|u; C^{0,\lambda}(\overline{\Omega})\| &= \|u; C(\overline{\Omega})\| + \sup_{\substack{x,y \in \Omega \\ x \ne y}} \frac{|u(x) - u(y)|}{|x-y|^\lambda} \\
&\le A_1 A_2 + B_1 B_2 \\
&\le \{A_1{}^p + B_1{}^p\}^{1/p} \{A_2^{p'} + B_2^{p'}\}^{1/p'} \\
&\le \|u; C^{0,\mu}(\overline{\Omega})\|^{\lambda/\mu} (3\|u; C(\overline{\Omega})\|)^{1-\lambda/\mu}
\end{aligned}$$

as required. ∎

**6.24 THEOREM** Let $\Omega$ satisfy the hypothesis of Theorem 6.13. Then the following imbeddings are compact:

$$W_0^{j+m,p}(\Omega) \to C^j(\overline{\Omega}) \qquad \text{if} \quad mp > n \tag{29}$$

$$W_0^{j+m,p}(\Omega) \to C^{j,\lambda}(\overline{\Omega}) \qquad \text{if} \quad mp > n \ge (m-1)p \quad \text{and} \quad 0 < \lambda < m - (n/p). \tag{30}$$

PROOF It is sufficient to deal with the case $j = 0$. If $mp > n$, let $j^*$ be the non-negative integer satisfying $(m-j^*)p > n \geq (m-j^*-1)p$. Then we have the chain

$$W_0^{m,p}(\Omega) \to W_0^{m-j^*,p}(\Omega) \to C^{0,\mu}(\overline{\Omega}) \to C(\overline{\Omega}),$$

where $0 < \mu < m-j^*-(n/p)$. If $\{u_i\}$ is a sequence bounded in $W_0^{m,p}(\Omega)$, then $\{u_i\}$ is also bounded in $C^{0,\mu}(\overline{\Omega})$. By Theorem 6.13, $\{u_i\}$ has a subsequence $\{u_i'\}$ converging in $L^p(\Omega)$. By (27), which applies by completion to the functions $u_i$, $\{u_i'\}$ is a Cauchy sequence in $C(\overline{\Omega})$ and so converges there. Hence (29) is compact for $j = 0$. Furthermore, if $mp > n \geq (m-1)p$ (that is, $j^* = 0$) and $0 < \lambda < \mu$, then by (28), $\{u_i'\}$ is also a Cauchy sequence in $C^{0,\lambda}(\overline{\Omega})$ whence (30) is also compact. ∎

## An Equivalent Norm for $W_0^{m,p}(\Omega)$

**6.25** Closely related to the problem of determining for which unbounded domains $\Omega$ the imbedding $W_0^{m,p}(\Omega) \to L^p(\Omega)$ is compact, is that concerned with determining for which domains $\Omega$ the seminorm $|\cdot|_{m,p,\Omega}$ defined by

$$|u|_{m,p,\Omega} = \left\{ \sum_{|\alpha|=m} \|D^\alpha u\|_{0,p,\Omega}^p \right\}^{1/p}$$

is actually a norm on $W_0^{m,p}(\Omega)$, equivalent to the given norm $\|\cdot\|_{m,p,\Omega}$. Such is certainly the case for any bounded domain as we now show.

**6.26** A domain $\Omega \subset \mathbb{R}^n$ is said to have *finite width* if it lies between two parallel hyperplanes. Let $\Omega$ be such a domain and suppose, without loss of generality, that $\Omega$ lies between the hyperplanes $x_n = 0$ and $x_n = d$. Letting $x = (x', x_n)$ where $x' = (x_1, \ldots, x_{n-1})$, we have for any $\phi \in C_0^\infty(\Omega)$

$$\phi(x) = \int_0^{x_n} \frac{d}{dt} \phi(x', t)\, dt$$

so that

$$\begin{aligned} \|\phi\|_{0,p,\Omega}^p &= \int_{\mathbb{R}^{n-1}} dx' \int_0^d |\phi(x)|^p\, dx_n \\ &\leq \int_{\mathbb{R}^{n-1}} dx' \int_0^d x_n^{p-1}\, dx_n \int_0^d |D_n \phi(x', t)|^p\, dt \\ &\leq (d^p/p)\, |\phi|_{1,p,\Omega}^p \end{aligned} \tag{31}$$

and

$$|\phi|_{1,p,\Omega}^p \leq \|\phi\|_{1,p,\Omega}^p = \|\phi\|_{0,p,\Omega}^p + |\phi|_{1,p,\Omega}^p \leq (1+(d^p/p))\, |\phi|_{1,p,\Omega}^p.$$

Successive application of the above inequality to derivatives $D^\alpha\phi$, $|\alpha| \le m-1$, then yields

$$|\phi|_{m,p,\Omega} \le \|\phi\|_{m,p,\Omega} \le K|\phi|_{m,p,\Omega} \tag{32}$$

and by completion (32) holds for all $u \in W_0^{m,p}(\Omega)$. Inequality (31) is often called *Poincaré's inequality*.

**6.27** An unbounded domain $\Omega$ in $\mathbb{R}^n$ is called *quasicylindrical* provided

$$\limsup_{x \in \Omega, |x| \to \infty} \operatorname{dist}(x, \operatorname{bdry} \Omega) < \infty.$$

Evidently every quasibounded domain is quasicylindrical, as is every (unbounded) domain of finite width. We leave to the reader the construction of a suitable counterexample to show that if $\Omega$ is not quasicylindrical, then $|\cdot|_{m,p,\Omega}$ is not an equivalent norm to $\|\cdot\|_{m,p,\Omega}$ on $W_0^{m,p}(\Omega)$.

The following theorem is clearly an analog of Theorem 6.13.

**6.28 THEOREM** Suppose there exist constants $K$, $R$, $h$, and $\nu$ with $0 < K \le 1$, $0 \le R < \infty$, $0 < h < \infty$, and $1 \le \nu \le n$, $\nu$ an integer, such that either $\nu < p$ or $\nu = p = 1$, and such that for every cube $H$ in $\mathbb{R}^n$, having edge length $h$ and having nonempty intersection with $\Omega_R = \{x \in \Omega : |x| \ge R\}$ we have

$$\mu_{n-\nu}(H, \Omega)/h^{n-\nu} \ge K,$$

where $\mu_{n-\nu}(H, \Omega)$ is as defined in the statement of Theorem 6.13. Then $|\cdot|_{m,p,\Omega}$ and $\|\cdot\|_{m,p,\Omega}$ are equivalent norms for $W_0^{m,p}(\Omega)$.

PROOF As noted in Section 6.26 it is sufficient to prove that $\|u\|_{0,p,\Omega} \le K_1 |u|_{1,p,\Omega}$ for $u \in C_0^\infty(\Omega)$. Let $H$ be a cube of edge length $h$ having nonempty intersection with $\Omega_R$. Since $\nu < p$ (or $\nu = p = 1$) the proof of Theorem 6.13 (Section 6.21) shows that

$$C^{1,p}(H, H \sim \Omega) \ge \mu_{n-\nu}(H, \Omega)/K_2 h^{n-\nu} \ge K/K_2$$

for all $u \in C_0^\infty(H)$, $K_2$ being independent of $u$. Hence

$$\|u\|_{0,p,H}^p \le (K_2/K)\, I_H^{1,p}(u) = K_3 |u|_{1,p,H}^p. \tag{33}$$

By summing (33) over the cubes $H$ comprising a tesselation of some neighborhood of $\Omega_R$, we obtain

$$\|u\|_{0,p,\Omega_R}^p \le K_3 |u|_{1,p,\Omega}^p. \tag{34}$$

It remains to be proven that

$$\|u\|_{0,p,B_R}^p \le K_4 |u|_{1,p,\Omega}^p,$$

where $B_R = \{x \in \mathbb{R}^n : |x| < R\}$. Let $(\rho, \phi)$ denote spherical polar coordinates of the point $x$ in $\mathbb{R}^n$ $(\rho \geq 0, \phi \in \Sigma)$ and denote the volume element by $dx = \rho^{n-1}\omega(\phi)\, d\rho\, d\phi$. For any $u \in C^\infty(\mathbb{R}^n)$ we have

$$u(\rho, \phi) = u(\rho + R, \phi) - \int_\rho^{R+\rho} \frac{d}{dt} u(t, \phi)\, dt$$

so that

$$|u(\rho, \phi)|^p \leq 2^{p-1} |u(\rho + R, \phi)|^p + 2^{p-1} R^{p-1} \rho^{1-n} \int_\rho^{R+\rho} |\operatorname{grad} u(t, \phi)|^p t^{n-1}\, dt.$$

Hence

$$\begin{aligned} \|u\|_{0,p,B_R}^p &= \int_\Sigma \omega(\phi)\, d\phi \int_0^R |u(\rho, \phi)|^p \rho^{n-1}\, d\rho \\ &\leq 2^{p-1} \int_\Sigma \omega(\phi)\, d\phi \int_0^R |u(\rho + R, \phi)|^p (\rho + R)^{n-1}\, d\rho \\ &\quad + 2^{p-1} R^p \int_\Sigma \omega(\phi)\, d\phi \int_0^{2R} |\operatorname{grad} u(t, \phi)|^p t^{n-1}\, dt. \end{aligned}$$

Therefore we have for $u \in C_0^\infty(\Omega)$

$$\begin{aligned} \|u\|_{0,p,B_R}^p &\leq 2^{p-1} \|u\|_{0,p,B_{2R} \sim B_R}^p + 2^{p-1} R^p |u|_{1,p,B_{2R}}^p \\ &\leq 2^{p-1} \|u\|_{0,p,\Omega_R}^p + 2^{p-1} R^p |u|_{1,p,\Omega}^p \leq K_4 |u|_{1,p,\Omega}^p \end{aligned}$$

by (34). ∎

## Unbounded Domains—Decay at Infinity

**6.29** The vanishing, in a generalized sense, on the boundary of $\Omega$ of elements of $W_0^{m,p}(\Omega)$ played a critical role in our earlier establishment of the compactness of the imbedding

$$W_0^{m,p}(\Omega) \to L^p(\Omega) \tag{35}$$

for certain unbounded domains. For elements of $W^{m,p}(\Omega)$ we no longer have this vanishing and the question remains: when, if ever, is the imbedding

$$W^{m,p}(\Omega) \to L^p(\Omega) \tag{36}$$

compact for unbounded $\Omega$, or even for bounded $\Omega$ which are sufficiently irregular that no imbedding of the form

$$W^{m,p}(\Omega) \to L^q(\Omega) \tag{37}$$

exists for any $q > p$? Note that if $\Omega$ has finite volume, the existence of (37) for some $q > p$ implies the compactness of (36) by the method of the first part of Section 6.7. By Theorem 5.30 imbedding (37) cannot, however, exist if $q > p$ and $\Omega$ is unbounded but has finite volume.

**6.30 EXAMPLE** For $j = 1, 2, \ldots$ let $B_j$ be an open ball in $\mathbb{R}^n$ having radius $r_j$, and suppose $\bar{B}_j \cap \bar{B}_k$ is empty if $j \neq k$. Let $\Omega = \bigcup_{j=1}^{\infty} B_j$; $\Omega$ may be bounded or unbounded. The sequence $\{u_j\}$ defined by

$$u_j(x) = \begin{cases} (\operatorname{vol} B_j)^{-1/p} & \text{if} \quad x \in \bar{B}_j \\ 0 & \text{if} \quad x \notin \bar{B}_j \end{cases}$$

is clearly bounded in $W^{m,p}(\Omega)$ but not precompact in $L^p(\Omega)$ no matter how rapidly $r_j \to 0$ as $j$ tends to infinity. Hence (36) is not compact. [Note that (35) is compact by Theorem 6.13 provided $\lim_{j\to\infty} r_j = 0$.] If $\Omega$ is bounded, imbedding (37) cannot exist for any $q > p$.

**6.31** There do exist unbounded domains $\Omega$ for which imbedding (36) is compact (see Section 6.48). An example of such a domain was given by Adams and Fournier [3] and it provided a basis for an investigation of the general problem by the same authors [4]. The approach of this latter paper is used in the following sections. First we concern ourselves with necessary conditions for the compactness of (37) ($q \geq p$). These conditions involve rapid decay at infinity for any unbounded domain (see Theorem 6.40). The techniques involved in the proof also yield a strengthened version of Theorem 5.30 (viz. Theorem 6.36) and a converse of the assertion [see Remark 5.5(6)] that $W^{m,p}(\Omega) \to L^p(\Omega)$ for $1 \leq q < p$ if $\Omega$ has finite volume.

A sufficient condition for the compactness of (36) is given in Theorem 6.47. It applies to many domains, bounded and unbounded, to which neither the Rellich–Kondrachov theorem, nor any generalizations of that theorem obtained by the same techniques can be applied (e.g., exponential cusps—see Example 6.49).

**6.32** Let $T$ be a tesselation of $\mathbb{R}^n$ by closed $n$-cubes of edge length $h$. If $H$ is one of the cubes in $T$, let $N(H)$ denote the cube of side $3h$ concentric with $H$ and having faces parallel to those of $H$. The $N(H)$ will be called the *neighborhood* of $H$. Clearly $N(H)$ is the union of the $3^n$ cubes in $T$ which intersect $H$. By the *fringe* of $H$ we shall mean the shell $F(H) = N(H) \sim H$.

Let $\Omega$ be a given domain in $\mathbb{R}^n$ and $T$ a given tesselation as above. Let $\lambda > 0$. A cube $H \in T$ will be called *$\lambda$-fat* (with respect to $\Omega$) if

$$\mu(H \cap \Omega) > \lambda\mu(F(H) \cap \Omega),$$

where $\mu$ denotes $n$-dimensional Lebesgue measure in $\mathbb{R}^n$. (We use "$\mu$" instead of "vol" for notational simplicity in the following discussion where the symbol must be used many times.) If $H$ is not $\lambda$-fat, it is called *$\lambda$-thin*.

**6.33 THEOREM** Suppose that there exists a compact imbedding of the form

$$W^{m,p}(\Omega) \to L^q(\Omega) \tag{38}$$

for some $q \geq p$. Then for every $\lambda > 0$ and every tesselation $T$ of $\mathbb{R}^n$ by cubes of fixed size, $T$ has only finitely many $\lambda$-fat cubes.

PROOF Suppose, to the contrary, that for some $\lambda > 0$ there exists a tesselation $T$ of $\mathbb{R}^n$ by cubes of edge length $h$ containing a sequence $\{H_j\}_{j=1}^{\infty}$ of $\lambda$-fat cubes. Passing to a subsequence if necessary we may assume that $N(H_j) \cap N(H_k) = \varnothing$ if $j \neq k$. For each $j$ there exists $\phi_j \in C_0^{\infty}(N(H_j))$ such that

(i) $|\phi_j(x)| \leq 1$ for all $x \in \mathbb{R}^n$,
(ii) $\phi_j(x) = 1$ if $x \in H_j$,
(iii) $|D^\alpha \phi_j(x)| \leq M$ for all $x \in \mathbb{R}^n$ and $0 \leq |\alpha| \leq m$,

where $M = M(n, m, h)$ is independent of $j$. Let $\psi_j = c_j \phi_j$, where the positive constant $c_j$ is so chosen that

$$\|\psi_j\|_{0,q,\Omega}^q \geq c_j^q \int_{H_j \cap \Omega} |\phi_j(x)|^q \, dx = c_j^q \mu(H_j \cap \Omega) = 1.$$

But then

$$\begin{aligned}
\|\psi_j\|_{m,p,\Omega}^p &= c_j^p \sum_{0 \leq |\alpha| \leq m} \int_{N(H_j) \cap \Omega} |D^\alpha \phi_j(x)|^p \, dx \\
&\leq M^p c_j^p \mu(N(H_j) \cap \Omega) \\
&< M^p c_j^p \mu(H_j \cap \Omega)[1 + (1/\lambda)] = M^p[1 + (1/\lambda)] c_j^{p-q},
\end{aligned}$$

since $H_j$ is $\lambda$-fat. Now $\mu(H_j \cap \Omega) \leq \mu(H_j) = h^n$ so $c_j \geq h^{-n/q}$. Since $p - q \leq 0$, $\{\psi_j\}$ is bounded in $W^{m,p}(\Omega)$. Since the functions $\psi_j$ have disjoint supports, $\{\psi_j\}$ cannot be precompact in $L^q(\Omega)$, contradicting the compactness of (38). Thus $T$ can possess only finitely many $\lambda$-fat cubes. ▮

**6.34 COROLLARY** Suppose there exists an imbedding (38) for some $q > p$. If $T$ is a tesselation of $\mathbb{R}^n$ by cubes of fixed edge length $h$, and if $\lambda > 0$ is given, then there exists $\varepsilon > 0$ such that $\mu(H \cap \Omega) \geq \varepsilon$ for every $\lambda$-fat $H \in T$.

PROOF Suppose, to the contrary, there exists a sequence $\{H_j\}$ of $\lambda$-fat cubes with $\lim_{j\to\infty} \mu(H_j \cap \Omega) = 0$. If $c_j$ is defined as in the above proof, we have $\lim_{j\to\infty} c_j = \infty$. But then $\lim_{j\to\infty} \|\psi_j\|_{m,p,\Omega} = 0$ since $p < q$. Since $\{\psi_j\}$ is bounded away from 0 in $L^q(\Omega)$ we have contradicted the continuity of imbedding (38). ▮

**6.35** Let us consider the implications of the above corollary. If imbedding (38) exists for some $q > p$, then one of the following alternatives must hold:

(a) There exists $\varepsilon > 0$ and a tesselation $T$ of $\mathbb{R}^n$ consisting of cubes of fixed size such that $\mu(H \cap \Omega) \geq \varepsilon$ for infinitely many cubes $H \in T$.

(b) For every $\lambda > 0$ and every tesselation $T$ of $\mathbb{R}^n$ by cubes of fixed size, $T$ contains only finitely many $\lambda$-fat cubes.

We shall show in Theorem 6.37 that (b) implies that $\Omega$ has finite volume. By Theorem 5.30, (b) is therefore inconsistent with the existence of (38) for $q > p$. On the other hand, (a) implies that $\mu(\{x \in \Omega : N \leq |x| \leq N+1\})$ does not tend to zero as $N$ tends to infinity. We have therefore proved the following theorem strengthening Theorem 5.30.

**6.36 THEOREM** Let $\Omega$ be an unbounded domain in $\mathbb{R}^n$ satisfying

$$\limsup_{N \to \infty} \operatorname{vol}\{x \in \Omega : N \leq |x| \leq N+1\} = 0.$$

Then there can be no imbedding of the form (38) for any $q > p$.

**6.37 THEOREM** Suppose that imbedding (38) is compact for some $q \geq p$. Then $\Omega$ has finite volume.

PROOF Let $T$ be a tesselation of $\mathbb{R}^n$ by cubes of unit edge, and let $\lambda = 1/[2(3^n - 1)]$. Let $P$ be the union of the finitely many $\lambda$-fat cubes in $T$. Clearly $\mu(P \cap \Omega) \leq \mu(P) < \infty$. Let $H$ be a $\lambda$-thin cube. Let $H_1$ be one of the $(3^n - 1)$ cubes of $T$ contained in the fringe $F(H)$ and selected so that $\mu(H_1 \cap G)$ is maximal. Thus

$$\mu(H \cap G) \leq \lambda\mu(F(H) \cap G) \leq \lambda(3^n - 1)\mu(H_1 \cap G) = \tfrac{1}{2}\mu(H_1 \cap G).$$

If $H_1$ is also $\lambda$-thin, we may select $H_2 \in T$, $H_2 \subset F(H_1)$ such that $\mu(H_1 \cap \Omega) \leq \frac{1}{2}\mu(H_2 \cap \Omega)$.

Suppose an infinite chain $\{H, H_1, H_2, \ldots\}$ of $\lambda$-thin cubes can be constructed in the above manner. Then

$$\mu(H \cap \Omega) \leq \tfrac{1}{2}\mu(H_1 \cap \Omega) \leq \cdots \leq (1/2^j)\mu(H_j \cap \Omega) \leq 1/2^j$$

for each $j$, since $\mu(H_j \cap \Omega) \leq \mu(H_j) = 1$. Hence $\mu(H \cap \Omega) = 0$. Denoting by $P_\infty$ the union of $\lambda$-thin cubes $H \in T$ for which such an infinite chain can be constructed, we have $\mu(P_\infty \cap \Omega) = 0$.

Let $P_j$ denote the union of the $\lambda$-thin cubes $H \in T$ for which some such chain ends on the $j$th step (that is, $H_j$ is $\lambda$-fat). Any particular $\lambda$-fat cube $H'$ can occur as the end $H_j$ of a chain beginning at $H$ only if $H$ is contained in the cube of edge $2j+1$ centered on $H'$. Hence there are at most $(2j+1)^n$ such cubes

$H \subset P_j$ having $H'$ as chain endpoint. Thus

$$\begin{aligned}\mu(P_j \cap \Omega) &= \sum_{H \subset P_j} \mu(H \cap \Omega) \\ &\le (1/2^j) \sum_{H \subset P_j} \mu(H_j \cap \Omega) \\ &\le [(2j+1)^n/2^j] \sum_{H' \subset P} \mu(H' \cap \Omega) = [(2j+1)^n/2^j]\, \mu(P \cap \Omega),\end{aligned}$$

so that $\sum_{j=1}^{\infty} \mu(P_j \cap \Omega) < \infty$. Since $\mathbb{R}^n = P \cup P_\infty \cup P_1 \cup P_2 \cup \cdots$ we have $\mu(\Omega) < \infty$. ∎

Suppose $1 \le q < p$. By Theorem 2.8 the imbedding

$$W^{m,p}(\Omega) \to L^q(\Omega) \tag{39}$$

exists if $\operatorname{vol} \Omega < \infty$. We are now in a position to prove the converse.

**6.38 THEOREM** Suppose imbedding (39) exists for some $p, q$ such that $1 \le q < p$. Then $\Omega$ has finite volume.

PROOF Let $T, \lambda$ be as in the proof of the previous theorem. Once again let $P$ denote the union of the $\lambda$-fat cubes in $T$. If we can show that $\mu(P \cap \Omega)$ is finite, it will follow by the same argument used in the above theorem that $\operatorname{vol} \Omega$ is finite.

Accordingly, suppose $\mu(P \cap \Omega)$ is not finite. Then there exists a sequence $\{H_j\}_{j=1}^{\infty}$ of $\lambda$-fat cubes in $T$ such that $\sum_{j=1}^{\infty} \mu(H_j \cap \Omega) = \infty$. If $L$ is the lattice of centers of the cubes in $T$, we may break up $L$ into $3^n$ mutually disjoint sublattices $\{L_i\}_{i=1}^{3^n}$ each having period 3 in each coordinate direction. For each $i$ let $T_i$ be the set of all cubes in $T$ with centers in $L_i$. For some $i$ we must have $\sum_{\lambda\text{-fat},\, H \in T_i} \mu(H \cap \Omega) = \infty$. Thus we may assume the cubes of the sequence $\{H_j\}$ all belong to $T_i$ for some fixed $i$, so that $N(H_j) \cap N(H_k)$ do not overlap.

Let the integer $j_1$ be chosen so that

$$2 \le \sum_{j=1}^{j_1} \mu(H_j \cap \Omega) < 4.$$

Let $\phi_j$ be as in the proof of Theorem 6.33, and let

$$\psi_1(x) = 2^{-1/p} \sum_{j=1}^{j_1} \phi_j(x).$$

We have, since the supports of the functions $\phi_j$ are disjoint, and since the cubes $H_j$ are $\lambda$-fat, for $|\alpha| \le m$,

$$\|D^\alpha \psi_1\|_{0,p,\Omega}^p = \tfrac{1}{2} \sum_{j=1}^{j_1} \int_\Omega |D^\alpha \phi_j(x)|^p \, dx$$

$$\le \tfrac{1}{2} M^p \sum_{j=1}^{j_1} \mu(N(H_j) \cap \Omega)$$

$$< \tfrac{1}{2} M^p (1 + (1/\lambda)) \sum_{j=1}^{j_1} \mu(H_j \cap \Omega) < 2M^p(1 + (1/\lambda)).$$

On the other hand,

$$\|\psi_1\|_{0,q,\Omega}^q \ge 2^{-q/p} \sum_{j=1}^{j_1} \mu(H_j \cap \Omega) \ge 2^{1-q/p}.$$

Having so defined $j_1$ and $\psi_1$, we may now define $j_2, j_3, \ldots$ and $\psi_2, \psi_3, \ldots$ inductively so that

$$2^k \le \sum_{j=j_{k-1}}^{j_k} \mu(H_j \cap \Omega) < 2^{k+1}$$

and

$$\psi_k(x) = 2^{-k/p} k^{-2/p} \sum_{j=j_{k-1}}^{j_k} \phi_j(x).$$

As above we have for $|\alpha| \le m$,

$$\|D^\alpha \psi_k\|_{0,p,\Omega}^p < (2/k^2) M^p (1 + (1/\lambda))$$

and

$$\|\psi_k\|_{0,q,\Omega}^q \ge 2^{k(1-q/p)} (1/k)^{2q/p}.$$

Thus $\psi = \sum_{k=1}^\infty \psi_k$ belongs to $W^{m,p}(\Omega)$ but not $L^q(\Omega)$, contradicting (39). Hence $\mu(P \cap \Omega) < \infty$ as required. ∎

**6.39** If there exists a compact imbedding of the form (38) for some $q \ge p$, then, as we have shown, $\Omega$ has finite volume. In fact, considerably more is true: $\mu(\{x \in \Omega : |x| \ge R\})$ must tend very rapidly to zero as $R \to \infty$, as we now show.

If $Q$ is a union of cubes $H$ in some tesselation of $\mathbb{R}^n$, we extend the notions of neighborhood and fringe to $Q$ in an obvious manner:

$$N(Q) = \bigcup_{H \subset Q} N(H), \qquad F(Q) = N(Q) \sim Q.$$

Given $\delta > 0$, let $\lambda = \delta/3^n(1+\delta)$. If all of the cubes $H \subset Q$ are $\lambda$-thin, then $Q$ is itself $\delta$-thin in the sense that

$$\mu(Q \cap \Omega) \le \delta \mu(F(Q) \cap \Omega). \tag{40}$$

To see this note that as $H$ runs through the cubes comprising $Q$, $F(H)$ covers

$N(Q)$ at most $3^n$ times. Hence

$$\mu(Q \cap \Omega) = \sum_{H \subset Q} \mu(H \cap \Omega) \leq \lambda \sum_{H \subset Q} \mu(F(H) \cap \Omega) \leq 3^n \lambda \mu(N(Q) \cap \Omega)$$
$$= 3^n \lambda [\mu(Q \cap \Omega) + \mu(F(Q) \cap \Omega)]$$

from which (40) follows by transposition (permissible since $\mu(\Omega) < \infty$) and since $3^n \lambda/(1-3^n \lambda) = \delta$.

For any measurable set $S \subset \mathbb{R}^n$ let $Q$ be the union of all cubes $H$ of our tesselation whose interiors intersect $S$, and define $F(S) = F(Q)$. If $S$ is at a positive distance from the finitely many $\lambda$-fat cubes in the tesselation, then $Q$ consists of $\lambda$-thin cubes and we obtain from (40),

$$\mu(S \cap \Omega) \leq \mu(Q \cap \Omega) \leq \delta \mu(F(S) \cap \Omega). \tag{41}$$

**6.40 THEOREM** Suppose there exists a compact imbedding of the form (38) for some $q \geq p$. For each $r \geq 0$ let $\Omega_r = \{x \in \Omega : |x| > r\}$, let $S_r = \{x \in \Omega : |x| = r\}$, and let $A_r$ denote the surface area [$(n-1)$-measure] of $S_r$. Then:

(a) For given $\varepsilon, \delta > 0$ there exists $R$ such that if $r \geq R$, then

$$\mu(\Omega_r) \leq \delta \mu(\{x \in \Omega : r - \varepsilon \leq |x| \leq r\}).$$

(b) If $A_r$ is positive and ultimately nonincreasing as $r$ tends to infinity, then for each $\varepsilon > 0$

$$\lim_{r \to \infty} \frac{A_{r+\varepsilon}}{A_r} = 0.$$

PROOF Given $\varepsilon > 0$ let $T$ be a tesselation of $\mathbb{R}^n$ by cubes of edge length $\varepsilon/2\sqrt{n}$. Then any cube $H \in T$ whose interior intersects $\Omega_r$ is contained in $\Omega_{r-\varepsilon/2}$ and

$$F(\Omega_r) \subset \{x \in \Omega : r - \varepsilon \leq |x| \leq r\}.$$

For given $\delta > 0$ let $\lambda$ be as given in Section 6.39. Let $R$ be large enough that the finitely many $\lambda$-fat cubes in $T$ are all contained in the ball of radius $R - \varepsilon/2$ centered at the origin. Then for any $r \geq R$ all the cubes in $T$ whose interiors intersect $\Omega_r$ are $\lambda$-thin and (a) follows from (41).

For (b) choose $R_0$ so that $A_r$ is nonincreasing on $[R_0, \infty)$. Fix $\varepsilon', \delta > 0$, and let $\varepsilon = \varepsilon'/2$. Let $R$ be as in (a). If $r \geq \max(R_0 + \varepsilon', R)$, then

$$A_{r+\varepsilon'} \leq (1/\varepsilon) \int_{r+\varepsilon}^{r+2\varepsilon} A_s \, ds \leq (1/\varepsilon) \mu(\Omega_{r+\varepsilon}) \leq (\delta/\varepsilon) \mu\{x \in \Omega : r \leq |x| \leq r + \varepsilon\}$$
$$= (\delta/\varepsilon) \int_r^{r+\varepsilon} A_s \, ds \leq \delta A_r.$$

Since $\varepsilon'$ and $\delta$ are arbitrary, (b) follows. ∎

**6.41 COROLLARY** If there exists a compact imbedding of the form (38) for some $q \geq p$, then for every $k$ we have

$$\lim_{r\to\infty} e^{kr}\mu(\Omega_r) = 0.$$

PROOF Fix $k$ and let $\delta = e^{-(k+1)}$. From conclusion (a) of Theorem 6.40 there exists $R$ such that $r \geq R$ implies $\mu(\Omega_{r+1}) \leq \delta\mu(\Omega_r)$. Thus if $j$ is a positive integer and $0 \leq t < 1$, we have

$$\begin{aligned} e^{k(R+j+t)}\mu(\Omega_{R+j+t}) &< e^{k(R+j+1)}\mu(\Omega_{R+j}) \leq e^{k(R+1)}e^{kj}\delta^j\mu(\Omega_R) \\ &= e^{k(R+1)}\mu(\Omega_R)e^{-j}. \end{aligned}$$

The last term tends to zero as $j$ tends to infinity. ∎

**6.42 REMARKS** (1) The argument used in the proof of Theorem 6.40(a) works for any norm $\rho$ on $\mathbb{R}^n$ in place of the usual norm $\rho(x) = |x|$. The same holds for (b) provided $A_r$ is well defined (with respect to the norm $\rho$) and provided

$$\mu(\{x \in \Omega : r \leq \rho(x) \leq r+\varepsilon\}) = \int_r^{r+\varepsilon} A_s\, ds.$$

This is true, for example, if $\rho(x) = \max_{1\leqslant i\leqslant n}|x_i|$.

(2) For the proof of (b) it is sufficient that $A_r$ have an equivalent, positive, nonincreasing majorant, that is, there should exist a positive, nonincreasing function $f(r)$ and a constant $M > 0$ such that for all sufficiently large $r$,

$$A_r \leq f(r) \leq MA_r.$$

(3) Theorem 6.33 is sharper than Theorem 6.40, because the conclusions of the latter theorem are global whereas the compactness of (38) evidently depends on local properties of $\Omega$. We illustrate this by means of two examples.

**6.43 EXAMPLE** Let $f \in C^1([0,\infty))$ be positive and nonincreasing with bounded derivative $f'$. We consider the planar domain (Fig. 6a)

$$\Omega = \{(x,y) \in \mathbb{R}^2 : x > 0,\ 0 < y < f(x)\}. \tag{42}$$

With respect to the supremum norm on $\mathbb{R}^2$, that is, $\rho(x,y) = \max(|x|,|y|)$, we have $A_s = f(s)$ for sufficiently large $s$. Hence $\Omega$ satisfies conclusion (b) of Theorem 6.40 [and since $f$ is monotonic conclusion (a) as well] if and only if

$$\lim_{s\to\infty}\frac{f(s+\varepsilon)}{f(s)} = 0 \tag{43}$$

holds for every $\varepsilon > 0$. For example, $f(x) = \exp(-x^2)$ satisfies (43) but

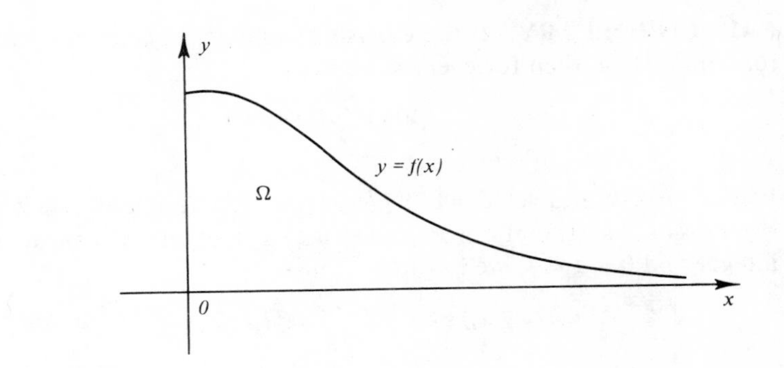


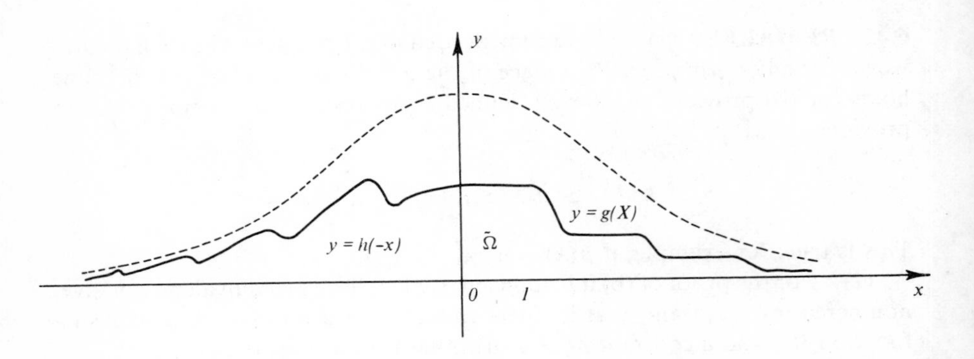


FIG. 6 Domains: (a) $\Omega$ of Example 6.43 and (b) $\tilde{\Omega}$ of Example 6.44

$f(x) = e^{-x}$ does not. We shall see (Section 6.48) that the imbedding

$$W^{m,p}(\Omega) \to L^p(\Omega) \tag{44}$$

is compact if (43) holds. Thus (43) is necessary and sufficient for the compactness of (38) for domains of the type (42).

**6.44 EXAMPLE** Let $f$ be as in Example 6.43 and assume also that $f'(0) = 0$. Let $g$ be a positive, nonincreasing function in $C^1([0, \infty))$ satisfying

(i) $g(0) = \frac{1}{2}f(0)$, $g'(0) = 0$,
(ii) $g(x) < f(x)$ for all $x \geq 0$,
(iii) $g(x)$ is constant on infinitely many disjoint intervals of unit length.

Let $h(x) = f(x) - g(x)$ and consider the domain (Fig. 6b)

$$\tilde{\Omega} = \{(x, y) \in \mathbb{R}^2 : 0 < y < g(x) \text{ if } x \geq 0,\ 0 < y < h(-x) \text{ if } x < 0\}.$$

Again we have $A_s = f(s)$ for sufficiently large $s$ so $\tilde{\Omega}$ satisfies the conclusions of Theorem 6.40 if (43) holds.

If, however, $T$ is a tesselation of $\mathbb{R}^2$ by squares of edge $\frac{1}{4}$ having edges parallel to the coordinate axes, and if one square of $T$ has center at the origin, then $T$ has infinitely many $\frac{1}{3}$-fat squares with centers on the positive $x$-axis. By Theorem 6.33, imbedding (44) cannot be compact.

## Unbounded Domains—Compact Imbeddings of $W^{m,p}(\Omega)$

**6.45** The above examples suggests that any sufficient condition for the compactness of the imbedding

$$W^{m,p}(\Omega) \to L^p(\Omega) \tag{45}$$

for unbounded domains $\Omega$ must involve the rapid decay of volume locally in each branch of $\Omega_r$ as $r$ tends to infinity. A convenient way to express such local decay is in terms of flows on $\Omega$.

By a *flow* on $\Omega$ we mean a continuously differentiable map $\Phi: U \to \Omega$, where $U$ is an open set in $\Omega \times \mathbb{R}$ containing $\Omega \times \{0\}$, and where $\Phi(x,0) = x$ for every $x \in \Omega$.

For fixed $x \in \Omega$ the curve $t \to \Phi(x,t)$ is called a *streamline* of the flow. For fixed $t$ the map $\Phi_t : x \to \Phi(x,t)$ sends a subset of $\Omega$ into $\Omega$. We shall be concerned with the Jacobian of this map:

$$\det \Phi_t'(x) = \left. \frac{\partial(\Phi_1, \ldots, \Phi_n)}{\partial(x_1, \ldots, x_n)} \right|_{(x,t)}.$$

It is sometimes required of a flow $\Phi$ that $\Phi_{s+t} = \Phi_s \circ \Phi_t$ but we do not need this property and so do not assume it.

**6.46 EXAMPLE** Let $\Omega$ be the domain given by (42). Define the flow

$$\Phi(x,y,t) = (x-t, [f(x-t)/f(x)]\, y), \qquad 0 < t < x.$$

The flow is toward the line $x = 0$ and the streamlines diverge as the domain widens (see Fig. 7). $\Phi_t$ is a local magnification for $t > 0$:

$$\det \Phi_t'(x,y) = f(x-t)/f(x).$$

Note that $\lim_{x \to \infty} \det \Phi_t'(x,y) = \infty$ if $f$ satisfies (43).

For $N = 1, 2, \ldots$ let $\Omega_N{}^* = \{(x,y) \in \Omega : 0 < x < N\}$. $\Omega_N{}^*$ is bounded and has the cone property, so the imbedding

$$W^{1,p}(\Omega_N{}^*) \to L^p(\Omega_N{}^*)$$

is compact. This compactness, together with properties of the flow $\Phi$ are sufficient to force the compactness of (45) as we now show.

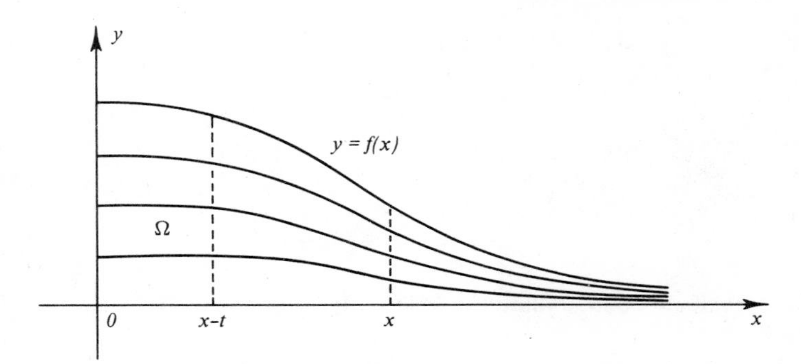


FIG. 7 Streamlines of the flow $\Phi$ given in Example 6.46

**6.47 THEOREM** Let $\Omega$ be an open set in $\mathbb{R}^n$ having the following properties:

(a) There exists a sequence $\{\Omega_N^*\}_{N=1}^{\infty}$ of open subsets of $\Omega$ such that $\Omega_N^* \subset \Omega_{N+1}^*$ and such that for each $N$ the imbedding

$$W^{1,p}(\Omega_N^*) \to L^p(\Omega_N^*)$$

is compact.

(b) There exists a flow $\Phi: U \to \Omega$, such that if $\Omega_N = \Omega \sim \Omega_N^*$, then

(i) $\Omega_N \times [0, 1] \subset U$ for each $N$,
(ii) $\Phi_t$ is one-to-one for all $t$,
(iii) $|(\partial/\partial t)\Phi(x, t)| \le M$ (const) for all $(x, t) \in U$.

(c) The functions $d_N(t) = \sup_{x \in \Omega_N} |\det \Phi_t'(x)|^{-1}$ satisfy

(i) $\lim_{N\to\infty} d_N(1) = 0$,
(ii) $\lim_{N\to\infty} \int_0^1 d_N(t)\, dt = 0$.

Then imbedding (45) is compact.

PROOF Since we have $W^{m,p}(\Omega) \to W^{1,p}(\Omega) \to L^p(\Omega)$ it is sufficient to prove that the latter imbedding is compact. Let $u \in C^1(\Omega)$. For each $x \in \Omega_N$ we have

$$u(x) = u(\Phi_1(x)) - \int_0^1 \frac{\partial}{\partial t} u(\Phi_t(x))\, dt.$$

Now

$$\begin{aligned}\int_{\Omega_N} |u(\Phi_1(x))|\, dx &\le d_N(1) \int_{\Omega_N} |u(\Phi_1(x))|\,|\det \Phi_1'(x)|\, dx \\ &= d_N(1) \int_{\Phi_1(\Omega_N)} |u(y)|\, dy \\ &\le d_N(1) \int_{\Omega} |u(y)|\, dy.\end{aligned}$$

Also

$$\int_{\Omega_N}\left|\int_0^1 \frac{\partial}{\partial t}u(\Phi_t(x))\,dt\right| dx \le \int_{\Omega_N} dx \int_0^1 |\operatorname{grad} u(\Phi_t(x))| \left|\frac{\partial}{\partial t}\Phi_t(x)\right| dt$$
$$\le M\int_0^1 d_N(t)\,dt \int_{\Omega_N} |\operatorname{grad} u(\Phi_t(x))|\,|\det \Phi_t{}'(x)|\,dx$$
$$\le M\left\{\int_0^1 d_N(t)\,dt\right\}\left\{\int_\Omega |\operatorname{grad} u(y)|\,dy\right\}.$$

Putting $\delta_N = \max(d_N(1), M\int_0^1 d_N(t)\,dt)$, we have

$$\int_{\Omega_N} |u(x)|\,dx \le \delta_N \int_\Omega (|u(y)| + |\operatorname{grad} u(y)|)\,dy \le \delta_N \|u\|_{1,1,\Omega} \tag{46}$$

and $\lim_{N\to\infty} \delta_N = 0$.

Now suppose $u$ is real valued and belongs to $C^1(\Omega) \cap W^{1,p}(\Omega)$. By Hölder's inequality the distributional derivatives of $|u|^p$,

$$D_j(|u|^p) = p \cdot |u|^{p-1} \cdot \operatorname{sgn} u \cdot D_j u,$$

satisfy

$$\int_\Omega |D_j(|u(x)|^p)|\,dx \le p\,\|D_j u\|_{0,p,\Omega} \|u\|_{0,p,\Omega}^{p-1} \le p\,\|u\|_{1,p,\Omega}^p.$$

Thus $|u|^p \in W^{1,1}(\Omega)$ and by Theorem 3.16 there is a sequence $\phi_j$ of functions in $C^1(\Omega) \cap W^{1,1}(\Omega)$ such that $\lim_{j\to\infty} \|\phi_j - |u|^p\|_{1,1,\Omega} = 0$. Thus by (46)

$$\int_{\Omega_N} |u(x)|^p\,dx = \lim_{j\to\infty} \int_{\Omega_N} \phi_j(x)\,dx \le \limsup_{j\to\infty} \delta_N \|\phi_j\|_{1,1,\Omega}$$
$$= \delta_N \| |u|^p \|_{1,1,\Omega} \le K\delta_N \|u\|_{1,p,\Omega}^p, \tag{47}$$

where $K = K(n,p)$. Inequality (47) holds for arbitrary (complex-valued) $u \in C^1(\Omega) \cap W^{1,p}(\Omega)$ by virtue of its separate application to the real and imaginary parts of $u$. (The constant $K$ may be changed.)

If $S$ is a bounded set in $W^{1,p}(\Omega)$ and $\varepsilon > 0$, we may, by (47), select $N$ so that for all $u \in S$

$$\int_{\Omega_N} |u(x)|^p\,dx < \varepsilon.$$

Since $W^{1,p}(\Omega \sim \Omega_N) \to L^p(\Omega \sim \Omega_N)$ is compact, the precompactness of $S$ in $L^p(\Omega)$ follows by Theorem 2.22. Hence $W^{1,p}(\Omega) \to L^p(\Omega)$ is compact. ∎

**6.48 EXAMPLE** Consider again the domain $\Omega$ of Examples 6.43 and 6.46 and the flow $\Phi$ given in the latter example. We have

$$d_N(t) = \sup_{x\ge N} \frac{f(x)}{f(x-t)} \le 1 \qquad \text{if} \quad 0 \le t \le 1$$

and by (43)

$$\lim_{N\to\infty} d_N(t) = 0 \quad \text{if} \quad t > 0.$$

Thus by dominated convergence

$$\lim_{N\to\infty} \int_0^1 d_N(t)\, dt = 0.$$

The assumption that $f'$ is bounded guarantees that the speed $|(\partial/\partial t)\,\Phi(x,y,t)|$ is bounded on $U$. Thus $\Omega$ satisfies the hypotheses of Theorem 6.47 and (45) is compact for this domain.

**6.49 EXAMPLE** Theorem 6.47 can also be used to show the compactness of (45) for some bounded domains to which neither the Rellich–Kondrachov theorem nor the techniques used in its proof can be applied. For example, we consider

$$\Omega = \{(x,y) \in \mathbb{R}^2 : 0 < x < 2,\ 0 < y < f(x)\},$$

where $f \in C^1(0,2)$ is positive, nondecreasing, has bounded derivative $f'$, and satisfies $\lim_{x\to 0+} f(x) = 0$. Let $U = \{(x,y,t) \in \mathbb{R}^3 : (x,y) \in \Omega,\ -x < t < 2-x\}$ and define the flow $\Phi\colon U \to \Omega$ by

$$\Phi(x,y,t) = \left(x+t,\ \frac{f(x+t)}{f(x)}\, y\right).$$

Thus $\det \Phi_t'(x,y) = f(x+t)/f(x)$. If $\Omega_N{}^* = \{(x,y) \in \Omega : x > 1/N\}$, then

$$d_N(t) = \sup_{0<x\leq 1/N} \left|\frac{f(x)}{f(x+t)}\right|$$

satisfies $d_N(t) \leq 1$ for $0 \leq t \leq 1$, and $\lim_{N\to\infty} d_N(t) = 0$ if $t > 0$. Hence also $\lim_{N\to\infty} \int_0^1 d_N(t)\, dt = 0$ by dominated convergence. Since $\Omega_N{}^*$ is bounded and has the cone property, and since the boundedness of $\partial\Phi/\partial t$ is assured by that of $f'$, we have by Theorem 6.47 the compactness of

$$W^{m,p}(\Omega) \to L^p(\Omega). \tag{48}$$

Suppose $\lim_{x\to 0+} f(x)/x^k = 0$ for every $k$. [For example, let $f(x) = e^{-1/x}$.] Then $\Omega$ has an exponential cusp at the origin, and so by Theorem 5.32 there exists no imbedding of the form

$$W^{m,p}(\Omega) \to L^q(\Omega)$$

for any $q > p$. Hence the method of Section 6.7 cannot be used to show the compactness of (48).

**6.50 REMARKS** (1) It is easy to imagine domains more general than those in the above examples to which Theorem 6.47 can be applied, although it may be difficult to specify a suitable flow. A domain with many (perhaps infinitely many) unbounded branches can, if connected, admit a suitable flow provided the volume decays sufficiently rapidly and regularly in each branch, a condition not fulfilled by the domain $\tilde{\Omega}$ of Example 6.44. For unbounded domains in which the volume decays monotonically in each branch Theorem 6.40 is essentially a converse of Theorem 6.47 in that the proof of Theorem 6.40 can be applied separately to show the volume decays in each branch in the required way.

(2) Since the only unbounded domains $\Omega$ for which $W^{m,p}(\Omega)$ imbeds compactly into $L^p(\Omega)$ have finite volume there can be no extension of Theorem 6.47 to give compact imbeddings into $L^q(\Omega)$ $(q > p)$, $C_B(\Omega)$, etc.—there do not exist such imbeddings.

## Hilbert–Schmidt Imbeddings

**6.51** A *complete orthonormal system* in a separable Hilbert space $X$ is a sequence $\{e_i\}_{i=1}^{\infty}$ of elements satisfying

$$(e_i, e_j)_X = \begin{cases} 1 & \text{if } \; i = j \\ 0 & \text{if } \; i \neq j \end{cases}$$

[where $(\cdot, \cdot)_X$ is the inner product in $X$], and such that for each $x \in X$ we have

$$\lim_{k\to\infty} \|x - \sum_{i=1}^{k} (x, e_i)_X\, e_i; X\| = 0. \tag{49}$$

Thus $x = \sum_{i=1}^{\infty} (x, e_i)_X\, e_i$, the series converging with respect to the norm in $X$. It is well known that every separable Hilbert space possesses such complete orthonormal systems. There follows from (49) the Parseval identity

$$\|x; X\|^2 = \sum_{i=1}^{\infty} |(x, e_i)_X|^2.$$

Let $X$ and $Y$ be two separable Hilbert spaces and let $\{e_i\}_{i=1}^{\infty}$ and $\{f_i\}_{i=1}^{\infty}$ be given complete orthonormal systems in $X$ and $Y$, respectively. Let $A$ be a bounded linear operator with domain $X$ taking values in $Y$, and let $A^*$ be the adjoint of $A$ taking $Y$ into $X$ and defined by

$$(x, A^*y)_X = (Ax, y)_Y, \qquad x \in X, \quad y \in Y.$$

Define

$$|||A|||^2 = \sum_{i=1}^{\infty} \|Ae_i; Y\|^2, \qquad |||A^*|||^2 = \sum_{i=1}^{\infty} \|A^*f_i; X\|^2.$$

If $|||A|||$ is finite, $A$ is called a *Hilbert–Schmidt operator* and we call $|||A|||$ its *Hilbert–Schmidt norm.* (Recall that the operator norm of $A$ is given by

$$\|A\| = \sup_{x \neq 0} \frac{\|Ax; Y\|}{\|x; X\|}.)$$

We must justify the above definition.

**6.52 LEMMA** The norms $|||A|||$ and $|||A^*|||$ are independent of the particular orthonormal systems $\{e_i\}$, $\{f_i\}$ used. Moreover,

$$|||A||| = |||A^*||| \geq \|A\|.$$

PROOF By Parseval's identity

$$\begin{aligned} |||A|||^2 &= \sum_{i=1}^{\infty} \|Ae_i; Y\|^2 = \sum_{i=1}^{\infty} \sum_{j=1}^{\infty} |(Ae_i, f_j)_Y|^2 \\ &= \sum_{j=1}^{\infty} \sum_{i=1}^{\infty} |(e_i, A^*f_j)_X|^2 = \sum_{j=1}^{\infty} \|A^*f_j; X\|^2 = |||A^*|||. \end{aligned}$$

Hence each expression is independent of $\{e_i\}$ and $\{f_j\}$. For any $x \in X$ we have

$$\begin{aligned} \|Ax; Y\|^2 &= \left\| \sum_{i=1}^{\infty} (x, e_i)_X Ae_i; Y \right\|^2 \leq \left( \sum_{i=1}^{\infty} |(x, e_i)_X| \, \|Ae_i; Y\| \right)^2 \\ &\leq \left( \sum_{i=1}^{\infty} |(x, e_i)_X|^2 \right) \left( \sum_{j=1}^{\infty} \|Ae_j; Y\|^2 \right) = \|x; X\|^2 \, |||A|||^2. \end{aligned}$$

Hence $\|A\| \leq |||A|||$ as required. ∎

We leave to the reader the task of verifying the following assertions:

(a) If $X$, $Y$, and $Z$ are separable Hilbert spaces and $A, B$ bounded linear operators from $X$ into $Y$ and $Y$ into $Z$, respectively, then $B \circ A$, which takes $X$ into $Z$, is a Hilbert–Schmidt operator if either $A$ or $B$ is. (If $A$ is Hilbert–Schmidt, then $|||B \circ A||| \leq \|B\| \, |||A|||$.)

(b) Every Hilbert–Schmidt operator is compact.

The following theorem, due to Maurin [43] has far-reaching implications for eigenfunction expansions corresponding to differential operators.

**6.53 THEOREM** Let $\Omega$ be a bounded set having the cone property in $\mathbb{R}^n$. Let $m, k$ be nonnegative integers with $k > n/2$. Then the imbedding mappings

$$W^{m+k,2}(\Omega) \to W^{m,2}(\Omega) \qquad (50)$$

are Hilbert–Schmidt operators. Similarly, the imbeddings

$$W_0^{m+k,2}(\Omega) \to W_0^{m,2}(\Omega) \tag{51}$$

are Hilbert–Schmidt operators for any bounded domain $\Omega$.

PROOF Given $y \in \Omega$ and $\alpha$ with $|\alpha| \le m$ we define a linear functional $T_y^\alpha$ on $W^{m+k,2}(\Omega)$ by

$$T_y^\alpha(u) = D^\alpha u(y).$$

Since $k > n/2$ the Sobolev imbedding Theorem 5.4 implies that $T_y^\alpha$ is bounded on $W^{m+k,2}(\Omega)$ and has norm bounded by a constant $K$ independent of $y$ and $\alpha$:

$$|T_y^\alpha(u)| \le \max_{0 \le |\alpha| \le m} \sup_{x \in \Omega} |D^\alpha u(x)| \le K \|u\|_{m+k,2,\Omega}. \tag{52}$$

By the Riesz representation theorem for Hilbert spaces there exists $v_y^\alpha \in W^{m+k,2}(\Omega)$ such that

$$D^\alpha u(y) = T_y^\alpha(u) = (u, v_y^\alpha)_{m+k}, \tag{53}$$

where $(\cdot, \cdot)_{m+k}$ is the inner product in $W^{m+k,2}(\Omega)$, and moreover

$$\|v_y^\alpha\|_{m+k,2,\Omega} = \|T_y^\alpha; [W^{m+k,2}(\Omega)]'\| \le K. \tag{54}$$

If $\{e_i\}_{i=1}^\infty$ is a complete orthonormal system in $W^{m+k,2}(\Omega)$, then

$$\|v_y^\alpha\|_{m+k,2,\Omega}^2 = \sum_{i=1}^\infty |(e_i, v_y^\alpha)_{m+k}|^2 = \sum_{i=1}^\infty |D^\alpha e_i(y)|^2.$$

Consequently,

$$\sum_{i=1}^\infty \|e_i\|_{m,2,\Omega}^2 \le \sum_{|\alpha| \le m} \int_\Omega \|v_y^\alpha\|_{m+k,2,\Omega}^2 \, dy \le \sum_{|\alpha| \le m} K \operatorname{vol} \Omega < \infty. \tag{55}$$

Hence imbedding (50) is Hilbert–Schmidt. The same proof holds for (51) without regularity needed for the application of Theorem 5.4. ∎

The following generalization of Maurin's theorem is due to Clark [17].

**6.54 THEOREM** Let $\mu$ be a nonnegative, measurable function defined on the domain $\Omega$ in $\mathbb{R}^n$. Let $W_0^{m,2;\mu}(\Omega)$ be the Hilbert space obtained by completing $C_0^\infty(\Omega)$ with respect to the weighted norm

$$\|u\|_{m,2;\mu} = \left\{ \sum_{|\alpha| \le m} \int_\Omega |D^\alpha u(x)|^2 \mu(x) \, dx \right\}^{1/2}.$$

For $y \in \Omega$ let $\tau(y) = \operatorname{dist}(y, \operatorname{bdry} \Omega)$. Let $\nu$ be a nonnegative integer such that

$$\int_\Omega [\tau(y)]^{2\nu} \mu(y) \, dy < \infty. \tag{56}$$

If $k > \nu + n/2$, then the imbedding

$$W_0^{m+k,2}(\Omega) \to W_0^{m,2;\mu}(\Omega) \tag{57}$$

(exists and) is Hilbert–Schmidt.

PROOF Let $\{e_i\}$, $T_y^\alpha$, and $v_y^\alpha$ be defined as in the previous theorem. If $y \in \Omega$, let $y_0$ be chosen in bdry $\Omega$ such that $\tau(y) = |y - y_0|$. If $\nu$ is a positive integer and $u \in C_0^\infty(\Omega)$, we have by Taylor's formula with remainder

$$D^\alpha u(y) = \sum_{|\beta| = \nu} (1/\beta!)\, D^{\alpha+\beta} u(y_\beta)(y - y_\beta)^\beta$$

for some points $y_\beta$ satisfying $|y - y_\beta| \le \tau(y)$. If $|\alpha| \le m$ and $k > \nu + n/2$, we obtain from the Sobolev imbedding theorem [as in (52)]

$$|D^\alpha u(y)| \le K \|u\|_{m+k,2} [\tau(y)]^\nu. \tag{58}$$

Inequality (58) holds, by completion, for any $u \in W_0^{m+k,2}(\Omega)$, and also holds if $\nu = 0$, directly from (52). Hence by (53) and (54)

$$\|v_y^\alpha\|_{m+k,2} = \sup_{\|u\|_{m+k,2} = 1} |D^\alpha u(y)| \le K[\tau(y)]^\nu.$$

It finally follows as in (55) that

$$\begin{aligned} \sum_{i=1}^\infty \|e_i\|_{m,2;\mu}^2 &\le \sum_{|\alpha| \le m} \int_\Omega \|v_y^\alpha\|_{m+k,2}^2 \mu(y)\, dy \\ &\le K^2 \sum_{|\alpha| \le m} \int_\Omega [\tau(y)]^{2\nu} \mu(y)\, dy < \infty \end{aligned}$$

by (56). Hence imbedding (57) is Hilbert–Schmidt. ∎

**6.55 REMARK** Various choices of $\mu$ and $\nu$ lead to generalizations of Maurin's theorem for imbeddings of the sort (51). If $\mu(x) = 1$ and $\nu = 0$, we obtain the obvious generalization to unbounded domains of finite volume. If $\mu(x) \equiv 1$ and $\nu > 0$, $\Omega$ may be unbounded and even have infinite volume, but by (56) it must be quasibounded. [Of course quasiboundedness may not be sufficient to guarantee (56).] If $\mu$ is the characteristic function of a bounded subdomain $\Omega_0$ of $\Omega$, and $\nu = 0$, we obtain the Hilbert–Schmidt imbedding

$$W_0^{m+k,2}(\Omega) \to W^{m,2}(\Omega_0), \qquad k > n/2.$$

# VII

# Fractional Order Spaces

## Outline

**7.1** In this chapter we are concerned with the problem of extending the notion of Sobolev space to allow for nonintegral values of $m$. There is no unique method for doing this, and different approaches may or may not lead to different families of spaces. The major families of spaces that arise are the following:

(i) The spaces $W^{s,p}(\Omega)$—which can be defined by a "real interpolation" method but can also be characterized in terms of an intrinsically defined norm involving first-order differences of the highest-order derivatives involved.

(ii) The spaces $L^{s,p}(\Omega)$—which can be defined by a "complex interpolation" method but, if $\Omega = \mathbb{R}^n$, can also be characterized in terms of Fourier transforms.

(iii) The Besov spaces $B^{s,p}(\Omega)$—defined in terms of an intrinsic norm similar to that of $W^{s,p}(\Omega)$ but involving second rather than first differences.

(iv) The Nikol'skii spaces $H^{s,p}(\Omega)$—having norm involving Hölder conditions in the $L^p$-metric.

Only the spaces $W^{s,p}(\Omega)$ and $L^{s,p}(\Omega)$ coincide with $W^{m,p}(\Omega)$ when $s = m$, an integer. $B^{s,p}(\Omega)$ coincides with $W^{s,p}(\Omega)$ for all $s$ if $p = 2$ but otherwise only for nonintegral $s$. The space $H^{s,p}(\Omega)$ is always larger than (but close to)

$W^{s,p}(\Omega)$. From the point of view of imbeddings, the simplest and most complete results obtain for the spaces $B^{s,p}$ and $H^{s,p}$ (see Theorem 7.70 and Section 7.73). However, it is in terms of the $W^{s,p}$ spaces that the problem, mentioned in Section 5.20, of characterizing the traces on smooth manifolds of functions in $W^{m,p}(\Omega)$ has its solution (Theorem 7.53). For this reason we concentrate our effort in this chapter on elucidating the properties of the spaces $W^{s,p}(\Omega)$ and give only brief descriptions of the other classes.

About half of the chapter is concerned with developing the "trace interpolation method" of Lions, on which we base our study of the spaces $W^{s,p}(\Omega)$. These latter spaces are introduced in Section 7.36. The trace interpolation method is one of several essentially equivalent real interpolation methods for Banach spaces concerning which there is now a considerable literature. Descriptions of these methods may be found in the work of Butzer and Berens [13] and Stein and Weiss [65], and the interested reader is referred to the work of Peetre [56] and Grisvard [28] for some applications in the direction of fractional order Sobolev spaces. A treatment of these spaces is also given in Stein [64a]. Most of the material in this chapter follows that of Lions [37, 38] and Lions and Magenes [40].

## The Bochner Integral

**7.2** In this chapter we shall need to make frequent use of the notion of integral of a Banach space-valued function $f$ defined on an interval of $\mathbb{R}$. We begin therefore with a brief discussion of the Bochner integral, referring the reader to the text by Yosida [69], for instance, for further details and proofs of our assertions.

Let $B$ be a Banach space with norm denoted by $\|\cdot\|_B$. Let $\{A_1, A_2, \dots, A_m\}$ be a finite collection of mutually disjoint, measurable subsets of $\mathbb{R}$, each having finite measure, and let $\{b_1, b_2, \dots, b_m\}$ be a corresponding collection of points of $B$. The function $f$ on $\mathbb{R}$ into $B$ defined by

$$f(t) = \sum_{j=1}^{m} \chi_{A_j}(t) b_j,$$

$\chi_A$ being the characteristic function of $A$, is called a *simple function*. For simple functions we define, obviously,

$$\int_{\mathbb{R}} f(t)\, dt = \sum_{j=1}^{m} \mu(A_j) b_j,$$

where $\mu(A)$ denotes the (Lebesgue) measure of $A$.

Let $A$ be a measurable set in $\mathbb{R}$ and $f$ an arbitrary function defined a.e. on $A$ into $B$. The function $f$ is called (*strongly*) *measurable* on $A$ if there exists a sequence $\{f_n\}$ of simple functions with supports in $A$ such that

$$\lim_{n\to\infty} \|f_n(t)-f(t)\|_B = 0 \qquad \text{a.e. in } A. \tag{1}$$

Let $\langle \cdot, \cdot \rangle$ denote the pairing between $B$ and its dual space $B'$ (i.e., $b'(b) = \langle b, b' \rangle$, $b \in B$, $b' \in B'$). It can be shown that any function $f$ whose range is separable is measurable provided the scalar-valued function $\langle f(\cdot), b' \rangle$ is measurable on $A$ for each $b \in B'$.

Suppose that a sequence of simple functions $f_n$ satisfying (1) can be chosen in such a way that

$$\lim_{n\to\infty} \int_A \|f_n(t)-f(t)\|_B \, dt = 0.$$

Then $f$ is called *(Bochner) integrable* on $A$ and we define

$$\int_A f(t)\, dt = \lim_{n\to\infty} \int_{\mathbb{R}} f_n(t)\, dt. \tag{2}$$

[The integrals on the right side of (2) do converge in (the norm topology of) $B$ to a limit which is independent of the choice of approximating sequence $\{f_n\}$.]

A measurable function $f$ is integrable on $A$ if and only if $\|f(\cdot)\|_B$ is (Lebesgue) integrable on $A$. In fact,

$$\left\| \int_A f(t)\, dt \right\|_B \le \int_A \|f(t)\|_B \, dt.$$

**7.3** Let $-\infty \le a < b \le \infty$. We denote by $L^p(a,b;B)$ the vector space of (equivalence classes of) functions $f$ measurable on $(a,b)$ into $B$ such that $\|f(\cdot)\|_B \in L^p(a,b)$. The space $L^p(a,b;B)$ is a Banach space with respect to the norm

$$\|f; L^p(a,b;B)\| = \begin{cases} \left\{ \int_a^b \|f(t)\|_B^p \, dt \right\}^{1/p} & \text{if } 1 \le p < \infty \\ \operatorname*{ess\,sup}_{t\in(a,b)} \|f(t)\|_B & \text{if } p = \infty. \end{cases}$$

Similarly, if $f \in L^p(c,d;B)$ for every $c, d$ with $a < c < d < b$, then we write $f \in L^p_{\text{loc}}(a,b;B)$, and, in case $p = 1$, call $f$ locally integrable.

A locally integrable function $g$ on $(a,b)$ is called the $j$th distributional derivative of the locally integrable function $f$ provided

$$\int_a^b \phi^{(j)}(t) f(t)\, dt = (-1)^j \int_a^b \phi(t) g(t)\, dt$$

for every (scalar-valued) testing function $\phi \in \mathscr{D}(a,b) = C_0^\infty(a,b)$.

## Semigroups of Operators and the Abstract Cauchy Problem

**7.4** The following sections present a discussion of those aspects of the theory of semigroups of operators in Banach spaces which we shall require in the subsequent development of "trace spaces" and the fractional order spaces $W^{s,p}(\Omega)$. In this treatment we follow the work of Zaidman [70].

**7.5** Let $B$ be a Banach space, and $L(B)$ the Banach space of bounded linear operators with domain $B$ and range in $B$. We denote the norms in $B$ and $L(B)$ by $\|\cdot\|_B$ and $\|\cdot\|_{L(B)}$, respectively.

A function $G$ with domain the interval $[0, \infty)$ and range in $L(B)$ is called a *(strongly) continuous semigroup* on $B$ provided

(i) $G(0) = I$, the identity operator on $B$,
(ii) $G(s)\,G(t) = G(s+t)$ for all $s, t \geq 0$,
(iii) for each $b \in B$ the function $G(\cdot)b$ is continuous from $[0, \infty)$ to (the norm topology of) $B$.

We note that (ii) implies that the operators $G(s)$ and $G(t)$ commute. Also, (iii) implies that for each $t_0 \geq 0$ the set $\{t : \|G(t)\|_{L(B)} > t_0\}$ is open in $\mathbb{R}$ and hence measurable. If $0 \leq t_0 < t_1 < \infty$, then for each $b \in B$, $G(\cdot)b$ is uniformly continuous on $[t_0, t_1]$ and hence there exists a constant $K_b$ such that $\|G(t)b\|_B \leq K_b$ if $t_0 \leq t \leq t_1$. By the uniform boundedness principle of functional analysis, there exists a constant $K$ such that $\|G(t)\|_{L(B)} \leq K$ for all $b \in B$ and all $t \in [t_0, t_1]$. Thus $\|G(\cdot)\|_{L(B)} \in L^{\infty}_{\text{loc}}(0, \infty)$. We amplify this result in the following lemma.

**7.6 LEMMA** (a) The limit

$$\lim_{t\to\infty} (1/t) \log \|G(t)\|_{L(B)} = \delta_0$$

exists and is finite.

(b) For each $\delta > \delta_0$ there exists a constant $M_\delta$ such that for every $t \geq 0$

$$\|G(t)\|_{L(B)} \leq M_\delta e^{\delta t}.$$

PROOF Let $N(t) = \log \|G(t)\|_{L(B)}$. Since

$$\|G(s+t)\|_{L(B)} \leq \|G(s)\|_{L(B)} \|G(t)\|_{L(B)}$$

we have that $N$ is subadditive;

$$N(s+t) \leq N(s) + N(t).$$

Let $\delta_0 = \inf_{t>0} N(t)/t$. Clearly $0 \leq \delta_0 < \infty$. Let $\varepsilon > 0$ be given and choose $r > 0$ so that $N(r)/r < \delta_0 + \varepsilon$. If $t \geq 2r$, let $k$ be the integer such that

$(k+1)r \le t < (k+2)r$. Then

$$\delta_0 \le \frac{N(t)}{t} \le \frac{N(kr)+N(t-kr)}{t} \le \frac{k}{t}N(r) + \frac{1}{t}N(t-kr).$$

Now $t-kr \in [r, 2r)$ so, as noted above, $N(t-kr)$ is bounded, say by $K$. Thus

$$\delta_0 \le \frac{N(t)}{t} \le \frac{kr}{t}(\delta_0+\varepsilon) + \frac{K}{t} \le \left(1 - \frac{r}{t}\right)(\delta_0+\varepsilon) + \frac{K}{t}.$$

The right side tends to $\delta_0+\varepsilon$ as $t \to \infty$, and (a) follows since $\varepsilon$ is arbitrary.

If $\delta > \delta_0$, there exists $t_\delta$ such that if $t \ge t_\delta$, then $N(t) \le \delta t$, or equivalently $\|G(t)\|_{L(B)} \le e^{\delta t}$. Conclusion (b) now follows with

$$M_\delta = \max(1, \sup_{0 \le t \le t_\delta} \|G(t)\|_{L(B)}). \quad ∎$$

**7.7** For given $b \in B$ the quotient $(G(t)b-b)/t$ may or may not converge (strongly) in $B$ as $t \to 0+$. Let $D(\Lambda)$ be the set of all those elements $b$ for which the limit exists, and for $b \in D(\Lambda)$ set

$$\Lambda b = \lim_{t\to 0+} \frac{G(t)b-b}{t} = \lim_{t\to 0+} \frac{G(t)b-G(0)b}{t}.$$

Clearly $D(\Lambda)$ is a linear subspace of $B$ and $\Lambda$ a linear operator from $D(\Lambda)$ into $B$. We call $\Lambda$ the *infinitesimal generator* of the semigroup $G$. Note that $\Lambda$ commutes with $G(t)$ $(t \ge 0)$ on $D(\Lambda)$.

**7.8 LEMMA** (a) For each $b \in B$

$$\lim_{t\to 0+} (1/t) \int_0^t G(\tau)b \, d\tau = b.$$

(b) For each $b \in B$ and $t > 0$ we have

$$\int_0^t G(\tau)b \, d\tau \in D(\Lambda) \qquad \text{and} \qquad \Lambda \int_0^t G(\tau)b \, d\tau = G(t)b - b.$$

(c) For each $b \in D(\Lambda)$ and $t > 0$ we have

$$\int_0^t G(\tau)\Lambda b \, d\tau = G(t)b - b.$$

(d) $D(\Lambda)$ is dense in $B$.

(e) $\Lambda$ is a closed operator in $B$, that is, the graph $\{(b, \Lambda b) : b \in D(\Lambda)\}$ is a closed subspace of $B \times B$.

PROOF Let $b \in B$. By the continuity of $G(\cdot)b$ we have $\lim_{\tau\to 0+} \|G(\tau)b-b\|_B = \lim_{\tau\to 0+} \|G(\tau)b-G(0)b\|_B = 0$. Conclusion (a) follows since $b = (1/t)\int_0^t b \, d\tau$.

For fixed $t$ we have

$$\begin{aligned}\lim_{s\to 0+}\frac{G(s)-G(0)}{s}\int_0^t G(\tau)b\,d\tau &= \lim_{s\to 0+}\frac{1}{s}\int_0^t [G(s+\tau)-G(\tau)]b\,d\tau \\ &= \lim_{s\to 0+}\left(\frac{1}{s}\int_s^{s+t} G(\tau)b\,d\tau - \frac{1}{s}\int_0^t G(\tau)b\,d\tau\right) \\ &= \lim_{s\to 0+}\left(\frac{1}{s}\int_t^{s+t} G(\tau)b\,d\tau - \frac{1}{s}\int_0^s G(\tau)b\,d\tau\right) \\ &= \lim_{s\to 0+}\frac{1}{s}\int_0^s G(\tau)G(t)b\,d\tau - b = G(t)b - b.\end{aligned}$$

This proves (b). If $b \in D(\Lambda)$, then

$$\left\|\int_0^t G(\tau)\left(\frac{G(s)b-b}{s} - \Lambda b\right)d\tau\right\|_B \le t \sup_{0<\tau<t}\|G(\tau)\|_{L(B)}\left\|\frac{G(s)b-b}{s} - \Lambda b\right\|_B$$
$$\to 0 \quad \text{as} \quad s \to 0+.$$

Thus

$$\Lambda\int_0^t G(\tau)b\,d\tau = \lim_{s\to 0+}\int_0^t G(\tau)\frac{G(s)b-b}{s}\,d\tau = \int_0^t G(\tau)\Lambda b\,d\tau,$$

which proves (c). Part (d) is an immediate consequence of (a) and (b).

If $b_n \in D(\Lambda)$, $b_n \to b$ and $\Lambda b_n \to b_0$ in $B$, then by (c)

$$G(t)b_n - b_n = \int_0^t G(\tau)\Lambda b_n\,d\tau.$$

We may let $n \to \infty$, justifying the interchange of limit and integral in the same way as was done in the proof of (c), and obtain

$$G(t)b - b = \int_0^t G(\tau)b_0\,d\tau.$$

Thus by (a)

$$\lim_{t\to 0+}\frac{G(t)b-b}{t} = \lim_{t\to 0+}\frac{1}{t}\int_0^t G(\tau)b_0\,d\tau = b_0,$$

whence $b \in D(\Lambda)$ and $\Lambda b = b_0$. This proves (e). ∎

**7.9 REMARK** It follows from the closedness of $\Lambda$ that $D(\Lambda)$ is a Banach space with respect to the norm $\|b; D(\Lambda)\| = \|b\|_B + \|\Lambda b\|_B$. We have, moreover, the obvious imbedding $D(\Lambda) \to B$.

The following theorem is concerned with an abstract setting of a typical Cauchy initial value problem for a first-order differential operator.

**7.10 THEOREM** Let $\Lambda$ be the infinitesimal generator of a continuous semigroup $G$ on the Banach space $B$. Let $a \in D(\Lambda)$ and let $f$ be a continuously differentiable function on $[0, \infty)$ into $B$. Then there exists a unique function $u$, continuous on $[0, \infty)$ into $D(\Lambda)$, and having derivative $u'$ continuous on $[0, \infty)$ into $B$, such that

$$\begin{aligned} u'(t) - \Lambda u(t) &= f(t), \qquad t \geq 0 \\ u(0) &= a. \end{aligned} \tag{3}$$

In fact, $u$ is given by

$$u(t) = G(t)a + \int_0^t G(t-\tau) f(\tau)\, d\tau. \tag{4}$$

PROOF (*Uniqueness*) We must show that the only solution of (3) for $f(t) \equiv 0$ and $a = 0$ is $u(t) \equiv 0$. If $u$ is any solution and $t > \tau$, we have

$$\begin{aligned} \frac{\partial}{\partial \tau} G(t-\tau)u(\tau) &= \lim_{s \to 0} \frac{G(t-\tau-s)u(\tau+s) - G(t-\tau)u(\tau)}{s} \\ &= \lim_{s \to 0} \frac{G(t-\tau-s) - G(t-\tau)}{s} u(s) \\ &\quad + \lim_{s \to 0} G(t-\tau-s) \frac{u(\tau+s) - u(\tau)}{s} \\ &= -G(t-\tau)\Lambda u(\tau) + G(t-\tau)u'(\tau) = 0. \end{aligned}$$

Thus $G(t-\tau)u(\tau) = G(t)u(0) = 0$ for all $t > \tau$. Letting $t \to \tau+$, we obtain $u(\tau) = G(0)u(\tau) = 0$ for all $\tau \geq 0$.

(*Existence*) We verify that $u$ given by (4) satisfies (3). First note that $u(0) = a$ and $(d/dt)G(t)a = \Lambda G(t)a$. Hence it is sufficient to show that the function

$$g(t) = \int_0^t G(t-\tau) f(\tau)\, d\tau$$

is continuously differentiable on $[0, \infty)$ into $B$, takes values in $D(\Lambda)$, and satisfies $g'(t) = \Lambda g(t) + f(t)$.

Now

$$\begin{aligned}\frac{g(t+s)-g(t)}{s} &= \frac{1}{s}\int_0^{t+s} G(t+s-\tau)\,f(\tau)\,d\tau - \frac{1}{s}\int_0^t G(t-\tau)\,f(\tau)\,d\tau \\ &= \frac{1}{s}\int_{-s}^{t} G(t-\tau)\,f(\tau+s)\,d\tau - \frac{1}{s}\int_0^t G(t-\tau)\,f(\tau)\,d\tau \\ &= \int_0^t G(t-\tau)\frac{f(\tau+s)-f(\tau)}{s} + \frac{1}{s}\int_t^{t+s} G(\tau)\,f(t+s-\tau)\,d\tau.\end{aligned}$$

Since $f$ is continuously differentiable on $[0,\infty)$ into $B$ we have that

$$g'(t) = \int_0^t G(t-\tau)\,f'(\tau)\,d\tau + G(t)\,f(0)$$

exists and is continuous on $[0,\infty)$ into $B$. On the other hand,

$$\begin{aligned}\frac{g(t+s)-g(t)}{s} &= \int_0^t \frac{G(s)-G(0)}{s}\,G(t-\tau)\,f(\tau)\,d\tau + \frac{1}{s}\int_t^{t+s} G(t+s-\tau)\,f(\tau)\,d\tau \\ &= \frac{G(s)-G(0)}{s}\,g(t) + \frac{1}{s}\int_0^s G(s-\tau)\,f(t+\tau)\,d\tau.\end{aligned}$$

By Lemma 7.8(a) and the continuity of $f$ the latter integral converges to $f(t)$ as $s\to 0+$. This, together with the existence of $g'(t)$ guarantees that $\lim_{s\to 0+}[G(s)-G(0)]\,g(t)/s$ exists, that is, $g(t)\in D(\Lambda)$. Thus $g'(t)=\Lambda g(t)+f(t)$ as required. ∎

## The Trace Spaces of Lions

**7.11** Let $B_1$ and $B_2$ be two Banach spaces with norms $\|\cdot\|_{B_1}$ and $\|\cdot\|_{B_2}$, respectively, and let $X$ be a topological vector space in which $B_1$ and $B_2$ are each imbedded continuously (i.e., $B_i \cap U$ is open in $B_i$, $i=1,2$, for every $U$ open in $X$). Then the vector sum of $B_1$ and $B_2$

$$B_1 + B_2 = \{b_1+b_2 \in X : b_1\in B_1,\ b_2\in B_2\}$$

is itself a Banach space with respect to the norm:

$$\|u; B_1+B_2\| = \inf_{\substack{b_i\in B_i\\ b_1+b_2=u}} (\|b_1\|_{B_1} + \|b_2\|_{B_2}).$$

Let $1\le p\le\infty$, and for each real number $\nu$ let $t^\nu$ denote the real-valued function defined on $[0,\infty)$ by $t^\nu(t)=t^\nu$, $0\le t<\infty$.

We designate by $W(p,\nu;B_1,B_2)$, or when confusion is not likely to occur simply by $W$, the vector space of (equivalence classes of) measurable functions

$f$ on $[0, \infty)$ into $B_1 + B_2$ such that

$$t^\nu f \in L^p(0, \infty; B_1) \qquad \text{and} \qquad t^\nu f' \in L^p(0, \infty; B_2),$$

$f'$ denoting the distributional derivative of $f$. The space $W$ is a Banach space with respect to the norm

$$\begin{aligned} \|f\|_W &= \|f; W(p, \infty; B_1, B_2)\| \\ &= \max(\|t^\nu f; L^p(0, \infty; B_1)\|, \|t^\nu f'; L^p(0, \infty; B_2)\|). \end{aligned}$$

As an example of this construction the reader may verify that $W(p, 0; W^{1,p}(\mathbb{R}^n), L^p(\mathbb{R}^n))$ is isomorphic to the Sobolev space $W^{1,p}(\Omega)$ where $\Omega = \{(x, t) = (x_1, \ldots, x_n, t) \in \mathbb{R}^{n+1} : t > 0\}$.

We shall show that for certain values of $p$ and $\nu$, functions $f$ in $W$ possess "traces" $f(0)$ in $B_1 + B_2$.

**7.12 LEMMA** Let $f \in W$. There exists $b \in B_1 + B_2$ such that

$$f(t) = b + \int_1^t f'(\tau)\, d\tau \qquad \text{a.e. in } (0, \infty). \tag{5}$$

Hence $f$ is almost everywhere equal to a continuous function on $(0, \infty)$ into $B_1 + B_2$.

PROOF Since $t^\nu f \in L^p(0, \infty; B_1)$, therefore $f \in L^p_{\text{loc}}(0, \infty; B_1)$. Similarly $f' \in L^p_{\text{loc}}(0, \infty; B_2)$, and the function $v$ defined a.e. on $(0, \infty)$ into $B_1 + B_2$ by

$$v(t) = f(t) - \int_1^t f'(\tau)\, d\tau$$

belongs to $L^p_{\text{loc}}(0, \infty; B_1 + B_2)$. It follows that the scalar-valued function $\langle v(\cdot), b' \rangle$ belongs to $L^p_{\text{loc}}(0, \infty)$ for each $b' \in (B_1 + B_2)'$. Thus for every $\phi \in C_0^\infty(0, \infty)$ we have

$$\begin{aligned} \int_0^\infty \frac{d}{dt} \langle v(t), b' \rangle \phi(t)\, dt &= -\int_0^\infty \langle v(t), b' \rangle \phi'(t)\, dt \\ &= -\left\langle \int_0^\infty v(t) \phi'(t)\, dt, b' \right\rangle \\ &= -\left\langle \int_0^\infty f(t) \phi'(t)\, dt - \int_0^\infty \phi'(t)\, dt \int_1^t f'(\tau)\, d\tau, b' \right\rangle \\ &= \left\langle \int_0^\infty f'(t) \phi(t)\, dt - \int_0^\infty f'(t) \phi(t)\, dt, b' \right\rangle = 0. \end{aligned}$$

(The change of order of integration and pairing $\langle \cdot, \cdot \rangle$ above is justified since $v$ is integrable on the support of $\phi$ and so can be approximated there by simple functions for which the interchange is clearly valid.) By Corollary 3.27, $\langle v(t), b' \rangle$ is constant a.e. on $(0, \infty)$ for each $b' \in (B_1 + B_2)'$. Thus $v(t) = b$, a fixed vector in $B_1 + B_2$ a.e. on $(0, \infty)$ and (5) follows at once. Clearly the integral in (5) is continuous on $(0, \infty)$ into $B_2$, and hence into $B_1 + B_2$. ∎

**7.13 LEMMA** Suppose $(1/p) + \nu < 1$. Then the right side of (5) converges in $B_1 + B_2$ as $t \to 0+$. The limit is defined to be the trace, $f(0)$, of $f$ at $t = 0$.

PROOF If $0 < s < t$, we have, for $1 < p < \infty$,

$$\left\| \int_s^t f'(\tau)\, d\tau \right\|_{B_2} \le \int_s^t \| \tau^\nu f'(\tau) \|_{B_2} \tau^{-\nu}\, d\tau$$

$$\le \| t^\nu f'; L^p(0, \infty; B_2) \| \left( \int_0^t \tau^{-\nu p/(p-1)}\, d\tau \right)^{(p-1)/p}.$$

The last factor tends to zero as $t \to 0+$ since $(1/p) + \nu < 1$. (The same argument works with the obvious modifications if $p = 1$ or $p = \infty$.) Thus $\int_1^t f'(\tau)\, d\tau$ converges in $B_2$ as $t \to 0+$, which proves the lemma. ∎

**7.14** Given real $p$ and $\nu$ with $1 \le p \le \infty$ and $\theta = (1/p) + \nu < 1$ we denote by $T(p, \nu; B_1, B_2)$, or simply by $T$, the space consisting of all traces $f(0)$ of functions $f$ in $W = W(p, \nu; B_1, B_2)$. Called the *trace space* of $W$, $T$ is a Banach space with respect to the norm

$$\|u\|_T = \inf_{\substack{u = f(0) \\ f \in W}} \|f\|_W.$$

$T$ is a subspace of $B_1 + B_2$ and lies topologically "between" $B_1$ and $B_2$ in a sense which will become more apparent later.

Before developing some properties of the trace space $T$ we prepare one further lemma concerning $W$ which will be needed later.

**7.15 LEMMA** The subspace of $W$ consisting of those functions $f \in W$ which are infinitely differentiable on $(0, \infty)$ into $B_1$ is dense in $W$ if $1 \le p < \infty$.

PROOF Under the transformation

$$t = e^\tau, \qquad f(e^\tau) = \tilde{f}(\tau),$$

we have that $f \in W$ if and only if

$$\int_{-\infty}^{\infty} \left( e^{\theta p \tau} \| \tilde{f}(\tau) \|_{B_1}^p + e^{(\theta - 1) p \tau} \| \tilde{f}'(\tau) \|_{B_2}^p \right) d\tau < \infty,$$

where $\theta = (1/p) + \nu$. If $J_\varepsilon$ is the mollifier of Section 2.17, then, just as in Lemma

2.18, $J_\varepsilon * \tilde{f}$ is infinitely differentiable on $\mathbb{R}$ into $B_1$ and

$$\lim_{\varepsilon\to 0+} \int_{-\infty}^{\infty} (e^{\theta p\tau} \|J_\varepsilon * \tilde{f}(\tau) - \tilde{f}(\tau)\|_{B_1}^p + e^{(\theta-1)p\tau} \|(J_\varepsilon * \tilde{f})'(\tau) - \tilde{f}'(\tau)\|_{B_2}^p)\, dt = 0.$$

Thus $f_n(t) = J_{1/n} * \tilde{f}(\log t)$ is infinitely differentiable for $0 < t < \infty$ with values in $B_1$. Since $f_n \to f$ in $W$ the lemma is proved. ∎

We now investigate interpolation properties of the space $T$.

**7.16 LEMMA** Let $\theta = (1/p) + v$ satisfy $0 < \theta < 1$. Then

(a) every $u \in T$ satisfies

$$\|u\|_T = \inf_{\substack{f\in W\\ f(0)=u}} \|t^v f; L^p(0,\infty; B_1)\|^{1-\theta} \|t^v f'; L^p(0,\infty; B_2)\|^\theta; \tag{6}$$

(b) if $u \in B_1 \cap B_2$, then $u \in T$ and

$$\|u\|_T \le K \|u\|_{B_1}^{1-\theta} \|u\|_{B_2}^\theta, \tag{7}$$

where $K$ is a constant independent of $u$.

PROOF (a) Fix $u \in T$ and $\varepsilon > 0$ and let $f \in W$ satisfy $f(0) = u$ and $\|f\|_W \le \|u\|_T + \varepsilon$. Let

$$R = \|t^v f; L^p(0,\infty; B_1)\|, \qquad S = \|t^v f'; L^p(0,\infty; B_2)\|.$$

For $\lambda > 0$ the function $f_\lambda$ defined by $f_\lambda(t) = f(\lambda t)$ also belongs to $W$ and satisfies $f_\lambda(0) = u$. Moreover,

$$\|t^v f_\lambda; L^p(0,\infty; B_1)\| = \lambda^{-\theta} R, \qquad \|t^v (f_\lambda)'; L^p(0,\infty; B_2)\| = \lambda^{1-\theta} S.$$

These two expressions are both equal to $R^{1-\theta}S^\theta$ provided we choose $\lambda = R/S$. Hence

$$\begin{aligned}\max(R,S) &= \|f\|_W \le \|u\|_T + \varepsilon \\ &\le \inf_{\lambda>0} \|f_\lambda\|_W + \varepsilon \\ &\le \inf_{\lambda>0} \max(\lambda^{-\theta}R, \lambda^{1-\theta}S) + \varepsilon \le R^{1-\theta}S^\theta + \varepsilon \le \max(R,S) + \varepsilon.\end{aligned}$$

Since $\varepsilon$ is arbitrary, (6) follows at once.

(b) Let $\phi \in C^\infty([0,\infty))$ satisfy $\phi(0) = 1$, $\phi(t) = 0$ if $t \ge 1$, $|\phi(t)| \le 1$, and $|\phi'(t)| \le K_1$ for all $t \ge 0$. If $u \in B_1 \cap B_2$, let $f(t) = \phi(t)u$ so that $u = f(0)$. Now

$$\|t^v f; L^p(0,\infty; B_1)\| \le K_2 \|u\|_{B_1},$$

where $K_2 = \{\int_0^1 t^{\nu p}\,dt\}^{1/p}$. Similarly,

$$\|t^\nu f'; L^p(0,\infty; B_2)\| \le K_1 K_2 \|u\|_{B_2}.$$

Hence $f \in W$ and (7) follows from (6). ∎

The above lemma suggests the sense in which $T$ lies "between" $B_1$ and $B_2$ provided $0 < \theta < 1$. The spaces $T$ corresponding to all such values of $\theta$ are sometimes described as constituting a *scale* of Banach spaces interpolated between $B_1$ and $B_2$. Many properties of $T$ can be deduced from corresponding properties of $B_1$ and $B_2$ via the following interpolation theorem.

**7.17 THEOREM** Suppose that $B_1 \cap B_2$ is dense in $B_1$ and $B_2$. Let $\tilde{B}_1$, $\tilde{B}_2$, and $\tilde{X}$ be three spaces having the same properties specified for $B_1$, $B_2$, and $X$ in Section 7.11. Let $L$ be a linear operator defined on $B_1 \cap B_2$ into $\tilde{B}_1 \cap \tilde{B}_2$ and suppose that for every $v \in B_1 \cap B_2$ we have

$$\|Lv\|_{\tilde{B}_1} \le K_1 \|v\|_{B_1}, \tag{8}$$

$$\|Lv\|_{\tilde{B}_2} \le K_2 \|v\|_{B_2}. \tag{9}$$

Thus $L$ possesses unique continuous extensions (also denoted $L$) to $B_1$ and $B_2$ satisfying (8) and (9), respectively, and hence to $B_1 + B_2$ satisfying

$$\|Lu\|_{\tilde{B}_1+\tilde{B}_2} \le \max(K_1, K_2)\,\|u\|_{B_1+B_2}.$$

If $0 < \theta = (1/p)+\nu < 1$, then for every $u \in T = T(p,\nu; B_1, B_2)$ we have $Lu \in \tilde{T} = T(p,\nu;\tilde{B}_1,\tilde{B}_2)$ and

$$\|Lu\|_{\tilde{T}} \le K_1^{1-\theta} K_2^{\theta}\,\|u\|_T. \tag{10}$$

PROOF $L$ is defined on $B_1 + B_2$ and hence on $T$. By (6) we have for $u \in T$

$$\|Lu\|_{\tilde{T}} = \inf_{\substack{\tilde{f}\in \tilde{W} \\ \tilde{f}(0)=Lu}} \|t^\nu \tilde{f}; L^p(0,\infty;\tilde{B}_1)\|^{1-\theta}\,\|t^\nu \tilde{f}'; L^p(0,\infty;\tilde{B}_2)\|^{\theta}.$$

Also, there exists, for any given $\varepsilon > 0$, an element $f \in W$ with $f(0) = u$ such that

$$\|t^\nu f; L^p(0,\infty;B_1)\|^{1-\theta}\,\|t^\nu f'; L^p(0,\infty;B_2)\|^{\theta} < \|u\|_T + \varepsilon.$$

For $t \ge 0$ let $\tilde{f}(t) = Lf(t)$ so that $\tilde{f}(0) = Lu$ and

$$\|t^\nu \tilde{f}; L^p(0,\infty;\tilde{B}_1)\|^{1-\theta}\,\|t^\nu \tilde{f}'; L^p(0,\infty;\tilde{B}_2)\|^{\theta}$$
$$\le K_1^{1-\theta}\,\|t^\nu f; L^p(0,\infty;B_1)\|^{1-\theta} K_2^{\theta}\,\|t^\nu f'; L^p(0,\infty;B_2)\|^{\theta}.$$

Hence

$$\|Lu\|_{\tilde{T}} < K_1^{1-\theta} K_2^{\theta}(\|u\|_T + \varepsilon).$$

Since $\varepsilon$ is arbitrary (10) follows. ∎

If we denote by $\|L\|_{L(S,\tilde{S})}$ the norm of $L$ as an element of the Banach space $L(S,\tilde{S})$ of continuous linear operators on $S$ into $\tilde{S}$ we may write (10) in the form

$$\|L\|_{L(T,\tilde{T})} \le \|L\|^{1-\theta}_{L(B_1,\tilde{B}_1)} \|L\|^{\theta}_{L(B_2,\tilde{B}_2)}.$$

**7.18 LEMMA** Suppose that $B_1 \cap B_2$ is dense in $B_1$ and $B_2$, and that there exists a sequence $\{P_j\}_{j=1}^{\infty}$ of linear operators belonging simultaneously to $L(B_1)$ and $L(B_2)$ and having range in $B_1 \cap B_2$. Suppose also that for each $b_i \in B_i$, $i = 1, 2$,

$$\lim_{j\to\infty} \|P_j b_i - b_i\|_{B_i} = 0.$$

Then for every $u \in T$ we have

$$\lim_{j\to\infty} \|P_j u - u\|_T = 0.$$

In particular, $B_1 \cap B_2$ is dense in $T$.

PROOF Fix $u \in T$ and choose $f \in W$ such that $f(0) = u$. Let $f_j(t) = P_j f(t)$. If $b_i \in B_i$, $i = 1, 2$, there exists an integer $j_0 = j_0(b_i)$ such that if $j \ge j_0$, then

$$\|P_j b_i - b_i\|_{B_i} \le 1.$$

Hence $\{P_j b_i\}$ is bounded in $B_i$, $i = 1, 2$, independently of $j$. By the uniform boundedness principle there exist constants $K_1$ and $K_2$ such that for every $j$

$$\|P_j\|_{L(B_i)} \le K_i.$$

It follows that

$$\|f_j(t)\|_{B_1} \le K_1 \|f(t)\|_{B_1}, \qquad \|f_j'(t)\|_{B_2} \le K_2 \|f'(t)\|_{B_2}.$$

Since for almost all $t > 0$, $f_j(t) \to f(t)$ in $B_1$ and $f_j'(t) \to f'(t)$ in $B_2$ as $j \to \infty$, we have by dominated convergence that $t^\nu f_j \to t^\nu f$ in $L^p(0,\infty; B_1)$ and $t^\nu f_j' \to t^\nu f'$ in $L^p(0,\infty; B_2)$. Hence $f_j \to f$ in $W$ and so $P_j u = P_j f(0) \to f(0) = u$ in $T$. Since $t^\nu f_j$ and $t^\nu f_j'$ take values in $B_1 \cap B_2$, $P_j u$ belongs to $B_1 \cap B_2$. ∎

We quote now a theorem characterizing the dual of a trace space as another trace space. The proof is rather long and will not be given here—the interested reader may find it in the work of Lions [37] where trace spaces somewhat more general than those introduced above are studied. (The following theorem is a special case of Theorem 1.1 of Ref. [37, Chap. II].)

**7.19 THEOREM** Suppose that $B_1$ and $B_2$ are reflexive and also satisfy the conditions of Lemma 7.18. If $1 < p < \infty$ and $(1/p) + \nu = \theta$ satisfies $0 < \theta < 1$, then $(1/p')-\nu = 1-(1/p)-\nu = 1-\theta$ and

$$[T(p,\nu;B_1,B_2)]' \cong T(p',-\nu;B_2',B_1').$$

In particular, $T(p,\nu;B_1,B_2)$ is reflexive.

We now prove an imbedding theorem for trace spaces between two $L^p$-spaces which will play a vital role in extending certain aspects of the Sobolev imbedding theorem to fractional order spaces (see Theorem 7.57). If $\Omega$ is a domain in $\mathbb{R}^n$, then $B_1 = L^q(\Omega)$ and $B_2 = L^p(\Omega)$ are both continuously imbedded in the topological vector space $X = L^1_{\text{loc}}(\Omega)$. (A subset $U \subset L^1_{\text{loc}}(\Omega)$ is open if for every $u \in U$ there exists $\varepsilon > 0$ and $K \subset\subset \Omega$ such that $\|v-u\|_{0,1,K} < \varepsilon$, $v \in L^1_{\text{loc}}(\Omega)$ implies $v \in U$.)

**7.20 THEOREM** Let $p, q, \theta$ satisfy $1 \le p \le q < \infty$, $0 < \theta < 1$, $\theta = (1/p)+\nu$. Then

$$T(p,\nu;L^q(\Omega),L^p(\Omega)) \to L^r(\Omega), \tag{11}$$

where

$$1/r = [(1-\theta)/q] + (\theta/p).$$

PROOF Suppose that $f \in C^\infty([0,\infty))$. From the identity

$$f(0) = f(t) - \int_0^t f'(\tau)\, d\tau$$

we may readily obtain

$$\begin{aligned}|f(0)| &\le \int_0^1 |f(t)|\, dt + \int_0^1 |f'(t)|\, dt \\ &\le \left\{\left(\int_0^\infty t^{\nu p}|f(t)|^p\, dt\right)^{1/p} + \left(\int_0^\infty t^{\nu p}|f'(t)|^p\, dt\right)^{1/p}\right\}\left(\int_0^1 t^{-\nu p'}\, dt\right)^{1/p'} \\ &= K_1(\|t^\nu f\|_{0,p,(0,\infty)} + \|t^\nu f'\|_{0,p,(0,\infty)}),\end{aligned}$$

where $K_1 < \infty$ since $\theta = (1/p)+\nu < 1$. By a homogeneity argument similar to that used in the proof of Lemma 7.16(a), we may now obtain

$$|f(0)| \le 2K_1 \|t^\nu f\|^{1-\theta}_{0,p,(0,\infty)} \|t^\nu f'\|^{\theta}_{0,p,(0,\infty)}. \tag{12}$$

Now suppose that $f \in W(p,\nu;L^q(\Omega),L^p(\Omega))$ and, for the moment, that $f$ is infinitely differentiable on $(0,\infty)$ into $L^q(\Omega)$. Let $\tilde{f}(x,t) = f(t)(x)$ for $0 \le t < \infty$, $x \in \Omega$. From (12) we have that $\tilde{f}(x,0) = \lim_{t\to 0+} \tilde{f}(x,t)$ satisfies

$$|\tilde{f}(x,0)|^r \le K_2\left(\int_0^\infty t^{\nu p}|\tilde{f}(x,t)|^p\, dt\right)^{(1-\theta)r/p}\left(\int_0^\infty t^{\nu p}\left|\frac{\partial}{\partial t}\tilde{f}(x,t)\right|^p dt\right)^{\theta r/p}$$

for almost all $x \in \Omega$. Thus, by Hölder's inequality,

$$\int_\Omega |\tilde{f}(x,0)|^r\,dx \le K_2\left(\int_\Omega\left(\int_0^\infty t^{\nu p}|\tilde{f}(x,t)|^p\,dt\right)^{(1-\theta)rs/p} dx\right)^{1/s}$$
$$\times\left(\int_\Omega\left(\int_0^\infty t^{\nu p}\left|\frac{\partial}{\partial t}\tilde{f}(x,t)\right|^p dt\right)^{\theta rs'/p} dx\right)^{1/s'},$$

where $(1/s)+(1/s')=1$. If we choose $s$ so that $(1-\theta)rs=q$, then also $\theta rs'=p$ and we have

$$\|\tilde{f}(\cdot,0)\|_{0,r,\Omega} \le K_3\left(\int_\Omega\left(\int_0^\infty t^{\nu p}|\tilde{f}(x,t)|^p\,dt\right)^{q/p} dx\right)^{(1-\theta)/q}$$
$$\times\left(\int_\Omega\int_0^\infty t^{\nu p}\left|\frac{\partial}{\partial t}\tilde{f}(x,t)\right|^p dt\,dx\right)^{\theta/p}$$
$$= K_3\left\|\int_0^\infty t^{\nu p}|f(t)|^p\,dt\right\|_{0,q/p,\Omega}^{(1-\theta)/p}\|t^\nu f';L^p(0,\infty;L^p(\Omega))\|^\theta$$
$$\le K_3\left(\int_0^\infty t^{\nu p}\,\||f(t)|^p\|_{0,q/p,\Omega}\,dt\right)^{(1-\theta)/p}\|t^\nu f';L^p(0,\infty;L^p(\Omega))\|^\theta$$
$$= K_3\,\|t^\nu f;L^p(0,\infty;L^q(\Omega))\|^{1-\theta}\,\|t^\nu f';L^p(0,\infty;L^p(\Omega))\|^\theta.$$

By the density of infinitely differentiable functions in $W$ (Lemma 7.15) the above inequality holds for any $f\in W$.

It now follows that if $u\in T(p,\nu;L^q(\Omega),L^p(\Omega))$, then

$$\|u\|_{0,r,\Omega} \le \inf_{\substack{f\in W\\ f(0)=u}} K_3\,\|t^\nu f;L^p(0,\infty;L^q(\Omega))\|^{1-\theta}\,\|t^\nu f';L^p(0,\infty;L^p(\Omega))\|^\theta$$
$$= K_3\,\|u\|_T$$

by Lemma 7.16(a). This establishes imbedding (11). ∎

**7.21 REMARK** With minor modifications in the proof, the above theorem extends to cover $q=\infty$ provided we use in place of $L^q(\Omega)$ the closure of $L^p(\Omega)\cap L^\infty(\Omega)$ in $L^\infty(\Omega)$.

## Semigroup Characterization of Trace Spaces

**7.22** Let $B$ be a Banach space and $G$ a continuous semigroup on $B$ which is uniformly bounded, that is, for which there exists a constant $M$ such that

$$\|G(t)\|_{L(B)} \le M, \qquad 0\le t<\infty.$$

Let $\Lambda$ be the infinitesimal generator of $G$ so that (see Remark 7.9) $D(\Lambda)$, the domain of $\Lambda$ in $B$, is a Banach space with respect to the graph-norm

$$\|u; D(\Lambda)\| = \|u\|_B + \|\Lambda u\|_B,$$

and is also a dense vector subspace of $B$. The spaces $B_1 = D(\Lambda)$ and $B_2 = X = B$ satisfy the conditions of Section 7.11, and we may accordingly construct the trace space $T = T(p, \nu; D(\Lambda), B)$ provided $\theta = (1/p) + \nu < 1$ Theorem 7.24 characterizes $T$ in terms of an explicit norm involving the semigroup $G$. First, however, we obtain an inequality of Hardy, Littlewood, and Pólya [28] which will be needed.

**7.23 LEMMA** Let $f$ be a scalar-valued function defined a.e. on $[0, \infty)$ and let

$$g(t) = (1/t)\int_0^t f(\xi)\, d\xi.$$

If $1 \le p < \infty$ and $(1/p) + \nu = \theta < 1$, then

$$\int_0^\infty t^{\nu p}|g(t)|^p\, dt \le [1/(1-\theta)^p]\int_0^\infty t^{\nu p}|f(t)|^p\, dt. \tag{13}$$

PROOF We may certainly suppose that the right side of (13) is finite. Under the transformation $t = e^\tau$, $f(e^\tau) = \tilde{f}(\tau)$, $\xi = e^\sigma$, $g(e^\tau) = \tilde{g}(\tau)$, (13) becomes

$$\int_{-\infty}^\infty e^{\theta p\tau}|\tilde{g}(\tau)|^p\, d\tau \le [1/(1-\theta)^p]\int_{-\infty}^\infty e^{\theta p\tau}|\tilde{f}(\tau)|^p\, d\tau. \tag{14}$$

Note that

$$\tilde{g}(\tau) = e^{-\tau}\int_{-\infty}^\tau \tilde{f}(\sigma)e^\sigma\, d\sigma.$$

Let $E(\tau) = e^{\theta\tau}$ and

$$F(\tau) = \begin{cases} e^{(\theta-1)\tau} & \text{if } \tau > 0 \\ 0 & \text{if } \tau \le 0. \end{cases}$$

Then $E\cdot\tilde{g} = F * (E\cdot\tilde{f})$, and so by Young's theorem 4.30,

$$\|E\cdot\tilde{g}\|_{0,p,\mathbb{R}} \le \|F\|_{0,1,\mathbb{R}}\|E\tilde{f}\|_{0,p,\mathbb{R}}.$$

This inequality is precisely (14) since $\int_{-\infty}^\infty |F(\tau)|\, d\tau = 1/(1-\theta)$. ∎

**7.24 THEOREM** Let $\Lambda$ be the infinitesimal generator of a uniformly bounded, continuous semigroup $G$ on the Banach space $B$. If $1 \le p < \infty$ and $0 < (1/p) + \nu < 1$, then $T = T(p, \nu; D(\Lambda), B)$ coincides with the space $T^0$ of all $u \in B$ for which the norm

$$\|u\|_{T^0} = \left( \|u\|_B^p + \int_0^\infty t^{(\nu-1)p} \|G(t)u - u\|_B^p \, dt \right)^{1/p} \tag{15}$$

is finite. The norms $\|\cdot\|_T$ and $\|\cdot\|_{T^0}$ are equivalent.

PROOF First let $u \in T$ and choose $f \in W$ such that $f(0) = u$. For the moment let us assume that $f$ is infinitely differentiable on $(0, \infty)$ into $D(\Lambda)$. Let $f'(t) - \Lambda f(t) = h(t)$. If $t \geq \varepsilon > 0$, we obtain by Theorem 7.10,

$$f(t) = G(t-\varepsilon) f(\varepsilon) + \int_\varepsilon^t G(t-\tau) h(\tau) \, d\tau.$$

Hence

$$G(t-\varepsilon) f(\varepsilon) - f(\varepsilon) = \int_\varepsilon^t f'(\tau) \, d\tau - \int_\varepsilon^t G(t-\tau) h(\tau) \, d\tau.$$

Letting $\varepsilon \to 0+$ in this identity and noting that $f(\varepsilon) \to f(0)$ by definition, we obtain

$$G(t) f(0) - f(0) = \int_0^t f'(\tau) \, d\tau - \int_0^t G(t-\tau) h(\tau) \, d\tau. \tag{16}$$

Now (16) holds for any $f \in W$ since, by Lemma 7.15, $f$ is the limit in $W$ of a sequence $\{f_n\}$ of infinitely differentiable functions on $(0, \infty)$ with values in $D(\Lambda)$. Hence for $u \in T$ and any $f \in W$ with $f(0) = u$, we have

$$G(t)u - u = \int_0^t f'(\tau) \, d\tau - \int_0^t G(t-\tau) h(\tau) \, d\tau, \qquad h(\tau) = f'(\tau) - \Lambda f(\tau).$$

Thus

$$\left\| \frac{G(t)u - u}{t} \right\|_B \leq \frac{1}{t} \int_0^t \|f'(\tau)\|_B \, d\tau + \frac{M}{t} \int_0^t \|h(\tau)\|_B \, d\tau,$$

where we have used the uniform boundedness of $G$. Applying Lemma 7.23, we obtain, putting $\theta = (1/p) + \nu$,

$$\begin{aligned}
&\int_0^\infty t^{(\nu-1)p} \|G(t)u - u\|_B^p \, dt \\
&\quad \leq \frac{1}{(1-\theta)^p} \int_0^\infty t^{\nu p} (\|f'(t)\|_B + M \|h(t)\|_B)^p \, dt \\
&\quad \leq \frac{2^{p-1}(M+1)^p}{(1-\theta)^p} (\|t^\nu f'; L^p(0,\infty;B)\|^p + \|t^\nu \Lambda f; L^p(0,\infty;B)\|^p) \\
&\quad \leq \frac{2^p (M+1)^p}{(1-\theta)^p} \|f\|_W^p.
\end{aligned}$$

Since this holds for any $f \in W$ with $f(0) = u$ we have

$$\int_0^\infty t^{(\nu-1)p} \|G(t)u-u\|_B^p \, dt \le \left(\frac{2M+2}{1-\theta}\right)^p \|u\|_T^p.$$

Also, the identity operator on $B$ provides imbeddings of $B$ into $B$ (trivially) and $D(\Lambda)$ into $B$, each imbedding having unit imbedding constant. By Theorem 7.17 we have as well, $T \to B$ and $\|u\|_B \le \|u\|_T$. Hence $u \in T$ implies $u \in T^0$ and

$$\|u\|_{T^0} \le \left(1 + \frac{2M+2}{1-\theta}\right) \|u\|_T.$$

Conversely, suppose that $u \in T^0$. Let $\phi \in C^\infty([0,\infty))$ satisfy $\phi(0) = 1$, $\phi(t) = 0$ for $t \ge 1$, $|\phi(t)| \le 1$ and $|\phi'(t)| \le K_1$ for $t \ge 0$. Let

$$f(t) = \phi(t) g(t),$$

where

$$g(t) = (1/t) \int_0^t G(\tau) u \, d\tau, \qquad t > 0. \tag{17}$$

In order to show that $u \in T$ and $\|u\|_T \le K_2 \|u\|_{T^0}$, it is sufficient to prove that $f \in W$ and

$$\|f\|_W \le K_2 \|u\|_{T^0} \tag{18}$$

[since $f(0) = \lim_{t\to 0+} \phi(t) g(t) = u$ by Lemma 7.8(a)] and this can be done by showing that $t^\nu g \in L^p(0,1; D(\Lambda))$ and $t^\nu g' \in L^p(0,1;B)$ with appropriate norms bounded by $K_3 \|u\|_{T^0}$.

By Lemma 7.8(b), $\int_0^t G(\tau) u \, d\tau \in D(\Lambda)$ and

$$\Lambda \int_0^t G(\tau) u \, d\tau = G(t)u - u.$$

Thus

$$\begin{aligned}
\int_0^1 & t^{\nu p} \|g(t); D(\Lambda)\|^p \, dt \\
&= \int_0^1 t^{(\nu-1)p} \left( \left\| \int_0^t G(\tau) u \, d\tau \right\|_B + \left\| \Lambda \int_0^t G(\tau) u \, d\tau \right\|_B \right)^p dt \\
&\le 2^{p-1} M^p \|u\|_B^p \int_0^1 t^{\nu p} \, dt + 2^{p-1} \int_0^\infty t^{(\nu-1)p} \|G(t)u-u\|_B^p \, dt \\
&\le 2^{p-1} \max(M^p \theta/p, 1) \|u\|_{T^0}^p.
\end{aligned}$$

Since

$$\begin{aligned}
g'(t) &= (1/t) G(t) u - (1/t^2) \int_0^t G(\tau) u \, d\tau \\
&= (1/t)(G(t)u-u) - (1/t^2) \int_0^t (G(\tau)u-u) \, d\tau,
\end{aligned}$$

and since

$$\int_0^1 t^{\nu p} \left\| \frac{G(t)u-u}{t} \right\|_B^p dt \le \|u\|_{T^0}^p$$

and, by Lemma 7.23 with $\nu-1$ replacing $\nu$,

$$\int_0^1 t^{\nu p} \left\| (1/t^2) \int_0^t (G(\tau)u-u)\, d\tau \right\|_B^p dt \le [1/(2-\theta)^p] \int_0^\infty t^{(\nu-1)p} \|G(t)u-u\|_B^p\, dt$$
$$\le [1/(2-\theta)^p]\, \|u\|_{T^0}^p$$

we therefore have

$$\int_0^1 t^{\nu p} \|g'(t)\|_B^p\, dt \le K_4 \|u\|_{T^0}^p.$$

Thus $g \in W$, (18) holds, and the proof is complete. ▮

**7.25** For our purposes we require a slight generalization of Theorem 7.24. Let $\Lambda_1, \Lambda_2, \ldots, \Lambda_n$ be a finite family of infinitesimal generators of commuting, uniformly bounded, continuous semigroups $G_1, G_2, \ldots, G_n$ on $B$. Thus

$$\|G_j(t)\|_{L(B)} \le M_j; \qquad 1 \le j \le n, \quad t \ge 0;$$
$$G_j(s)\, G_k(t) = G_k(t)\, G_j(s); \qquad 1 \le j, k \le n, \quad s, t \ge 0.$$

Let $B^n$ denote the product space $B \times B \times \cdots \times B$ ($n$ factors) which is a Banach space with norm

$$\|(b_1, b_2, \ldots, b_n)\|_{B^n} = \sum_{j=1}^n \|b_j\|_B.$$

The operator $\Lambda$ is defined on $D(\Lambda) = \bigcap_{j=1}^n D(\Lambda_j)$ into $B^n$ by

$$\Lambda u = (\Lambda_1 u, \Lambda_2 u, \ldots, \Lambda_n u).$$

We leave it to the reader to generalize Lemma 7.8 to show that $D(\Lambda)$ is dense in $B$, that $\Lambda$ is a closed operator, and hence that $D(\Lambda)$ is a Banach space with respect to the norm

$$\|u; D(\Lambda)\| = \|u\|_B + \|\Lambda u\|_{B^n} = \|u\|_B + \sum_{j=1}^n \|\Lambda_j u\|_B.$$

**7.26 THEOREM** If $0 < (1/p) + \nu < 1$, $1 \le p < \infty$, then $T = T(p, \nu; D(\Lambda), B)$ coincides with the space $T^0$ of all $u \in B$ for which the norm

$$\|u\|_{T^0} = \left( \|u\|_B^p + \sum_{j=1}^n \int_0^\infty t^{(\nu-1)p} \|G_j(t)u-u\|_B^p\, dt \right)^{1/p}$$

is finite. The norms $\|\cdot\|_T$ and $\|\cdot\|_{T^0}$ are equivalent.

PROOF The proof is nearly identical to that of Theorem 7.24 except that in place of the function $g(t)$ given by (17) we use

$$g(t) = (1/t^n) \int_0^t \int_0^t \cdots \int_0^t G_1(\tau_1) G_2(\tau_2) \cdots G_n(\tau_n) u \, d\tau_1 \, d\tau_2 \cdots d\tau_n.$$

The details are left to the reader. ▮

**7.27 EXAMPLE** Let $B = L^p(\mathbb{R}^n)$, $1 \le p < \infty$, and for $u \in B$ set

$$(G_j(t)u)(x) = u(x_1, \ldots, x_j + t, \ldots, x_n); \qquad j = 1, 2, \ldots, n.$$

Clearly the $G_j$ are commuting, uniformly bounded ($M_j = 1$), continuous semigroups on $L^p(\mathbb{R}^n)$. (In fact they are groups if we allow $t < 0$.) The corresponding infinitesimal generators satisfy

$$D(\Lambda_j) = \{u \in L^p(\mathbb{R}^n) : D_j u \in L^p(\mathbb{R}^n)\},$$

$$\Lambda_j u = D_j u, \qquad u \in D(\Lambda_j).$$

Accordingly, $D(\Lambda) = \bigcap_{j=1}^n D(\Lambda_j) = W^{1,p}(\mathbb{R}^n)$. By Theorem 7.26 the norm

$$\left( \|u\|_{0,p,\mathbb{R}^n}^p + \sum_{j=1}^n \int_0^t t^{(\nu-1)p} \int_{\mathbb{R}^n} |u(x_1, \ldots, x_j + t, \ldots, x_n) - u(x_1, \ldots, x_n)|^p \, dx \, dt \right)^{1/p}$$

is equivalent to $\|u\|_T$ on the space $T = T(p, \nu; W^{1,p}(\mathbb{R}^n), L^p(\mathbb{R}^n))$, $0 < (1/p) + \nu < 1$.

## Higher-Order Traces

**7.28** Up to this point we have considered only traces $f(0)$ of functions satisfying, with their first derivatives $f'$, suitable integrability conditions on $[0, \infty)$ into various Banach spaces. We now extend the notion of trace to obtain values for $f^{(j)}(0)$, $0 \le j \le m-1$, provided $f, f', \ldots, f^{(m)}$ satisfy such integrability conditions. As a result of this extension we will be able later to characterize the traces on the boundary of a regular domain $\Omega$, of functions belonging to $W^{m,p}(\Omega)$.

**7.29** Let $B$ be a Banach space and $\Lambda_1, \ldots, \Lambda_n$ infinitesimal generators of commuting, uniformly bounded, continuous semigroups $G_1, \ldots, G_n$ on $B$. For each multi-index $\alpha$ we define a subspace $D(\Lambda^\alpha)$ of $B$ and a corresponding linear operator $\Lambda^\alpha$ on $D(\Lambda^\alpha)$ into $B$ by induction on $|\alpha|$ as follows.

If $\alpha = (0, 0, \ldots, 0)$, then $D(\Lambda^\alpha) = B$ and $\Lambda^\alpha = I$, the identity on $B$.

If $\alpha = (0, \ldots, 1, \ldots, 0)$ (1 in the $j$th place), then $D(\Lambda^\alpha) = D(\Lambda_j)$ and $\Lambda^\alpha = \Lambda_j$.

If $D(\Lambda^\beta)$ and $\Lambda^\beta$ have been defined for all $\beta$ with $|\beta| \le r$, and if $|\alpha| = r + 1$, then

$$D(\Lambda^\alpha) = \{u : u \in D(\Lambda^\beta) \text{ and } \Lambda^{\alpha-\beta}u \in D(\Lambda^\beta) \text{ for all } \beta < \alpha\},$$
$$\Lambda^\alpha = \Lambda_1^{\alpha_1} \cdots \Lambda_n^{\alpha_n}.$$

If $k$ is a positive integer, let $D(\Lambda^k) = \bigcap_{|\alpha| \le k} D(\Lambda^\alpha)$. Once again we leave to the reader the task of verifying (say by induction on $k$) that $D(\Lambda^k)$ is dense in $B$, that $\Lambda^k = (\Lambda^\alpha)_{|\alpha| \le k}$ is a closed operator on $D(\Lambda^k)$ into $\prod_{|\alpha| \le k} B$, and hence that $D(\Lambda^k)$ is a Banach space with respect to the norm

$$\|u; D(\Lambda^k)\| = \sum_{|\alpha| \le k} \|\Lambda^\alpha u\|_B.$$

**7.30** For positive integers $m$ and real $p$, $1 \le p \le \infty$, let $W^m = W^m(p, \nu; \Lambda; B)$ denote the space of (equivalence classes of) measurable functions $f$ on $[0, \infty)$ into B such that

$$t^\nu f^{(k)} \in L^p(0, \infty; D(\Lambda^{m-k})) \qquad 0 \le k \le m,$$

$f^{(k)}$ being the distributional derivative $d^k f/dt^k$. The space $W^m$ is a Banach space with respect to the norm

$$\|f\|_{W^m} = \max_{0 \le k \le m} \|t^\nu f^{(k)}; L^p(0, \infty; D(\Lambda^{m-k}))\|.$$

Note that $W^1 = W(p, \nu; D(\Lambda), B)$ with $D(\Lambda)$ as in Section 7.25.

**7.31 LEMMA** Let $f \in W^m$, $m \ge 1$. If $0 \le k \le m-1$ and $\alpha$ is a multi-index such that $|\alpha| + k \le m-1$, then the function $f_{\alpha k} = \Lambda^\alpha f^{(k)} \in W^1$ and

$$\|f_{\alpha k}\|_{W^1} \le \|f\|_{W^m}.$$

PROOF We have (for $1 \le p < \infty$)

$$\begin{aligned}
\|t^\nu f_{\alpha k}; L^p(0, \infty; D(\Lambda))\|^p &= \int_0^\infty t^{\nu p}\left(\|\Lambda^\alpha f^{(k)}(t)\|_B + \sum_{j=1}^n \|\Lambda_j \Lambda^\alpha f^{(k)}(t)\|_B\right)^p dt \\
&\le \int_0^\infty t^{\nu p}\left(\sum_{|\beta| \le m-k} \|\Lambda^\beta f^{(k)}(t)\|_B\right)^p dt \\
&= \|t^\nu f^{(k)}; L^p(0, \infty; D(\Lambda^{m-k}))\|^p \le \|f\|_{W^m}^p.
\end{aligned}$$

Also,

$$\begin{aligned}
\|t^\nu f'_{\alpha k}; L^p(0, \infty; B)\|^p &= \int_0^\infty t^{\nu p} \|\Lambda^\alpha f^{(k+1)}(t)\|_B^p \, dt \\
&\le \int_0^\infty t^{\nu p}\left(\sum_{|\beta| \le m-k-1} \|\Lambda^\beta f^{(k+1)}(t)\|_B\right)^p dt \\
&= \|t^\nu f^{(k+1)}; L^p(0, \infty; D(\Lambda^{m-k-1}))\|^p \le \|f\|_{W^m}^p,
\end{aligned}$$

whence the lemma follows. ∎

**7.32** Let us assume hereafter that $0 < (1/p)+\nu < 1$ and denote by $T^0$ the trace space $T(p,\nu;D(\Lambda),B)$ corresponding to $W^1$. By Theorem 7.26 we may take the norm in $T^0$ to be

$$\|u\|_{T^0} = (\|u\|_B^p + \|u\|_G^p)^{1/p},$$

where

$$\|u\|_G = \left(\sum_{j=1}^{n} \int_0^\infty t^{(\nu-1)p} \|G_j(t)u - u\|_B^p\, dt\right)^{1/p}.$$

Higher-order trace spaces may now be defined as follows. For $k = 0, 1, 2, \dots$ we define $T^k = T^k(p,\nu;\Lambda;B)$ to be the space consisting of those elements $u \in D(\Lambda^k)$ for which $\Lambda^\alpha u \in T^0$ whenever $|\alpha| \le k$. The space $T^k$ is a Banach space with respect to the norm

$$\|u\|_{T^k} = \left(\|u;D(\Lambda^k)\|^p + \sum_{|\alpha|=k} \|\Lambda^\alpha u\|_G^p\right)^{1/p}.$$

It follows from Lemmas 7.31 and 7.13 that if $f \in W^m$ and $|\alpha| \le m-k-1$, then $\Lambda^\alpha f^{(k)}(0)$ exists in $T^0$ and

$$\|\Lambda^\alpha f^{(k)}(0)\|_{T^0} \le K_{\alpha k} \|f\|_{W^m},$$

where $K_{\alpha k}$ is a constant depending on $\alpha$ and $k$. Hence $f^{(k)}(0) \in T^{m-k-1}$ and

$$\|f^{(k)}(0)\|_{T^{m-k-1}} = \left(\|f^{(k)}(0); D(\Lambda^{m-k-1})\|^p + \sum_{|\beta|=m-k-1} \|\Lambda^\beta f^{(k)}(0)\|_G^p\right)^{1/p}$$
$$\le K_k \|f\|_{W^m}.$$

It follows that the linear mapping

$$f \to (f(0), f'(0), \dots, f^{(m-1)}(0)) \tag{19}$$

is continuous on $W^m$ into $T^{m-1} \times T^{m-2} \times \cdots \times T^0 = \prod_{k=0}^{m-1} T^{m-k-1}$; that is,

$$\sum_{k=0}^{m-1} \|f^{(k)}(0)\|_{T^{m-k-1}} \le K\|f\|_{W^m}.$$

We shall prove that this mapping is onto (see Lions [38]).

**7.33 THEOREM** The range of the mapping (19) is $\prod_{k=0}^{m-1} T^{m-k-1}$. If $(u_0, u_1, \dots, u_{m-1}) \in \prod_{k=0}^{m-1} T^{m-k-1}$, there exists $f \in W^m$ such that $f^{(k)}(0) = u_k$, $0 \le k \le m-1$, and

$$\|f\|_{W^m} \le K_0 \sum_{k=0}^{m-1} \|u_k\|_{T^{m-k-1}}.$$

PROOF The proof is similar to the second part of that of Theorem 7.24, but is rather more complicated. To achieve some simplification we shall deal with the special case $n = 1$ so that $\Lambda_j$ becomes just $\Lambda$ and $\Lambda^\alpha$ becomes $\Lambda^k$ ($|\alpha| = k$).

Suppose we have constructed for each $k$, $0 \le k \le m-1$, a function $f_k \in W^m$ such that $f_k^{(k)}(0) = u_k$ and

$$\|f_k\|_{W^m} \le K_k \|u_k\|_{T^{m-k-1}}.$$

Let $\lambda_{r,k}$, $0 \le r \le m-1$, satisfy the nonsingular system of equations

$$\sum_{r=0}^{m-1} r^j \lambda_{r,k} = \begin{cases} 1 & \text{if } \ j = k \\ 0 & \text{if } \ 0 \le j \le m-1, \ \ j \ne k. \end{cases}$$

Then the function

$$g_k(t) = \sum_{r=0}^{m-1} \lambda_{r,k} f_k(rt)$$

satisfies

$$g_k^{(k)}(0) = u_k, \qquad g_k^{(j)}(0) = 0, \qquad 0 \le j \le m-1, \ \ j \ne k.$$

Moreover, it is easily checked that

$$\|g_k\|_{W^m} \le \tilde{K}_k \|f_k\|_{W^m}.$$

Hence the function $f(t) = \sum_{k=0}^{m-1} g_k(t)$ has the properties required in the statement of the theorem. Thus we need only construct $f_k$.

In the rest of the proof we shall make extensive use of the convolution product of operator-valued functions on $[0, \infty)$. If for $t \ge 0$, $F_1(t)$ and $F_2(t)$ belong to $L(B)$, we define $F_1 * F_2$ on $[0, \infty)$ into $L(B)$ by

$$F_1 * F_2(t)b = \int_0^t F_1(t-\tau) F_2(\tau) b \, d\tau, \qquad b \in B.$$

(All the operators we use will commute.) If $F_1$ is continuously differentiable on $[0, \infty)$ into $L(B)$ so that $F_1'(t) \in L(B)$, we have evidently

$$(F_1 * F_2)'(t) = F_1' * F_2(t).$$

We denote by $F^{((m))}$ the convolution product $F * F * \cdots * F$ having $m$ factors; this is well defined since $*$ is associative for mutually commuting factors. If $I(t) = I$ denotes the identity on $B$, we have clearly

$$I^{((m))}(t) = [t^{m-1}/(m-1)!] I.$$

If $G$ is the continuous semigroup whose infinitesimal generator is $\Lambda$, then by Lemma 7.8 we have

$$\Lambda(I * G) = G - I$$

or, when both sides are restricted to elements of $D(\Lambda)$,

$$I * \Lambda G = G - I.$$

Given $u = u_k \in T^{m-k-1}$ we define

$$f_k(t) = [(m+k)!/k!]\,\phi(t)g(t)$$

where $\phi \in C([0,\infty))$ satisfies $\phi(t) = 1$ if $t \le \frac{1}{2}$, $\phi(t) = 0$ if $t \ge 1$, and $|\phi^{(j)}(t)| \le K_1$, $0 \le j \le m$, and where

$$g(t) = t^{-m} I^{((k+1))} * G^{((m))}(t)u. \tag{20}$$

[Note that this is the same function $g$ as defined in (17) in the proof of Theorem 7.24 if $m = 1$ and $k = 0$.] We must verify that

$$f_k^{(k)}(0) = u \tag{21}$$

and

$$\|f_k\|_{W^m} \le K_2 \|u\|_{T^{m-k-1}}. \tag{22}$$

In view of the constancy of $\phi$ for $t \le \frac{1}{2}$, (21) will follow if we can show that

$$g^{(k)}(0) = [k!/(m+k)!]u.$$

However,

$$g(t) = t^{-m} I^{((k+1))} * (G-I+I)^{((m))}(t)u$$
$$= t^{-m} \sum_{j=0}^{m} \binom{m}{j} I^{((k+1+m-j))} * (G-I)^{((j))}(t)u.$$

Since $t^{-m} I^{((k+1+m-j))}(t) = (t^{k-j}/(k+m-j)!)I$ has vanishing $k$th derivative if $j > 0$ we have

$$g^{(k)}(0) = \left(\frac{d}{dt}\right)^k \frac{t^k}{(k+m)!}\bigg|_{t=0} I * (G-I)^{((0))}(0)u = \frac{k!}{(k+m)!}u.$$

In order to establish (22) it is clearly sufficient to show that

$$\int_0^1 t^{\nu p} \|\Lambda^i g^{(j)}(t)\|_B^p \, dt \le K_3 \|u\|_{T^{m-k-1}}^p \tag{23}$$

holds for every $i, j$ such that $0 \le j \le m$ and $0 \le i \le m-j$. We distinguish three cases.

CASE 1 Suppose $0 \le j \le k$ and $m-k \le i \le m-j$. Let $w = \Lambda^{m-k-1}u$. Thus $w \in B$. Let $l = k+1+i-m$ so that $l > 1$ and $k+1-l \ge j$. Now

$$\Lambda^i g(t) = t^{-m} I^{((k+1-l))} * \Lambda^l I^{((l))} * G^{((l))} * G^{((m-l))}(t)w.$$

Since $\Lambda(I * G) = G - I$ we have

$$\Lambda^l I^{((l))} * G^{((l))} = (G-I)^{((l))},$$

and so

$$\Lambda^i g(t) = t^{-m} I^{((k+l-1))} * G^{((m-l))} * (G-I)^{((l))}(t)\,w.$$

Since $k+1-l \geq j$ we have, for $t > 0$,

$$\Lambda^i g^{(j)}(t) = \sum_{r=0}^{j} w_r(t),$$

where

$$w_r(t) = \tilde{K}_r t^{-m-r} I^{((k+1-l-j+r))} * G^{((m-l))} * (G-I)^{((l))}(t)\,w.$$

Now

$$\begin{aligned}\|I^{((k+1-l-j+r))} * G^{((m-l))} * G^{((l-1))}\|_{L(B)} &\leq K_4 t^{(k+1-l-j+r)+(m-l)+(l-1)} \\ &= K_4 t^{2m-i-j+r-2}\end{aligned}$$

Hence

$$\begin{aligned}\|w_r(t)\|_B &\leq K_5 t^{-m-r} \int_0^t (t-\tau)^{2m-i-j+r-2} \|G(\tau)w - w\|_B\, d\tau \\ &\leq K_5 t^{m-i-j-2} \int_0^t \|G(\tau)w - w\|_B\, d\tau.\end{aligned}$$

Since $i \leq m-j$ we therefore obtain for $0 < t \leq 1$

$$\|\Lambda^i g^{(j)}(t)\|_B \leq K_6 t^{-2} \int_0^t \|G(\tau)w - w\|_B\, d\tau.$$

It follows by Lemma 7.23 (with $\nu-1$ in place of $\nu$) that

$$\begin{aligned}\int_0^1 t^{\nu p} \|\Lambda^i g^{(j)}(t)\|_B^p\, dt &\leq K_6 \int_0^1 t^{(\nu-1)p} \left( (1/t) \int_0^t \|G(\tau)w - w\|_B\, d\tau \right)^p dt \\ &\leq K_7 \int_0^\infty t^{(\nu-1)p} \|G(t)w - w\|_B^p\, dt \\ &\leq K_7 \|w\|_{T^0}^p \leq K_7 \|u\|_{T^{m-k-l}}.\end{aligned}$$

CASE 2 Suppose $0 \leq j \leq k$ and $0 \leq i \leq m-k-1$. Then $w = \Lambda^i u \in B$ and

$$\Lambda^i g(t) = t^{-m} I^{((k+1))} * G^{((m))}(t)\,w.$$

Hence

$$\Lambda^i g^{(j)}(t) = \sum_{r=0}^{j} w_r(t) = \sum_{r=0}^{j} \tilde{K}_r t^{-m-r} I^{((k+1-j+r))} * G^{((m))}(t)\,w.$$

Now

$$\|w_r(t)\|_B \leq K_8 t^{-m-r+(k+1-j+r)-1+m} \|w\|_B = K_8 t^{k-j} \|w\|_B.$$

Thus

$$\|\Lambda^i g^{(j)}(t)\|_B \le K_9 t^{k-j} \|w\|_B$$

and

$$\int_0^1 t^{\nu p} \|\Lambda^i g^{(j)}(t)\|_B^p \, dt \le K_9{}^p \|w\|_B^p = K_9{}^p \|\Lambda^i u\|_B \le K_9{}^p \|u\|_{T^{m-k-1}}.$$

CASE 3 Suppose $k+1 \le j \le m$ and $1 \le i \le m-j$. Then $i \le m-k-1$ and $\tilde{u} = \Lambda^i u \in T^{m-k-1-i}$ with

$$\|\tilde{u}\|_{T^{m-k-1-i}} \le \|u\|_{T^{m-k-1}}.$$

Let $h(t) = \Lambda^i g(t)$. Thus

$$h(t) = t^{-m} I^{((k+1))} * G^{((m))}(t)\tilde{u}. \tag{24}$$

In order to prove (23) in this case it is sufficient to show that

$$\int_0^1 t^{\nu p} \|h^{(j)}(t)\|_B^p \, dt \le K_{10} \|\tilde{u}\|_{T^{m-k-1-i}}^p. \tag{25}$$

Now $G = \Lambda(I * G) + I$ so that

$$h(t) = t^{-m} I^{((k+2))} * G^{((m))}(t)\Lambda\tilde{u} + t^{-m} I^{((k+2))} * G^{((m-1))}(t)\tilde{u}.$$

Another $m-1$ repetitions of this argument yields

$$h(t) = \sum_{l=0}^{m-1} t^{-m} I^{((k+2+l))} * G^{((m-l))}(t)\Lambda\tilde{u} + t^{-m} I^{((k+1+m))}(t)\tilde{u}.$$

For purposes of proving (25) we may omit the term

$$t^{-m} I^{((k+1+m))}(t)\tilde{u} = [t^k/(k+m)!]\tilde{u}$$

since the $j$th derivative of this term vanishes for $t > 0$. Accordingly we consider

$$h(t) \sim \sum_{l=0}^{m-1} t^{-m} I^{((k+2+l))} * G^{((m-l))}(t)\Lambda\tilde{u}. \tag{26}$$

We repeat $m-k-i-2$ more times the preceding argument used in deriving (26) from (24), each time discarding terms which are polynomials of degree $\le j-1$ and so contribute nothing to $h^{(j)}$. This leads to

$$h(t) \sim \sum_{l=0}^{m-1} t^{-m} I^{((k+2+(m-k-i-2)+l))} * G^{((m-l))}(t)\Lambda^{m-k-1-i}\tilde{u}.$$

Let $w = \Lambda^{m-k-1-i}\tilde{u} = \Lambda^{m-k-1}u$. The terms of the above sum are of the form

$$w_l(t) = t^{-m} I^{((m+l-i))} * G^{((m-l))}(t)w,$$

where $0 \le l \le m-1$. Note that $m+l-i \ge j$. In order to prove (25) it is sufficient to show that

$$\int_0^1 t^{\nu p} \|w_l^{(j)}(t)\|_B^p \, dt \le K_{11} \|w\|_{T^0}^p. \tag{27}$$

At this point we must distinguish two subcases, $i \le m-j-1$ and $i = m-j$.

If $i \le m-j-1$, then $w_l^{(j)}$ is a linear combination of terms of the form

$$t^{-m-r} I^{((m+l-i-j+r))} * G^{((m-l))}(t) w$$

whose norms in $B$ are bounded by

$$K_{12} t^{-m-r+(m+l-i-j+r-1)+m-l} \|w\|_B \le K_{12} t^{m-j-i-1} \|w\|_{T^0}$$

and (27) follows at once.

If $i = m-j$, we have

$$\begin{aligned} w_l(t) &= t^{-m} I^{((j+l))} * G^{((m-l-1))} * (G-I+I)(t) w \\ &= t^{-m} I^{((j+l))} * G^{((m-l-1))} * (G-I)(t) w \\ &\quad + t^{-m} I^{((j+l+1))} * G^{((m-l-1))}(t) w. \end{aligned}$$

We repeat this procedure on the last term $m-l-1$ more times to obtain

$$w_l(t) = \sum_{s=0}^{m-l-1} t^{-m} I^{((j+l+s))} * G^{((m-l-s-1))} * (G-I)(t) w + t^{-m} I^{((j+m))}(t) w.$$

Again we may discard the last term which makes no contribution to (27). It is therefore sufficient to establish (27) with $w_l$ replaced by

$$w_{ls}(t) = t^{-m} I^{((j+l+s))} * G^{((m-l-s-1))} * (G-I)(t) w.$$

However, $w_{ls}^{(j)}(t)$ is a linear combination of terms of the form

$$t^{-m-r} I^{((l+s+r))} * G^{((m-l-s-1))} * (G-I)(t) w$$

for $0 \le r \le j$. It follows just as in Case 1 that for $0 < t \le 1$,

$$\|w_{ls}^{(j)}(t)\|_B \le K_{13} t^{-2} \int_0^t \|G(\tau) w - w\|_B \, d\tau$$

and hence, using Lemma 7.23 again, that (27) is satisfied. This completes the proof. ▮

We remark that the proof for general $n$ is essentially similar to that given for $n = 1$ above. In place of (20) one uses (a suitable multiple of)

$$g(t) = t^{-mn} I^{((k+1))} * G_1^{((m))} * \cdots * G_n^{((m))}(t) u.$$

**7.34 EXAMPLE** Let $B = L^p(\mathbb{R}^n)$ and $G_j$, $1 \le j \le n$, be as in Example 7.27 so that $\Lambda_j = D_j$. Evidently

$$D(\Lambda^k) = \{u \in L^p(\mathbb{R}^n) : D^\alpha u \in L^p(\mathbb{R}^n), |\alpha| \le k\} = W^{k,p}(\mathbb{R}^n).$$

For each $u \in L^p(\mathbb{R}^{n+1}_+)$ define $\tilde{u}$ a.e. on $[0, \infty)$ into $L^p(\mathbb{R}^n)$ by

$$\tilde{u}(t)(x_1, \ldots, x_n) = u(x_1, \ldots, x_n, t).$$

Then $u \in W^{m,p}(\mathbb{R}^{n+1}_+)$ provided $\tilde{u}^{(k)} \in L^p(0, \infty; W^{m-k,p}(\mathbb{R}^n))$, $0 \le k \le m$. Accordingly,

$$W^{m,p}(\mathbb{R}^{n+1}_+) \cong W^m(p, 0; \Lambda; L^p(\mathbb{R}^n))$$

with $\Lambda = (D_1, \ldots, D_n)$, If $1 < p < \infty$, the mapping $\gamma$

$$\gamma: u \to (u(\cdot, \ldots, \cdot, 0), D_{n+1} u(\cdot, \ldots, \cdot, 0), \ldots, D^{m-1}_{n+1} u(\cdot, \ldots, \cdot, 0))$$

is an isomorphism and a homeomorphism of $W^{m,p}(\mathbb{R}^{n+1}_+)/\ker \gamma$ onto the product $\prod_{k=0}^{m-1} T^{m-k-1}$ where

$$T^k = T^k(p, 0; \Lambda, L^p(\mathbb{R}^n)) = \{v \in W^{k,p}(\mathbb{R}^n) : D^\alpha v \in T^0, |\alpha| \le k\}$$

and

$$\|v\|_{T^k} = \Bigg\{ \sum_{|\alpha| \le k} \|D^\alpha v\|^p_{0,p,\mathbb{R}^n} + \sum_{|\alpha| = k} \sum_{j=1}^n \int_0^\infty t^{-p} \int_{\mathbb{R}^n} |D^\alpha v(x_1, \ldots, x_j + t, \ldots, x_n) - D^\alpha v(x_1, \ldots, x_n)|^p \, dx \, dt \Bigg\}^{1/p}.$$

## The Spaces $W^{s,p}(\Omega)$

**7.35** We now define spaces $W^{s,p}(\Omega)$ for arbitrary domains $\Omega$ in $\mathbb{R}^n$, arbitrary values of $s$, and $1 < p < \infty$. These spaces coincide for integer values of $s$ with the spaces $W^{m,p}(\Omega)$ and $W^{-m,p}(\Omega)$ defined in Chapter III. For $s \ge 0$ the definitions can be extended to $p = 1$ and $p = \infty$, but for the time being we ignore these limiting values.

The spaces $B_1 = W^{1,p}(\Omega)$ and $B_2 = X = L^p(\Omega)$ clearly satisfy the conditions laid down in Section 7.11. For $0 < \theta < 1$ let

$$T^{\theta,p}(\Omega) = T(p, \nu; W^{1,p}(\Omega), L^p(\Omega)),$$

where $\nu + (1/p) = \theta$. Denoting $W = W(p, \nu; W^{1,p}(\Omega), L^p(\Omega))$, we write the norm of $u$ in $T^{\theta,p}(\Omega)$ as

$$\|u; T^{\theta,p}(\Omega)\| = \inf_{\substack{f\in W \\ u=f(0)}} \max\left\{\left(\int_0^\infty t^{\nu p}\|f(t)\|_{1,p,\Omega}^p\,dt\right)^{1/p}, \left(\int_0^\infty t^{\nu p}\|f'(t)\|_{0,p,\Omega}^p\,dt\right)^{1/p}\right\}. \tag{28}$$

**7.36** Let $s \geq 0$ be arbitrary. If $s = m$, an integer, we define $W^{s,p}(\Omega) = W^{m,p}(\Omega)$. If $s$ is not an integer, we write $s = m+\sigma$ where $m$ is an integer and $0 < \sigma < 1$. The space $W^{s,p}(\Omega)$ is, in this case, defined to consist of those (equivalence classes of) functions $u \in W^{m,p}(\Omega)$ whose distributional derivatives $D^\alpha u$, $|\alpha| = m$, belong to $T^{1-\sigma,p}(\Omega)$. Then $W^{s,p}(\Omega)$ is a Banach space with respect to the norm

$$\|u\|_{s,p,\Omega} = \left\{\|u\|_{m,p,\Omega}^p + \sum_{|\alpha|=m}\|D^\alpha u; T^{1-\sigma,p}(\Omega)\|^p\right\}^{1/p}. \tag{29}$$

**7.37** The operator $P$ given by

$$Pu = (u, (D^\alpha u)_{|\alpha|=m})$$

(the multi-indices $\alpha$ with $|\alpha| = m$ being ordered in some conveient way) is an isometric isomorphism of $W^{s,p}(\Omega)$ onto a closed subspace of the (product) Banach space

$$S = W^{m,p}(\Omega) \times \prod_{|\alpha|=m} T^{1-\sigma,p}(\Omega)$$

having norm

$$\|(u, (v_\alpha)_{|\alpha|=m}); S\| = \left\{\|u\|_{m,p,\Omega}^p + \sum_{|\alpha|=m}\|v_\alpha; T^{1-\sigma,p}(\Omega)\|^p\right\}^{1/p}.$$

Since $W^{m,p}(\Omega)$ and $T^{1-\sigma,p}(\Omega)$ are reflexive (Theorems 3.5 and 7.19) it follows by Theorems 1.21 and 1.22 that $W^{s,p}(\Omega)$ is reflexive.

**7.38 THEOREM** For any $s \geq 0$, $C_0^\infty(\mathbb{R}^n)$ is dense in $W^{s,p}(\mathbb{R}^n)$.

PROOF This result has already been proved for $s = 0, 1, 2, \ldots$ (Theorems 2.19 and 3.18) and in particular $W^{1,p}(\mathbb{R}^n)$ is a dense subset of $L^p(\mathbb{R}^n)$ [i.e., dense with respect to the topology of $L^p(\mathbb{R}^n)$]. If $s = m+\sigma > 0$, where $m$ is an integer and $0 < \sigma < 1$, the theorem may be proved as follows.

Let $\psi \in C^\infty(\mathbb{R})$ satisfy $\psi(t) = 1$ if $t \leq 0$ and $\psi(t) = 0$ if $t \geq 1$. For $j = 1, 2, 3, \ldots$ let $\psi_j \in C_0^\infty(\mathbb{R}^n)$ be defined by

$$\psi_j(x) = \psi(|x|-j).$$

Let $J_\varepsilon$ be the mollifier introduced in Section 2.17. If $u$ is a function defined (a.e.) on $\mathbb{R}^n$, set

$$P_j u = J_{1/j} * (\psi_j \cdot u), \qquad n = 1, 2, \ldots.$$

Evidently $P_j$ is a bounded linear operator on $W^{m,p}(\mathbb{R}^n)$ into $W^{m,p}(\mathbb{R}^n)$ for $m = 0, 1, 2, \ldots$, and has range in $C_0^\infty(\mathbb{R}^n)$ in each case. We can deduce from Lemmas 2.18 and 3.15 that if $u \in W^{m,p}(\mathbb{R}^n)$, then

$$\lim_{j\to\infty} \|P_j u - u\|_{m,p,\mathbb{R}^n} = 0.$$

It follows by Lemma 7.18 that if $0 < \theta < 1$ and $u \in T^{\theta,p}(\mathbb{R}^n)$, then

$$\lim_{j\to\infty} \|P_j u - u; T^{\theta,p}(\mathbb{R}^n)\| = 0.$$

Since

$$D^\alpha P_j u = J_{1/j} * D^\alpha(\psi_j \cdot u) = P_j D^\alpha u + J_{1/j} * \omega_j,$$

where

$$\omega_j = \sum_{\beta<\alpha} \binom{\alpha}{\beta} D^{\alpha-\beta}\psi_j D^\beta u,$$

and since

$$\lim_{j\to\infty} \|\omega_j\|_{1,p,\mathbb{R}^n} = 0$$

provided $u \in W^{|\alpha|,p}(\mathbb{R}^n)$, it follows that for any $u \in W^{s,p}(\mathbb{R}^n)$ we have, taking $|\alpha| = m$,

$$\lim_{j\to\infty} \|D^\alpha P_j u - D^\alpha u; T^{1-\sigma,p}(\mathbb{R}^n)\| = 0.$$

Hence

$$\lim_{j\to\infty} \|P_j u - u\|_{s,p,\Omega} = 0$$

and the proof is complete. ∎

**7.39** Let $W_0^{s,p}(\Omega)$ denote the closure of $C_0^\infty(\Omega)$ in the space $W^{s,p}(\Omega)$ $(s \geq 0)$. By the above theorem $W_0^{s,p}(\mathbb{R}^n) = W^{s,p}(\mathbb{R}^n)$. For $s < 0$ we define

$$W^{s,p}(\Omega) = [W_0^{-s,p'}(\Omega)]', \qquad (1/p) + (1/p') = 1.$$

It follows by reflexivity that for every real $s$

$$[W^{s,p}(\mathbb{R}^n)]' \cong W^{-s,p'}(\mathbb{R}^n).$$

We shall not comment further on the structure of the spaces $W^{s,p}(\Omega)$ for $s < 0$ except to note that, being duals of spaces having $\mathscr{D}(\Omega) = C_0^\infty(\Omega)$ as dense subsets, they are spaces of distributions on $\Omega$.

Many properties of the spaces $W^{s,p}(\Omega)$ are conveniently proven only for $\Omega = \mathbb{R}^n$, and must then be deduced for more general $\Omega$ with the aid of an extension operator extending functions defined on $\Omega$ to $\mathbb{R}^n$ with preservation of differential properties (see Section 4.24). For fractional $s = m+\sigma$ suitable extensions are obtained by interpolation. Thus the existence of a strong $(m+1)$-extension operator for $\Omega$ will normally be required. Such is, for example, assured if $\Omega$ satisfies the hypotheses of Theorem 4.26 (see also Section 4.29).

**7.40 THEOREM** If $s = m+\sigma$ where $m$ is an integer and $0 < \sigma < 1$, and if there exists a strong $(m+1)$-extension operator $E$ for $\Omega$, a domain in $\mathbb{R}^n$, then the set of restrictions to $\Omega$ of functions in $C_0^\infty(\mathbb{R}^n)$ is dense in $W^{s,p}(\Omega)$.

PROOF (Recall that the conclusion holds for $W^{m,p}(\Omega)$ under the assumption only that $\Omega$ has the segment property.) The proof follows the same lines as that of Theorem 7.38 except that in place of the operator $P_j$ we use the operator $\tilde{P}_j$ defined by

$$\tilde{P}_j u = R_\Omega P_j Eu, \qquad u \text{ defined on } \Omega,$$

where $R_\Omega$ is the operator restricting to $\Omega$ functions defined on $\mathbb{R}^n$. The details are left to the reader. ∎

The following localization theorem requires, in addition to the existence of a strong $(m+1)$-extension operator $E$ for $\Omega$, a representation for the derivatives $D^\alpha Eu(x)$ in terms of the derivatives of $u$ such as is provided by Theorem 4.26. Thus the hypotheses of the theorem below will certainly be satisfied by any domain satisfying the hypotheses of Theorem 4.26.

**7.41 THEOREM** Let $\Omega$ be a domain in $\mathbb{R}^n$ for which there exists a strong $(m+1)$-extension operator $E$ and, for $|\gamma| \le |\alpha| = m$, linear operators $E_{\alpha\gamma}$ continuous from $W^{1,p}(\Omega)$ into $W^{1,p}(\mathbb{R}^n)$ and from $L^p(\Omega)$ into $L^p(\mathbb{R}^n)$ such that if $u \in W^{m,p}(\Omega)$, then

$$D^\alpha Eu(x) = \sum_{|\gamma| \le m} E_{\alpha\gamma} D^\gamma u(x) \qquad \text{a.e. in } \mathbb{R}^n. \tag{30}$$

If $s = m+\sigma > 0$, $0 \le \sigma < 1$, then $W^{s,p}(\Omega)$ coincides with the set of restrictions to $\Omega$ of funcitons in $W^{s,p}(\mathbb{R}^n)$.

PROOF If $\sigma = 0$, the result is an immediate consequence of the existence of a strong $m$-extension operator for $\Omega$. Suppose $0 < \sigma < 1$. If $u \in W^{s,p}(\Omega)$, then $u \in W^{m,p}(\Omega)$ and $Eu \in W^{m,p}(\mathbb{R}_n)$; also

$$\|Eu\|_{m,p,\mathbb{R}^n} \le K_1 \|u\|_{m,p,\Omega} \le K_1 \|u\|_{s,p,\Omega}. \tag{31}$$

If $|\gamma| \le m$, then $D^\gamma u \in T^{1-\sigma, p}(\Omega)$ and

$$\|D^\gamma u; T^{1-\sigma, p}(\Omega)\| \le K_2 \|u\|_{s, p, \Omega}. \tag{32}$$

(This holds by definition if $|\gamma| = m$ and via Lemma 7.16 if $|\gamma| < m$.) Since $E_{\alpha\gamma}$ is linear and continuous on $W^{1,p}(\Omega)$ into $W^{1,p}(\mathbb{R}^n)$ and on $L^p(\Omega)$ into $L^p(\mathbb{R}^n)$, by Theorem 7.17 it is also continuous on $T^{1-\sigma, p}(\Omega)$ into $T^{1-\sigma, p}(\mathbb{R}^n)$. By (30) and (32) we have for $|\alpha| = m$

$$\|D^\alpha Eu; T^{1-\sigma}(\mathbb{R}^n)\| \le K_3 \|u\|_{s, p, \Omega}.$$

Combining this with (31), we obtain

$$\|Eu\|_{s, p, \mathbb{R}^n} \le K_4 \|u\|_{s, p, \Omega}$$

and $u$ is the restriction to $\Omega$ of $Eu \in W^{s,p}(\mathbb{R}^n)$.

Conversely, the operator $R_\Omega$ of restriction to $\Omega$, being continuous from $W^{m,p}(\mathbb{R}^n)$ into $W^{m,p}(\Omega)$ for any $m$, is also continuous by Theorem 7.17 from $W^{s,p}(\mathbb{R}^n)$ into $W^{s,p}(\Omega)$ so the restriction $R_\Omega u$ of $u \in W^{s,p}(\mathbb{R}^n)$ belongs to $W^{s,p}(\Omega)$. ∎

We remark that under the conditions of the theorem the extension operator $E$ is continuous from $W^{s,p}(\Omega)$ into $W^{s,p}(\mathbb{R}^n)$ for any $s$, $0 \le s \le m+1$.

## An Intrinsic Norm for $W^{s,p}(\Omega)$

**7.42** We now investigate the possibility of constructing a new norm for $W^{s,p}(\Omega)$, $s \ge 0$, which is equivalent to the "trace norm" (29) ($s$ not an integer) but which is expressed "intrinsically" in terms of properties of the element involved. In view of Example 7.27 it is most convenient to begin with the case $\Omega = \mathbb{R}^n$. Following Lions and Magenes [34] we define new spaces $\tilde{W}^{s,p}(\Omega)$ with intrinsically defined norm and then show, at least for suitably regular domains $\Omega$, $\tilde{W}^{s,p}(\Omega)$ coincides with $W^{s,p}(\Omega)$.

**7.43** For $0 < \theta < 1$ and $1 \le p < \infty$ let $\tilde{T}^{\theta,p}(\Omega)$ denote the space of (equivalence classes of) functions $u \in L^p(\Omega)$ for which the norm

$$\|u; \tilde{T}^{\theta,p}(\Omega)\| = \left\{ \|u\|_{0,p,\Omega}^p + \int_\Omega \int_\Omega \frac{|u(x)-u(y)|^p}{|x-y|^{n-1+(1-\nu)p}}\, dx\, dy \right\}^{1/p} \tag{33}$$

is finite, where $\nu + (1/p) = \theta$.

**7.44 LEMMA** The space $\tilde{T}^{\theta,p}(\mathbb{R}^n)$ coincides with the Banach space $T^{\theta,p}(\mathbb{R}^n)$, the norms in the two spaces being equivalent.

PROOF The norm of an element $u$ in $T^{\theta,p}(\mathbb{R}^n)$ was defined [in (28)] to be its norm in the trace space $T(p, \nu; W^{1,p}(\mathbb{R}^n), L^p(\mathbb{R}^n))$. By Example 7.27 we may take the norm to be

$$\|u\|_T = \left\{ \|u\|_{0,p,\mathbb{R}^n}^p + \sum_{j=1}^n \int_0^t t^{(\nu-1)p} \int_{\mathbb{R}^n} |u(x_1, \ldots, x_j + t, \ldots, x_n) - u(x_1, \ldots, x_n)|^p \, dx \, dt \right\}^{1/p}.$$

Let us denote the norm given by (33) as $\|u\|_{\tilde{T}}$.

Let $u \in T^{\theta,p}(\mathbb{R}^n)$. Putting $\lambda = \frac{1}{2}[n-1+(1-\nu)p]$ and writing $u(x)-u(y)$ in the form

$$\sum_{j=1}^n [u(y_1, \ldots, y_{j-1}, x_j, x_{j+1}, \ldots, x_n) - u(y_1, \ldots, y_{j-1}, y_j, x_{j+1}, \ldots, x_n)],$$

we have

$$\int_{\mathbb{R}^n} \int_{\mathbb{R}^n} \frac{|u(x)-u(y)|^p}{|x-y|^{n-1+(1-\nu)p}} \, dx \, dy \le K_1 \sum_{j=1}^n Q_j,$$

where

$$Q_j = \int_{\mathbb{R}^n} \int_{\mathbb{R}^n} \frac{|u(y_1, \ldots, y_{j-1}, x_j, \ldots, x_n) - u(y_1, \ldots, y_j, x_{j+1}, \ldots, x_n)|^p}{(\sum_{k=1}^n (x_k - y_k)^2)^\lambda} \, dx \, dy.$$

Thus

$$\begin{aligned} Q_j = \int_{\mathbb{R}^j} dy_1 \cdots dy_j & \\ \times \int_{\mathbb{R}^{n+1-j}} dx_j \cdots dx_n & |u(y_1, \ldots, y_{j-1}, x_j, \ldots, x_n) \\ & - u(y_1, \ldots, y_j, x_{j+1}, \ldots, x_n)|^p R_j \end{aligned} \tag{34}$$

where

$$R_j = \int_{\mathbb{R}^{n-j}} \int_{\mathbb{R}^{j-1}} \frac{dx_1 \cdots dx_{j-1} \, dy_{j+1} \cdots dy_n}{(\sum_{k=1}^n (x_k - y_k)^2)^\lambda}.$$

Let $\rho^2 = (x_1 - y_1)^2 + \cdots + (x_{j-1} - y_{j-1})^2 + (x_{j+1} - y_{j+1})^2 + \cdots + (x_n - y_n)^2$. Then

$$\begin{aligned} R_j &= K_2 \left( \int_0^{|x_j - y_j|} + \int_{|x_j - y_j|}^\infty \right) \frac{\rho^{n-2}}{[\rho^2 + (x_j - y_j)^2]^\lambda} \, d\rho \\ &\le \frac{K_2}{|x_j - y_j|^{2\lambda}} \int_0^{|x_j - y_j|} \rho^{n-2} \, d\rho + K_2 \int_{|x_j - y_j|}^\infty \rho^{n-2-2\lambda} \, d\rho \\ &= K_3 |x_j - y_j|^{(\nu-1)p} \end{aligned} \tag{35}$$

since $\lambda > 0$ and $n-1-2\lambda < 0$. Setting

$$y_i = z_i, \quad 1 \le i \le j, \qquad x_j = t + y_j, \qquad x_i = z_i, \quad j+1 \le i \le n$$

in (35), we obtain, using (34),

$$Q_j \le 2K_3 \int_0^\infty t^{(\nu-1)p}\, dt \int_{\mathbb{R}^n} |u(z_1, \ldots, z_j + t, \ldots, z_n) - u(z_1, \ldots, z_n)|^p\, dz.$$

Thus $u \in \tilde{T}^{\theta, p}(\mathbb{R}^n)$ and $\|u\|_{\tilde{T}} \le K_4 \|u\|_T$.

Conversely, suppose $u \in \tilde{T}^{\theta, p}(\mathbb{R}^n)$. Let $x' = (x_2, \ldots, x_n)$ and $z' = (z_2, \ldots, z_n)$ and integrate the inequality

$$|u(x_1+t, x') - u(x, x')|^p \le K_5(|u(x_1+t, x') - u(x_1+\tfrac{1}{2}t, z')|^p + |u(x_1+\tfrac{1}{2}t, z') - u(x, x')|^p)$$

with respect to $z'$ over the disk $D(t, x')$ centered at $x' \in \mathbb{R}^{n-1}$ and having radius $\frac{1}{2}t$, thus obtaining

$$|u(x_1+t, x') - u(x_1, x')|^p \le (K_6/t^{n-1})[I_t(t, x) + I_t(0, x)],$$

where

$$I_t(s, x) = \int_{D(t, x')} |u(x_1+s, x') - u(x_1+\tfrac{1}{2}t, z')|^p\, dz'$$

for $s = t$ or $s = 0$. Now

$$\begin{aligned}
&\int_0^\infty t^{(\nu-1)p}\, dt \int_{\mathbb{R}^n} \frac{1}{t^{n-1}} I_t(t, x)\, dx \\
&\quad = \int_{\mathbb{R}^{n-1}} dx' \int_0^\infty \frac{1}{t^{2\lambda}}\, dt \int_{D(t, x')} dz' \int_{-\infty}^{\infty} |u(x_1+t, x') - u(x_1+\tfrac{1}{2}t, z')|^p\, dx_1 \\
&\quad = \int_{\mathbb{R}^{n-1}} dx' \int_0^\infty \frac{1}{t^{2\lambda}}\, dt \int_{D(t, x')} dz' \int_{-\infty}^{\infty} |u(x_1, x') - u(x_1-\tfrac{1}{2}t, z')|^p\, dx_1 \\
&\quad = \int_{\mathbb{R}^n} dx \int_{\mathbb{R}^{n-1}} dz' \int_{2|z'-x'|}^{\infty} \frac{|u(x_1, x') - u(x_1-\tfrac{1}{2}t, z')|^p}{t^{2\lambda}}\, dt \\
&\quad = \int_{\mathbb{R}^n} dx \int_{\mathbb{R}^{n-1}} dz' \int_{-\infty}^{x_1-|z'-x'|} \frac{|u(x)-u(z)|^p}{[2(x_1-z_1)]^{2\lambda}}\, dz_1 \\
&\quad \le \frac{1}{2^\lambda} \int_{\mathbb{R}^n} dx \int_{\mathbb{R}^n} \frac{|u(x)-u(z)|^p}{|x-z|^\lambda}\, dz,
\end{aligned}$$

where we have put $z_1 = x_1 - \frac{1}{2}t$, $dz = -\frac{1}{2}\, dt$ and used the fact that in the inner integral in the second last line $|x_1 - z_1| \ge |x' - z'|$ so that $|x_1 - z_1| \ge |x-z|/\sqrt{2}$.

A similar inequality holds for $I_t(0, x)$. Thus

$$\int_0^\infty t^{(\nu-1)p} \int_{\mathbb{R}^n} |u(x_1+t, x_2, \dots, x_n) - u(x_1, \dots, x_n)|^p \, dx \, dt$$

$$\leq K_7 \int_{\mathbb{R}^n} \int_{\mathbb{R}^n} \frac{|u(x)-u(z)|^p}{|x-z|^{n-1+(1-\nu)p}} \, dx \, dz.$$

Similar inequalities hold for differences in the other variables $x_2, \dots, x_n$ and combining these we obtain $\|u\|_T \leq K_8 \|u\|_{\tilde{T}}$. ∎

In order to extend the above lemma to more general domains $\Omega$ we require the following extension lemma.

**7.45 LEMMA** Let $\Omega$ be a half-space in $\mathbb{R}^n$ or a domain in $\mathbb{R}^n$ having the uniform 1-smooth regularity property and a bounded boundary. Then there exists a linear operator $E$ mapping $L^p(\Omega)$ into $L^p(\mathbb{R}^n)$ such that if $u \in L^p(\Omega)$, then

$$Eu(x) = u(x) \qquad \text{a.e. in } \Omega,$$

and if $0 < \theta < 1$ and $u \in \tilde{T}^{\theta,p}(\Omega)$, then $Eu \in \tilde{T}^{\theta,p}(\mathbb{R}^n)$ and

$$\|Eu; \tilde{T}^{\theta,p}(\mathbb{R}^n)\| \leq K \|u; \tilde{T}^{\theta,p}(\Omega)\|$$

with $K$ independent of $u$.

PROOF The proof is quite similar to that of Theorem 4.26. We begin with the case $\Omega = \mathbb{R}_+{}^n = \{x \in \mathbb{R}^n : x_n > 0\}$. Let us denote by $x' = (x_1, \dots, x_{n-1})$ and for $u \in L^p(\Omega)$ set

$$Eu(x) = \begin{cases} u(x) & \text{a.e. in } \mathbb{R}_+{}^n \\ u(x', -x_n) & \text{a.e. in } \mathbb{R}^n \sim \mathbb{R}_+{}^n. \end{cases}$$

Then

$$\|Eu\|_{0,p,\mathbb{R}^n}^p = \int_{\mathbb{R}^{n-1}} dx' \left\{ \int_0^\infty |u(x)|^p \, dx_n + \int_{-\infty}^0 |u(x', -x_n)|^p \, dx_n \right\}$$

$$= 2 \|u\|_{0,p,\mathbb{R}^n_+}^p.$$

Also, setting $2\lambda = n-1+(1-\nu)p = n+(1-\theta)p > 0$, we have

$$\int_{\mathbb{R}^n} \int_{\mathbb{R}^n} \frac{|Eu(x)-Eu(y)|^p}{|x-y|^{2\lambda}} \, dx \, dy = I_{++} + I_{+-} + I_{-+} + I_{--},$$

where

$$I_{++} = \int_{\mathbb{R}_+{}^n} \int_{\mathbb{R}_+{}^n} \frac{|u(x)-u(y)|^p}{|x-y|^{2\lambda}}\, dx\, dy,$$

$$I_{+-} = \int_{\mathbb{R}^{n-1}} dx' \int_{\mathbb{R}^{n-1}} dy' \int_0^\infty dx_n \int_{-\infty}^0 \frac{|u(x)-u(y', -y_n)|^p}{[|x'-y'|^2+(x_n-y_n)^2]^\lambda}\, dy_n$$

$$= \int_{\mathbb{R}^{n-1}} dx' \int_{\mathbb{R}^{n-1}} dy' \int_0^\infty dx_n \int_0^\infty \frac{|u(x)-u(y)|^p}{[|x'-y'|^2+(x_n+y_n)^2]^\lambda}\, dy_n$$

$$\leq \int_{\mathbb{R}_+{}^n} \int_{\mathbb{R}_+{}^n} \frac{|u(x)-u(y)|^p}{|x-y|^{2\lambda}}\, dx\, dy$$

[since $(x_n+y_n)^2 \geq (x_n-y_n)^2$ when $x_n \geq 0$ and $y_n \geq 0$], and similar inequalities hold for $I_{-+}$ and $I_{--}$. Thus

$$\|Eu; \tilde{T}^{\theta,p}(\mathbb{R}^n)\| \leq 4^{1/p} \|u; \tilde{T}^{\theta,p}(\mathbb{R}_+{}^n)\|.$$

Now suppose that $\Omega$ is uniformly $C^1$-regular and has a bounded boundary. Then the open cover $\{U_j\}$ of bdry $\Omega$ and the corresponding collection $\{\Phi_j\}$ of 1-smooth maps of $U_j$ onto $B = \{y \in \mathbb{R}^n : |y| < 1\}$ referred to in Section 4.6 are both finite collections, say $1 \leq j \leq N$. We may also assume that the sets $U_j$ are bounded. Let $U_0$ be an open subset of $\Omega$, bounded away from bdry $\Omega$, such that $\Omega \subset \bigcup_{j=0}^N U_j$. Let $\{\omega_j\}_{j=0}^\infty$ be a $C^\infty$-partition of unity for $\Omega$ subordinate to $\{U_j\}$. Given $u \in L^p(\Omega)$ let $u_j$ be defined a.e. in $\Omega$ by $u_j(x) = \omega_j(x)\,u(x)$. Clearly $u_j \in L^p(\Omega)$ and $\|u_j\|_{0,p,\Omega} \leq \|u\|_{0,p,\Omega}$. If $u \in \tilde{T}^{\theta,p}(\Omega)$, then for $1 \leq j \leq N$

$$\int_\Omega \int_\Omega \frac{|u_j(x)-u_j(y)|^p}{|x-y|^{2\lambda}}\, dx\, dy \leq K_1 \int_\Omega \int_\Omega \frac{|u(x)-u(y)|^p}{|x-y|^{2\lambda}}\, dx\, dy$$

$$+ K_1 \int_{\Omega \cap U_j} |u(y)|^p\, dy \int_{U_j} \frac{|\omega_j(x)-\omega_j(y)|^p}{|x-y|^{2\lambda}}\, dx.$$

But since $U_j$ is bounded we have for $y \in \Omega \cap U_j$ by Lemma 5.47,

$$\int_{U_j} \frac{|\omega_j(x)-\omega_j(y)|^p}{|x-y|^{2\lambda}}\, dx \leq K_2 \int_{U_j} |x-y|^{p+1-n}\, dx \leq K_3,$$

and $K_3$ may be chosen independent of the finitely many values of $j$ involved. Thus $u_j \in \tilde{T}^{\theta,p}(\Omega)$ and

$$\|u_j; \tilde{T}^{\theta,p}(\Omega)\| \leq K_4 \|u; \tilde{T}^{\theta,p}(\Omega)\|.$$

Since $\omega_0(x) = 1$ for all $x \in \Omega$ lying outside the bounded set $\bigcup_{j=1}^N U_j$, the above inequality also holds for $u_0$.

For $1 \le j \le N$ let $v_j$ be defined on $\mathbb{R}_+{}^n$ by

$$v_j(y) = \begin{cases} u_j \circ \Psi_j(y) & \text{if} \quad y \in B \cap \mathbb{R}_+{}^n \\ 0 & \text{if} \quad y \in \mathbb{R}_+{}^n \sim B, \end{cases}$$

where $\Psi_j = \Phi_j^{-1}$. Then $v_j \in \tilde{T}^{\theta,p}(\mathbb{R}_+{}^n)$. In fact, putting $y = \Phi_j(x)$, $\eta = \Phi_j(\xi)$, we have

$$\begin{aligned} \|v_j; \tilde{T}^{\theta,p}(\mathbb{R}_+{}^n)\|^p &= \int_{\mathbb{R}_+{}^n \cap B} |u_j(\Psi_j(y))|^p \, dy \\ &\quad + \int_{\mathbb{R}_+{}^n \cap B} \int_{\mathbb{R}_+{}^n \cap B} \frac{|u_j(\Psi_j(y)) - u_j(\Psi_j(\eta))|^p}{|y-\eta|^{2\lambda}} \, dy \, d\eta \\ &= \int_\Omega |u_j(x)|^p |\det \Phi_j'(x)| \, dx \\ &\quad + \int_\Omega \int_\Omega \frac{|u_j(x) - u_j(\xi)|^p}{|\Phi_j(x) - \Phi_j(y)|^{2\lambda}} |\det \Phi_j'(x)| |\det \Phi_j'(\xi)| \, dx \, d\xi \\ &\le K_5 \|u_j; \tilde{T}^{\theta,p}(\Omega)\|^p \end{aligned} \tag{36}$$

since $|\det \Phi_j'|$ is bounded and since, $\Psi_j$ being 1-smooth on $B$,

$$|x - \xi| = |\Psi_j(y) - \Psi_j(\eta)| \le K_6 |y - \eta| = K_6 |\Phi_j(x) - \Phi_j(\xi)|.$$

Now $Ev_j \in \tilde{T}^{\theta,p}(\mathbb{R}^n)$ and

$$\|Ev_j; \tilde{T}^{\theta,p}(\mathbb{R}^n)\| \le K_7 \|v_j; \tilde{T}^{\theta,p}(\mathbb{R}_+{}^n)\|.$$

Also $\operatorname{supp} Ev_j \subset\subset B$. We define $w_j$ a.e. on $\mathbb{R}^n$ by

$$w_j(x) = \begin{cases} Ev_j(\Phi(x)) & \text{if} \quad x \in U_j \\ 0 & \text{if} \quad x \in \mathbb{R}^n \sim U_j. \end{cases}$$

Then clearly $w_j(x) = u_j(x)$ a.e. in $\Omega$, $\operatorname{supp} w_j \subset\subset U_j$, and by a calculation similar to the one carried on in (36),

$$\|w_j; \tilde{T}^{\theta,p}(\mathbb{R}^n)\| \le K^8 \|Ev_j; \tilde{T}^{\theta,p}(\mathbb{R}^n)\|.$$

Finally, we set

$$E^* u(x) = u_0(x) + \sum_{j=1}^N w_j(x).$$

It is clear that $E^*$ has all the properties required of $u$ in the statement of the lemma. ∎

It should be remarked that the comments made in Section 4.29 concerning weakening the hypotheses for the extension theorems 4.26 and 4.28 apply as well to the above lemma.

**7.46 COROLLARY** Under the conditions of Lemma 7.45 the spaces $\tilde{T}^{\theta,p}(\Omega)$ and $T^{\theta,p}(\Omega)$ coincide, and their norms are equivalent.

PROOF The coincidence of the two vector spaces follows from the fact that they coincide with restrictions to $\Omega$ of functions in the coincident spaces $\tilde{T}^{\theta,p}(\mathbb{R}^n)$ and $T^{\theta,p}(\mathbb{R}^n)$. If $u \in \tilde{T}^{\theta,p}(\Omega)$ and $E$ is the extension operator constructed in the above lemma, we have

$$\|u; T^{\theta,p}(\Omega)\| \le \|Eu; T^{\theta,p}(\mathbb{R}^n)\| \le K_1 \|Eu; \tilde{T}^{\theta,p}(\mathbb{R}^n)\| \le K_2 \|u; \tilde{T}^{\theta,p}(\Omega)\|.$$

The reverse inequality follows in the same way, using instead of $E$ the strong 1-extension operator constructed in Theorem 4.6 (the case $m = 1$) which, as is implicit in the proof of Theorem 7.41, is an extension operator for $T^{\theta,p}(\Omega)$. ∎

**7.47** For $s \ge 0$ let $\tilde{W}^{s,p}(\Omega)$ be the space constructed in exactly the same way that $W^{s,p}(\Omega)$ was constructed in Section 7.36 except using the spaces $\tilde{T}^{1-\sigma,p}(\Omega)$ in place of $T^{1-\sigma,p}(\Omega)$. In view of Corollary 7.46 we have proved the following theorem.

**7.48 THEOREM** Let $\Omega$ be $\mathbb{R}^n$, or a half-space in $\mathbb{R}^n$, or a domain in $\mathbb{R}^n$ which is uniformly $C^1$-regular and has a bounded boundary. Then the spaces $\tilde{W}^{s,p}(\Omega)$ and $W^{s,p}(\Omega)$ coincide algebraically and topologically for each $s \ge 0$. In particular, if $s = m+\sigma$ where $m$ is an integer and $0 < \sigma < 1$, then the norm $\|\cdot\|^{\sim}_{s,p,\Omega}$ given by

$$\|u\|^{\sim}_{s,p,\Omega} = \left\{ \|u\|^p_{m,p,\Omega} + \sum_{|\alpha|=m} \int_\Omega \int_\Omega \frac{|D^\alpha u(x) - D^\alpha u(y)|^p}{|x-y|^{n+\sigma p}}\, dx\, dy \right\}^{1/p}$$

is equivalent to the original norm $\|\cdot\|_{s,p,\Omega}$ on $W^{s,p}(\Omega)$.

**7.49 REMARK** It is the spaces which we have above denoted $\tilde{W}^{s,p}(\Omega)$ which one encounters most frequently in the literature, and which are usually designated $W^{s,p}(\Omega)$. The space $\tilde{W}^{s,\infty}(\Omega)$ may obviously be defined in an analagous way. It consists of those $u \in W^{m,\infty}(\Omega)$ for which the norm

$$\|u\|_{s,\infty,\Omega} = \max\left( \|u\|_{m,\infty,\Omega},\ \max_{|\alpha|=m} \operatorname*{ess\,sup}_{\substack{x,y\in\Omega \\ x\ne y}} \frac{|D^\alpha u(x) - D^\alpha u(y)|}{|x-y|^\sigma} \right)$$

is finite.

## Imbedding Theorems

**7.50** As we have already seen (in Example 7.34), if $1 < p < \infty$, the linear mapping

$$u \to \gamma u = (\gamma_0 u, \ldots, \gamma_{m-1} u); \qquad \gamma_j u = D_n{}^j u(\cdot, \ldots, \cdot, 0)$$

establishes an isomorphism and a homeomorphism of $W^{m,p}(\mathbb{R}_+{}^n)/\ker\gamma$ onto $\prod_{k=0}^{m-1} T^{m-k-1}(p,0;\Lambda;L^p(\mathbb{R}^{n-1}))$, where $\Lambda=(D_1,\ldots,D_{n-1})$. Since $D(\Lambda^k)=W^{k,p}(\mathbb{R}^{n-1})$ and since $T^0(p,0;\Lambda;L^p(\mathbb{R}^{n-1}))=T^{1/p,p}(\mathbb{R}^{n-1})$, we have

$$T^k(p,0;\Lambda;L^p(\mathbb{R}^{n-1})) = W^{k+1-1/p,p}(\mathbb{R}^{n-1}).$$

Thus $\gamma$ is in fact an isomorphism and homeomorphism of $W^{m,p}(\mathbb{R}_+{}^n)/\ker\gamma$ onto $\prod_{k=0}^{m-1} W^{m-k-1/p,p}(\mathbb{R}^{n-1})$. In particular, the traces on $\mathbb{R}^{n-1}=\operatorname{bdry}\mathbb{R}_+{}^n$ of functions in $W^{m,p}(\mathbb{R}_+{}^n)$ belong to, and constitute the whole of the space $W^{m-1/p,p}(\operatorname{bdry}\mathbb{R}_+{}^n)$. [This phenomenon is sometimes described as the loss of $(1/p)$th of a derivative on the boundary.] The result can be extended to smoothly bounded domains.

**7.51** If $\Omega$ is a domain in $\mathbb{R}^n$ having the uniform $C^m$-regularity property and a bounded boundary, then the open cover $\{U_j\}$ of $\operatorname{bdry}\Omega$ and the associated collection $\{\overline{\Psi}_j\}$ of $m$-smooth maps from $B=\{y\in\mathbb{R}^n:|y|<1\}$ onto the sets $U_j$ (referred to in Section 4.6) are finite collections, say $1\le j\le r$. If $\{\omega_j\}$ is a partition of unity for $\operatorname{bdry}\Omega$ subordinate to $\{U_j\}$ and if $u$ is a function defined on $\operatorname{bdry}\Omega$, we define $\theta_j u$ on $\mathbb{R}^{n-1}$ by

$$\theta_j u(y') = \begin{cases} (\omega_j u)(\overline{\Psi}_j(y',0)) & \text{if } \ |y'|<1 \\ 0 & \text{otherwise,} \end{cases}$$

where $y'=(y_1,\ldots,y_{n-1})$.

For $s\ge 0$ and $1<p<\infty$ we define $W^{s,p}(\operatorname{bdry}\Omega)$ to be the class of functions $u\in L^p(\operatorname{bdry}\Omega)$ (see Section 5.21) such that $\theta_j u$ belongs to $W^{s,p}(\mathbb{R}^{n-1})$ for $1\le j\le r$. The space $W^{s,p}(\operatorname{bdry}\Omega)$ is a Banach space with respect to the norm

$$\|u\|_{s,p,\operatorname{bdry}\Omega} = \left\{\sum_{j=1}^{r} \|\theta_j u\|_{s,p,\mathbb{R}^{n-1}}^p\right\}^{1/p}.$$

As defined above, the space $W^{s,p}(\operatorname{bdry}\Omega)$ appears to depend on the particular cover $\{U_j\}$, the mappings $\{\Psi_j\}$, and the partition of unity $\{\omega_j\}$ used in the definition. It can be checked that the same space, with an equivalent norm, results if we carry out the construction for a different collection $\{\tilde{U}_j\}$, $\{\tilde{\Psi}_j\}$, and $\{\tilde{\omega}_j\}$. (We omit the details; see Lions and Magenes [40].) It can also be checked that $C^\infty(\operatorname{bdry}\Omega)$ is dense in $W^{s,p}(\operatorname{bdry}\Omega)$.

**7.52** Let $u\in C_0{}^\infty(\mathbb{R}^n)$. [The restrictions of such functions to $\Omega$ are dense in $W^{m,p}(\Omega)$.] Let $\gamma$ denote the linear mapping

$$u \to \gamma u = (\gamma_0 u,\ldots,\gamma_{m-1}u); \qquad \gamma_j u = \left.\frac{\partial^j u}{\partial n^j}\right|_{\operatorname{bdry}\Omega}, \tag{37}$$

where $\partial^j/\partial n^j$ denotes the $j$th directional derivative in the direction of the inward normal to bdry $\Omega$. Using a partition of unity for a neighborhood of bdry $\Omega$ subordinate to the open cover $\{u_j\}$, we can prove the following generalization (to $\Omega$) of the result of Section 7.50. (The details are left to the reader.)

**7.53 THEOREM** Let $1 < p < \infty$ and let $\Omega$ satisfy the conditions prescribed above. Then the mapping $\gamma$ given by (37) extends by continuity to an isomorphism and homeomorphism of $W^{m,p}(\Omega)/\ker\gamma$ onto

$$\prod_{k=0}^{m-1} W^{m-k-1/p,p}(\text{bdry}\,\Omega).$$

**7.54** It is an immediate consequence of the following theorem that the kernel $\ker\gamma$ of the mapping $\gamma$, that is, the class of $u \in W^{m,p}(\Omega)$ for which $\gamma u = 0$, is precisely the space $W_0^{m,p}(\Omega)$. We adopt again the notations of Section 7.30. Let $W_0{}^m$ denote the closure in $W^m = W^m(p, \nu; \Lambda; B)$ of the set of functions $f \in W^m$, each of which vanishes identically on an interval $[0, \varepsilon)$ for some positive $\varepsilon$ (which may depend on $f$).

**7.55 THEOREM** If $f \in W^m$ and if $f^{(k)}(0) = 0$ for $0 \le k \le m-1$, then $f \in W_0{}^m$. Thus $W_0{}^m$ is the kernel of the mapping

$$f \to (f(0), f'(0), \ldots, f^{(m-1)}(0))$$

of $W^m$ onto $\prod_{k=0}^{m-1} T^{m-k-1}$.

PROOF Let $f \in W^m$ satisfy $f(0) = \cdots = f^{(m-1)}(0) = 0$. Let $\psi \in C^\infty(\mathbb{R})$ satisfy $\psi(t) = 0$ for $t \le 1$, $\psi(t) = 1$ for $t \ge 2$, $0 \le \psi(t) \le 1$ and $|\psi^{(k)}(t)| \le K_1$ for all $t$, $0 \le k \le m$. Let $f_n(t) = \psi(nt) f(t)$. Clearly $f_n \in W_0{}^m$. We must show that $f(t) - f_n(t) = (1 - \psi(nt)) f(t) \to 0$ in $W^m$ as $n \to \infty$, that is, we must show that for each $k$, $0 \le k \le m$, and each multi-index $\alpha$, $|\alpha| \le m-k$, we have

$$\int_0^\infty t^{\nu p} \|\Lambda^\alpha (f - f_n)^{(k)}(t)\|_B^p \, dt \to 0 \qquad \text{as} \quad n \to \infty.$$

Now

$$\int_0^\infty t^{\nu p} \|(1 - \psi(nt)) \Lambda^\alpha f^{(k)}(t)\|_B^p \, dt \le \int_0^{2/n} t^{\nu p} \|\Lambda^\alpha f^{(k)}(t)\|_B^p \, dt \to 0$$

as $n \to \infty$ since $f \in W^m$. Hence we need only show that if $1 \le j \le k$, then

$$\int_0^\infty t^{\nu p} \left[ \left(\frac{d}{dt}\right)^j (1 - \psi(nt)) \right]^p \|\Lambda^\alpha f^{(k-j)}(t)\|_B^p \, dt \to 0 \tag{38}$$

as $n \to \infty$. But the left side of (38) does not exceed a constant times

$$n^{jp} \int_{1/n}^{2/n} t^{\nu p} \|\Lambda^\alpha f^{(k-j)}(t)\|_B^p \, dt. \tag{39}$$

Since $f(0) = f'(0) = \cdots = f^{(m-1)}(0) = 0$, and since $k-j \le m-1$ we have [where $p^{-1}+(p')^{-1} = 1$]

$$\begin{aligned}
\|\Lambda^\alpha f^{(k-j)}(t)\|_B^p &\le \left\{ \frac{1}{(j-1)!} \int_0^t (t-\tau)^{j-1} \|\Lambda^\alpha f^{(k)}(\tau)\|_B \, d\tau \right\}^p \\
&\le \frac{t^{(j-1)p}}{[(j-1)!]^p} \int_0^t \tau^{\nu p} \|\Lambda^\alpha f^{(k)}(\tau)\|_B^p \, d\tau \left\{ \int_0^t \tau^{-\nu p'} \, d\tau \right\}^{p/p'} \\
&\le K_2 t^{jp-\nu p-1} \int_0^t \tau^{\nu p} \|\Lambda^\alpha f^{(k)}(\tau)\|_B^p \, d\tau.
\end{aligned}$$

It follows that (39) does not exceed a constant times

$$n^{jp} \int_{1/n}^{2/n} t^{jp-1} \, dt \int_0^t \tau^{\nu p} \|\Lambda^\alpha f^{(k)}(\tau)\|_B^p \, d\tau \le (2^{jp}/jp) \int_0^{2/n} \tau^{\nu p} \|\Lambda^\alpha f^{(k)}(\tau)\|_B^p \, d\tau \to 0$$

as $n \to \infty$ since $f \in W^m$. ∎

**7.56** The characterization of traces on bdry $\Omega$ of functions in $W^{m,p}(\Omega)$ has important applications in the study of nonhomogeneous boundary value problems for differential operators defined on $\Omega$. Theorem 7.53 contains both "direct" and "converse" imbedding theorems for $W^{m,p}(\Omega)$ in the following sense: If $u \in W^{m,p}(\Omega)$, then the trace $v = u|_{\text{bdry}\,\Omega}$ belongs to $W^{m-1/p,p}(\text{bdry}\,\Omega)$ and

$$\|v\|_{m-1/p,p,\text{bdry}\,\Omega} \le K_1 \|u\|_{m,p,\Omega};$$

and conversely, if $v \in W^{m-1/p,p}(\text{bdry}\,\Omega)$, then there exists $u \in W^{m,p}(\Omega)$ with $v = u|_{\text{bdry}\,\Omega}$ and

$$\|u\|_{m,p,\Omega} \le K_2 \|v\|_{m-1/p,p,\text{bdry}\,\Omega}.$$

Before stating a very general imbedding theorem for the spaces $W^{s,p}(\Omega)$ we show how some (but not all) imbeddings for these spaces can be obtained from known cases for integral $s$ by the interpolation Theorem 7.17.

**7.57 THEOREM** Let $\Omega$ be a domain having the cone property in $\mathbb{R}^n$. Let $s > 0$ and $1 < p < n$.

(a) If $n > sp$, then $W^{s,p}(\Omega) \to L^r(\Omega)$ for $p \le r \le np/(n-sp)$.
(b) If $n = sp$, then $W^{s,p}(\Omega) \to L^r(\Omega)$ for $p \le r < \infty$.
(c) If $n < (s-j)p$ for some nonnegative integer $j$, then $W^{s,p}(\Omega) \to C_B^j(\Omega)$.

PROOF The results are already known for integer $s$ so we may assume $s$ is not an integer and write $s = m+\sigma$ where $m$ is an integer and $0 < \sigma < 1$. First let us suppose $m = 0$. Then

$$\begin{aligned} W^{\sigma,p}(\Omega) &= L^p(\Omega) \cap T^{1-\sigma,p}(\Omega) \\ &= L^p(\Omega) \cap T(p, 1-\sigma-(1/p); W^{1,p}(\Omega), L^p(\Omega)). \end{aligned}$$

Now the identity operator is continuous from $W^{1,p}(\Omega)$ into $L^{np/(n-p)}(\Omega)$ and (trivially) from $L^p(\Omega)$ into $L^p(\Omega)$. By Theorem 7.17 it is also continuous from $T(p, 1-\sigma-(1/p); W^{1,p}(\Omega), L^p(\Omega))$ into $T(p, 1-\sigma-(1/p); L^{np/(n-p)}(\Omega), L^p(\Omega))$. By Theorem 7.20 this latter space is imbedded in $L^{np/(n-\sigma p)}$ provided $n > \sigma p$. Hence

$$W^{\sigma,p}(\Omega) \to L^{np/(n-\sigma p)}(\Omega).$$

For general $m$, we argue as follows. Let $u \in W^{m+\sigma,p}(\Omega)$. If $|\alpha| = m$, then $D^\alpha u \in W^{\sigma,p}(\Omega) \to L^{np/(n-\sigma p)}(\Omega)$. If $|\alpha| \le m-1$, then $D^\alpha u \in W^{1,p}(\Omega) \to L^{np/(n-\sigma p)}(\Omega)$. Hence $W^{m+\sigma,p}(\Omega) \to W^{m,np/(n-\sigma p)}(\Omega)$. If $n > sp$, we have by Theorem 5.4 that $W^{m,np/(n-\sigma p)}(\Omega) \to L^{np/(n-sp)}(\Omega)$. Hence (a) is proved. If $n = sp$, then $W^{m,np/(n-\sigma p)}(\Omega) \to L^r(\Omega)$ for any $r$ such that $p \le r < \infty$, so (b) is proved. If $(s-j)p > n$, then $(m-j)np/(n-\sigma p) > n$ and so $W^{m,np/(n-\sigma p)}(\Omega) \to C_B^j(\Omega)$ and (c) is proved. ▮

The restriction $p < n$ in the above theorem is unnatural and was placed only for the purpose of achieving a very simple proof.

The following theorem contains all the imbedding results cited above as special cases. It comprises results obtained by several writers, in particular, Besov [9, 10], Uspenskii [67, 68], and Lizorkin [41]. The theorem is stated for $\mathbb{R}^n$ but can obviously be extended to domains with sufficient regularily, such as those satisfying the conditions of Theorem 7.41. We shall not attempt any proof.

**7.58 THEOREM** Let $s > 0$, $1 < p \le q < \infty$, and $1 \le k \le n$. Let $\chi = s-(n/p)+(k/q)$. If

(i) $\chi \ge 0$ and $p < q$, or
(ii) $\chi > 0$ and $\chi$ is not an integer, or
(iii) $\chi \ge 0$ and $1 < p \le 2$,

then (direct imbedding theorem)

$$W^{s,p}(\mathbb{R}^n) \to W^{\chi,q}(\mathbb{R}^k). \tag{40}$$

Imbedding (40) does not necessarily hold for $p = q > 2$ and $\chi$ a nonnegative integer. (In particular, one cannot in general strengthen Case A of Part I of Theorem 5.4 to allow $k = n-mp$.)

Conversely, if $p = q$ and if either

(iv) $\chi = s-(n-k)/p > 0$ and is not an integer, or
(v) $\chi \geq 0$ and $p \geq 2$,

then we have the reverse imbedding

$$W^{\chi,p}(\mathbb{R}^k) \to W^{s,p}(\mathbb{R}^n)$$

in the sense that each $u \in W^{\chi,p}(\mathbb{R}^k)$ is the trace on $\mathbb{R}^k$ (i.e., $u = w|_{\mathbb{R}^k}$) of a function $w \in W^{s,p}(\mathbb{R}^n)$ satisfying

$$\|w\|_{s,p,\mathbb{R}^n} \leq K\|u\|_{\chi,p,\mathbb{R}^k}$$

with $K$ independent of $u$. (The trace is understood in the sense of Section 5.2.)

## Bessel Potentials—The Spaces $L^{s,p}(\Omega)$

**7.59** We shall outline here, without proofs, another method of constructing fractional order spaces which originates in studies of Bessel potentials by Aronszajn and Smith [7] (and their collaborators—Adams *et al.* [5] and Aronszajn *et al.* [8]) and which is presented by Calderón [13] and Lions and Magenes [40]. The resulting spaces, denoted $L^{s,p}(\Omega)$ (or $H^{s,p}(\Omega)$ by Lions and Magenes—but not to be confused with the $H$-spaces of Nikol'skii defined in Section 7.73) coincide with the spaces $W^{s,p}(\Omega)$ for integer values of $s$ if $1 < p < \infty$, and for all $s$ when $p = 2$.

The space $L^{s,p}(\mathbb{R}^n)$ is constructed directly in terms of Fourier transforms of tempered distributions. It is shown then that for $1 < p < \infty$, $L^{s,p}(\mathbb{R}^n)$ and $W^{s,p}(\mathbb{R}^n)$ are isomorphic and homeomorphic when $s$ is an integer. For any values of $s_1, s, s_2$ with $s_1 \leq s \leq s_2$, the space $L^{s,p}(\mathbb{R}^n)$ can be identified as an intermediate space interpolated between $L^{s_1,p}(\mathbb{R}^n)$ and $L^{s_2,p}(\mathbb{R}^n)$ by a "complex" interpolation method (see Calderón [15] or Lions [36]) which is not identical to the trace method of Lions described earlier. This interpolation method then provides a means of defining $L^{s,p}(\Omega)$ for domain $\Omega \subset \mathbb{R}^n$ as an intermediate space between spaces of the form $W^{m,p}(\Omega)$ for integer values of $m$.

Proofs of assertions made in the discussion of the spaces $L^{s,p}(\Omega)$ and their relationship to the spaces $W^{s,p}(\Omega)$ can be found in one or another of the papers by Calderón, Lions, and Lions and Magenes cited above.

**7.60** First we introduce the notion of tempered distribution. We denote by $\mathcal{S}(\mathbb{R}^n)$ the space of rapidly decreasing functions in $\mathbb{R}^n$, that is, functions $\phi$ satisfying

$$\sup_{x \in \mathbb{R}^n} |x^\alpha D^\beta \phi(x)| < \infty$$

for all multi-indices $\alpha$ and $\beta$. The space $\mathscr{S}(\mathbb{R}^n)$ carries a locally convex topology characterized by the following notion of convergence: The sequence $\{\phi_j\}$ converges to 0 in $\mathscr{S}(\mathbb{R}^n)$ if for all $\alpha$ and $\beta$

$$\lim_{j\to\infty} x^\alpha D^\beta \phi_j(x) = 0 \quad \text{uniformly on } \mathbb{R}^n.$$

It may be readily verified that the Fourier transform

$$\mathscr{F}\phi(y) = (2\pi)^{-n/2} \int_{\mathbb{R}^n} e^{-ix\cdot y}\phi(x)\, dx$$

and the inverse Fourier transform

$$\mathscr{F}^{-1}\phi(x) = (2\pi)^{-n/2} \int_{\mathbb{R}^n} e^{ix\cdot y}\phi(y)\, dy$$

are each continuous on $\mathscr{S}(\mathbb{R}^n)$ into $\mathscr{S}(\mathbb{R}^n)$, and, since $\mathscr{F}^{-1}\mathscr{F}\phi = \mathscr{F}\mathscr{F}^{-1}\phi = \phi$, each is in fact an isomorphism and a homeomorphism of $\mathscr{S}(\mathbb{R}^n)$ onto $\mathscr{S}(\mathbb{R}^n)$.

It is clear from the definition of $\mathscr{S}(\mathbb{R}^n)$ that $\mathscr{D}(\mathbb{R}^n) \to \mathscr{S}(\mathbb{R}^n)$. Hence the dual space $\mathscr{S}'(\mathbb{R}^n)$ consists of those distributions $T \in \mathscr{D}'(\mathbb{R}^n)$ which possess continuous extensions to $\mathscr{S}(\mathbb{R}^n)$. For instance, if $1 \le p \le \infty$ and $f \in L^p(\mathbb{R}^n)$, then

$$T_f(\phi) = \int_{\mathbb{R}^n} f(x)\,\phi(x)\, dx$$

defines $T_f \in \mathscr{S}'(\mathbb{R}^n)$. The same holds for any function $f$ of "slow growth" at infinity, that is, for which for some finite $k$ we have $|f(x)| \le \text{const}\,|x|^k$ a.e. in some neighborhood of infinity. The elements of $\mathscr{S}'(\mathbb{R}^n)$ are therefore called *tempered distributions.* $\mathscr{S}'(\mathbb{R}^n)$ is given the weak-star topology as dual of $\mathscr{S}(\mathbb{R}^n)$ and is a locally convex topological vector space with this topology.

The direct and inverse Fourier transformations are extended to $\mathscr{S}'(\mathbb{R}^n)$ by

$$\mathscr{F}T(\phi) = T(\mathscr{F}\phi), \qquad \mathscr{F}^{-1}T(\phi) = T(\mathscr{F}^{-1}\phi).$$

Once again, each is an isomorphism and a homeomorphism of $\mathscr{S}'(\mathbb{R}^n)$ onto $\mathscr{S}'(\mathbb{R}^n)$ and $\mathscr{F}^{-1}\mathscr{F}T = \mathscr{F}\mathscr{F}^{-1}T = T$.

**7.61** Given a tempered distribution $u$ on $\mathbb{R}^n$ and a complex number $z$ the Bessel potential of order $z$ of $u$ is denoted $J^z u$ and defined by

$$J^z u = \mathscr{F}^{-1}((1 + |\cdot|^2)^{-z/2}\mathscr{F}u).$$

Evidently $J^z$ is one-to-one on $\mathscr{S}'(\mathbb{R}^n)$ into $\mathscr{S}'(\mathbb{R}^n)$. If $\operatorname{Re} z > 0$ and $1 \le p \le \infty$, or if $\operatorname{Re} z \geq 0$ and $1 < p < \infty$, then $J^z$ transforms $L^p(\mathbb{R}^n)$ continuously into $L^p(\mathbb{R}^n)$, and $D^\alpha J^{z+|\alpha|}$ does likewise.

**7.62** For real $s$ and $1 \leq p \leq \infty$ let $L^{s,p}(\mathbb{R}^n)$ denote the image of $L^p(\mathbb{R}^n)$ under the linear mapping $J^s$. Thus $L^{s,p}(\mathbb{R}^n) \to \mathscr{S}'(\mathbb{R}^n)$ for every $s$, and $L^{s,p}(\mathbb{R}^n) \to L^p(\mathbb{R}^n)$ for $s \geq 0$. If $u \in L^{s,p}(\mathbb{R}^n)$, then there exists unique $\tilde{u} \in L^p(\mathbb{R}^n)$ with $u = J^s \tilde{u}$. We define

$$\|u; L^{s,p}(\mathbb{R}^n)\| = \|\tilde{u}\|_{0,p,\mathbb{R}^n}.$$

With respect to this norm, $L^{s,p}(\mathbb{R}^n)$ is a Banach space. We summarize some of its properties.

**7.63 THEOREM** (a) If $s \geq 0$ and $1 \leq p < \infty$, then $\mathscr{D}(\mathbb{R}^n)$ is dense in $L^{s,p}(\mathbb{R}^n)$.

(b) If $1 < p < \infty$ and $p' = p/(p-1)$, then $[L^{s,p}(\mathbb{R}^n)]' \cong L^{-s,p'}(\mathbb{R}^n)$.

(c) It $t < s$, then $L^{s,p}(\mathbb{R}^n) \to L^{t,p}(\mathbb{R}^n)$.

(d) If $t \leq s$ and if either $1 < p \leq q \leq np/[n-(s-t)p] < \infty$ or $p = 1$ and $1 \leq q < n/(n-s+t)$, then $L^{s,p}(\mathbb{R}^n) \to L^{t,q}(\mathbb{R}^n)$.

(e) If $0 \leq \mu \leq s-(n/p) < 1$, then $L^{s,p}(\mathbb{R}^n) \to C^{0,\mu}(\mathbb{R}^n)$.

(f) If $s$ is a nonnegative integer and $1 < p < \infty$, then $L^{s,p}(\mathbb{R}^n)$ coincides with $W^{s,p}(\mathbb{R}^n)$, the norms in the two spaces being equivalent. This conclusion also holds for any $s$ if $p = 2$.

(g) If $1 < p < \infty$ and $\varepsilon > 0$, then for every $s$ we have

$$L^{s+\varepsilon,p}(\mathbb{R}^n) \to W^{s,p}(\mathbb{R}^n) \to L^{s-\varepsilon,p}(\mathbb{R}^n).$$

**7.64** We now describe a complex interpolation method of Calderón [15] and Lions [36] in which setting the spaces $L^{s,p}(\mathbb{R}^n)$ can also arise.

Let $B_0$ and $B_1$ be Banach spaces both imbedded in a topological vector space $X$ as in Section 7.11, and let the Banach space $B_0 + B_1$ be defined as in that section. We denote by $F(B_0, B_1)$ the space of functions $f$ of a complex variable $z = \sigma + i\tau$ taking values in $B_0 + B_1$ and satisfying

(i) $f$ is holomorphic on the strip $0 < \sigma < 1$,

(ii) $f$ is continuous and bounded on the strip $0 \leq \sigma \leq 1$,

(iii) $f(i\tau) \in B_0$ for $\tau \in \mathbb{R}$, the map $\tau \to f(i\tau)$ is continuous on $\mathbb{R}$ into $B_0$, and $\lim_{|\tau| \to \infty} f(i\tau) = 0$, and

(iv) $f(1+i\tau) \in B_1$ for $\tau \in \mathbb{R}$, the map $\tau \to f(1+i\tau)$ is continuous on $\mathbb{R}$ into $B_1$, and $\lim_{|\tau| \to \infty} f(1+i\tau) = 0$.

$F(B_0, B_1)$ is a Banach space with respect to the norm

$$\|f; F(B_0, B_1)\| = \max\left\{\sup_{\tau \in \mathbb{R}} \|f(i\tau)\|_{B_0}, \sup_{\tau \in \mathbb{R}} \|f(1+i\tau)\|_{B_1}\right\}.$$

For $0 \le \sigma \le 1$ set

$$B_\sigma = [B_0; B_1]_\sigma = \{u \in B_0 + B_1 : u = f(\sigma) \text{ for some } f \in F(B_0, B_1)\}.$$

With respect to the norm

$$\|u\|_{B_\sigma} = \|u; [B_0; B_1]_\sigma\| = \inf_{\substack{f \in F(B_0, B_1) \\ f(\sigma) = u}} \|f; F(B_0, B_1)\|,$$

$B_\sigma$ is a Banach space imbedded in $B_0 + B_1$.

The intermediate spaces $B_\sigma$ possess interpolation characteristics similar to those of the trace spaces of Lions. If $C_0$, $C_1$, and $Y$ are spaces having properties similar to those specified for $B_0$, $B_1$, and $X$, and if $L$ is a linear mapping from $B_0 + B_1$ into $C_0 + C_1$ satisfying

$$\|Lu\|_{C_0} \le K_0 \|u\|_{B_0}, \qquad \|Lu\|_{C_1} \le K_1 \|u\|_{B_1},$$

then for each $u \in B_\sigma$ we have $Lu \in C_\sigma$ and

$$\|Lu\|_{C_\sigma} \le K_0^{1-\sigma} K_1^{\sigma} \|u\|_{B_\sigma}.$$

The following theorem may be found in the papers by Calderón [15] or Lions [36].

**7.65 THEOREM** For any real $s_0$ and $s_1$, and for $0 \le \sigma \le 1$ we have

$$[L^{s_0, p}(\mathbb{R}^n); L^{s_1, p}(\mathbb{R}^n)]_\sigma = L^{(1-\sigma)s_0 + \sigma s_1, p}(\mathbb{R}^n).$$

We remark that the corresponding statement for intermediate spaces between $W^{s_0, p}(\mathbb{R}^n)$ and $W^{s_1, p}(\mathbb{R}^n)$ obtained by trace interpolation is not valid for all $s_0$ and $s_1$ though it is for certain values, in particular if $s_0$ and $s_1$ are consecutive nonnegative integers.

**7.66** The above theorem suggests how the spaces $L^{s,p}(\Omega)$ may be defined for arbitrary domains $\Omega \subset \mathbb{R}^n$. If $s \ge 0$, let $m$ be the integer satisfying $s \le m < s+1$ and define $L^{s,p}(\Omega) = [W^{m,p}(\Omega); L^p(\Omega)]_{(m-s)/m}$. If $\Omega$ is sufficiently regular to possess a strong $m$-extension operator, then an interpolation argument shows that $L^{s,p}(\Omega)$ coincides with the space of restrictions to $\Omega$ of functions in $L^{s,p}(\mathbb{R}^n)$. Also, Theorem 7.65 is valid for the spaces $L^{s,p}(\Omega)$ provided $0 \le s_0, s_1 \le m$.

The definition of $L^{s,p}(\Omega)$ for negative $s$ is carried out in the same manner as for the spaces $W^{s,p}(\Omega)$. One denotes by $L_0^{s,p}(\Omega)$ (where $s > 0$) the closure of $\mathscr{D}(\Omega)$ in $L^{s,p}(\Omega)$ and defines, for $1 < p < \infty$ and $s < 0$, the space $L^{s,p}(\Omega)$ to be $[L_0^{-s,p'}(\Omega)]'$, where $(1/p) + (1/p') = 1$.

All the properties stated for $L^{s,p}(\mathbb{R}^n)$ in Theorem 7.63 possess analogs for $L^{s,p}(\Omega)$ provided $\Omega$ is suitably regular.

## Other Fractional Order Spaces

**7.67** Certain gaps in the general imbedding theorem for the spaces $W^{s,p}(\mathbb{R}^n)$ (see Theorem 7.58) led to the construction by Besov [9, 10] of a family of spaces $B^{s,p}(\mathbb{R}^n)$ which differ from $W^{s,p}(\mathbb{R}^n)$ when $s$ is a positive integer and which naturally supplement these latter spaces in a sense to be made precise below.

$B^{s,p}(\mathbb{R}^n)$ is defined for $s > 0$ and $1 \le p \le \infty$ as follows. Let $s = m+\sigma$ where $m$ is a nonnegative integer and $0 < \sigma \le 1$. The space $B^{s,p}(\mathbb{R}^n)$ consists of those functions $u$ in $W^{m,p}(\mathbb{R}^n)$ for which the norm $\|u; B^{s,p}(\mathbb{R}^n)\|$ is finite. If $1 \le p < \infty$,

$$\|u; B^{s,p}(\mathbb{R}^n)\|$$

$$= \left\{ \|u\|_{m,p,\mathbb{R}^n}^p + \sum_{|\alpha|=m} \int_{\mathbb{R}^n} \int_{\mathbb{R}^n} \frac{|D^\alpha u(x) - 2D^\alpha u((x+y)/2) + D^\alpha u(y)|^p}{|x-y|^{n+\sigma p}} \, dx \, dy \right\}^{1/p}.$$

If $p = \infty$,

$$\|u; B^{s,\infty}(\mathbb{R}^n)\|$$

$$= \max \left\{ \|u\|_{m,\infty,\mathbb{R}^n}, \max_{|\alpha|=m} \operatorname*{ess\,sup}_{\substack{x,y\in\mathbb{R}^n \\ x\neq y}} \frac{|D^\alpha u(x) - 2D^\alpha u((x+y)/2) + D^\alpha u(y)|}{|x-y|^\sigma} \right\}.$$

$B^{s,p}(\mathbb{R}^n)$ is a Banach space with respect to the above norm. If $1 \le p < \infty$, $C_0^\infty(\mathbb{R}^n)$ is dense in $B^{s,p}(\mathbb{R}^n)$.

**7.68 LEMMA** If $1 \le p < \infty$ and $s > 0$ is not an integer, then the spaces $W^{s,p}(\mathbb{R}^n)$ and $B^{s,p}(\mathbb{R}^n)$ coincide, and have equivalent norms.

PROOF For functions $u$ defined on $\mathbb{R}^n$ we define the difference operator $\Delta_z$ by

$$\Delta_z u(x) = u(x+z) - u(x).$$

The second difference operator $\Delta_z^2$ is then given by

$$\Delta_z^2 u(x) = \Delta_z \Delta_z u(x) = u(x+2z) - 2u(x+z) + u(x).$$

The identity

$$\Delta_z u = (1/2^k)\Delta_{2^k z} u - \tfrac{1}{2}\sum_{j=0}^{k-1} (1/2^j)\Delta_{2^j z}^2 u \tag{41}$$

may readily be verified by expanding the sum on the right side.

Evidently, the norm of a function $u$ in $B^{s,p}(\mathbb{R}^n)$ is equivalent to

$$\left\{ \|u\|_{m,p,\mathbb{R}^n}^p + \sum_{|\alpha|=m} \int_{\mathbb{R}^n} |z|^{-n-\sigma p} \, dz \int_{\mathbb{R}^n} |\Delta_z^2 D^\alpha u(x)|^p \, dx \right\}^{1/p}, \tag{42}$$

while by Theorem 7.48 the norm of $u$ in $W^{s,p}(\mathbb{R}^n)$ can be expressed in the form

$$\left\{\|u\|_{m,p,\mathbb{R}^n}^p + \sum_{|\alpha|=m} \int_{\mathbb{R}^n} |z|^{-n-\sigma p}\, dz \int_{\mathbb{R}^n} |\Delta_z D^\alpha u(x)|^p\, dx\right\}^{1/p}. \tag{43}$$

It is clear that (42) is bounded by a constant times (43); we must prove the reverse assertion.

Suppose, therefore, that $u \in C_0^\infty(\mathbb{R}^n)$. We have, using (41),

$$\begin{aligned}
&\left\{\int_{\mathbb{R}^n} |z|^{-n-\sigma p}\, dz \int_{\mathbb{R}^n} |\Delta_z D^\alpha u(x)|^p\, dx\right\}^{1/p} \\
&\quad \le (1/2^k)\left\{\int_{\mathbb{R}^n} |z|^{-n-\sigma p}\, dz \int_{\mathbb{R}^n} |\Delta_{2^k z} D^\alpha u(x)|^p\, dx\right\}^{1/p} \\
&\qquad + \tfrac{1}{2}\sum_{j=0}^{k-1} (1/2^j)\left\{\int_{\mathbb{R}^n} |z|^{-n-\sigma p}\, dz \int_{\mathbb{R}^n} |\Delta_{2^j z}^2 D^\alpha u(x)|^p\, dx\right\}^{1/p} \\
&\quad = (1/2^{k(1-\sigma)})\left\{\int_{\mathbb{R}^n} |\rho|^{-n-\sigma p}\, d\rho \int_{\mathbb{R}^n} |\Delta_\rho D^\alpha u(x)|^p\, dx\right\}^{1/p} \\
&\qquad + \tfrac{1}{2}\sum_{j=0}^{k-1} (1/2^{j(1-\sigma)})\left\{\int_{\mathbb{R}^n} |\rho|^{-n-\sigma p}\, d\rho \int_{\mathbb{R}^n} |\Delta_\rho{}^2 D^\alpha u(x)|^p\, dx\right\}^{1/p}.
\end{aligned}$$

(We have substituted $\rho = 2^k z$ in the first integral, $\rho = 2^j z$ in the second.) Since $s$ is not an integer, we have $\sigma < 1$ and so $k$ may be chosen large enough that $k(1-\sigma) \ge 1$. It then follows that (43) is bounded by a constant times (42). Since $C_0^\infty(\mathbb{R}^n)$ is dense in $B^{s,p}(\mathbb{R}^n)$ the lemma follows. ▮

**7.69** If $s$ is a positive integer and $p = 2$, the spaces $W^{s,2}(\mathbb{R}^n)$ and $B^{s,2}(\mathbb{R}^n)$ coincide. For $p \ne 2$, $s$ an integer they are distinct but for any $\varepsilon > 0$ we have

$$W^{s+\varepsilon,p}(\mathbb{R}^n) \to B^{s,p}(\mathbb{R}^n) \to W^{s,p}(\mathbb{R}^n) \qquad \text{if} \quad 1 < p \le 2$$

$$B^{s+\varepsilon,p}(\mathbb{R}^n) \to W^{s,p}(\mathbb{R}^n) \to B^{s,p}(\mathbb{R}^n) \qquad \text{if} \quad p \ge 2.$$

The spaces $B^{s,p}(\mathbb{R}^n)$ are of interest primarily for their imbedding characteristics. They possess a "closed system" of imbeddings and at the same time fill gaps in the system of imbeddings for the spaces $W^{s,p}(\mathbb{R}^n)$.

**7.70 THEOREM** Let $s > 0$, $1 \le p \le q \le \infty$, and $1 \le k \le n$, $k$ an integer. Suppose

$$r = s - (n/p) + (k/q) > 0.$$

Then

$$B^{s,p}(\mathbb{R}^n) \to B^{r,q}(\mathbb{R}^k).$$

Conversely, if $p = q$ and $r = [s-(n-k)]/p > 0$, then the reverse imbedding

$$B^{r,p}(\mathbb{R}^k) \to B^{s,p}(\mathbb{R}^n)$$

holds, in the sense that each element $u$ in $B^{r,p}(\mathbb{R}^k)$ is the trace $u = v|_{\mathbb{R}^k}$ of some element $v$ in $B^{s,p}(\mathbb{R}^n)$ satisfying

$$\|v; B^{s,p}(\mathbb{R}^n)\| \le K\|u; B^{r,p}(\mathbb{R}^k)\|,$$

where $K$ is independent of $u$.

**7.71 THEOREM** If $s > 0$, $1 \le p \le \infty$, and $1 \le k \le n$, and if $r = [s-(n-k)]/p$, then

$$W^{s,p}(\mathbb{R}^n) \to B^{r,p}(\mathbb{R}^n)$$

and conversely

$$B^{r,p}(\mathbb{R}^k) \to W^{s,p}(\mathbb{R}^n).$$

**7.72** The definitions and theorems above can be extended to suitable domains $\Omega \subset \mathbb{R}^n$ and smooth manifolds $\Omega^k$ of dimension $k$ contained in $\overline{\Omega}$. For $1 \le p < \infty$ the norm in $B^{s,p}(\Omega)$ is

$$\|u; B^{s,p}(\Omega)\|$$
$$= \left\{\|u\|^p_{m,p,\Omega} + \sum_{|\alpha|=m} \int_\Omega \int_{\Omega_x} \frac{|D^\alpha u(x) - 2D^\alpha u((x+y)/2) + D^\alpha u(y)|^p}{|x-y|^{n+\sigma p}}\, dy\, dx\right\}^{1/p},$$

where $\Omega_x = \{y \in \Omega : (x+y)/2 \in \Omega\}$.

**7.73** A different class of spaces having imbedding properties similar to the Besov spaces are the spaces $H^{s,p}(\Omega)$ introduced by Nikol'skii [49–51]. These spaces, having norms involving Hölder conditions in the $L^p$-metric, were studied earlier than the (fractional order) $W$- or $B$-spaces and provided impetus for the latter.

Again we set $s = m+\sigma$ where $m \ge 0$ is an integer and $0 < \sigma \le 1$. For $1 \le p < \infty$ and $\Omega \subset \mathbb{R}^n$ a function $u$ belongs to $H^{s,p}(\Omega)$ provided the norm

$$\|u; H^{s,p}(\Omega)\| = \left\{\|u\|^p_{0,p,\Omega} + \sum_{|\alpha|=m} \sup_{\substack{h \in \mathbb{R}^n \\ \eta > 0 \\ 0<|h|<\eta}} \int_{\Omega_\eta} \frac{|\Delta_h{}^2 D^\alpha f(x)|^p}{|h|^{\sigma p}}\, dx\right\}^{1/p}$$

is finite, where $\Omega_\eta = \{x \in \Omega; \operatorname{dist}(x, \operatorname{bdry}\Omega) \ge 2\eta\}$. The obvious modification is made for $p = \infty$ so that in fact $H^{s,\infty}(\Omega) = B^{s,\infty}(\Omega)$. An argument similar to that of Lemma 7.68 shows that if $s$ is not an integer, the second difference $\Delta^2$

in the norm of $H^{s,p}(\Omega)$ can be replaced by the first difference $\Delta$ without changing the space.

The spaces $H^{s,p}(\Omega)$ are larger than the corresponding spaces $W^{s,p}(\Omega)$; but if $\varepsilon > 0$, we have

$$H^{s+\varepsilon,p}(\Omega) \to W^{s,p}(\Omega) \to H^{s,p}(\Omega).$$

The spaces $H^{s,p}(\mathbb{R}^n)$ possess a closed system of imbeddings identical to those of the Besov spaces, that is, Theorem 7.70 holds with $B$ everywhere replaced by $H$. Strong extension theorems can be proved for $H$-spaces over smoothly bounded domains so that the imbedding theorem can be extended to such domains and traces on smooth manifolds in them.

The imbedding theorems for the spaces $H^{s,p}(\mathbb{R}^n)$ and $B^{s,p}(\mathbb{R}^n)$ are proved by a technique involving approximation of functions in these spaces by entire functions of exponential type in several complex variables (see Nikol'skii [49], for example).

**7.74** Numerous generalizations of the above spaces have been made, partly for their own sake and partly to facilitate the solution of other problems in analysis. We mention two such directions of generalization. The first involves replacing ordinary $L^p$-norms by weighted norms. The second involves the use of different values of $s$ and $p$ in terms of the norm involving integration in different coordinate directions (anisotropic spaces). The interested reader is referred to two excellent survey articles (Nikol'skii [52] and Sobolev and Nikol'skii [64]), and their bibliographies for further information on the whole spectrum of spaces of differentiable functions of several real variables.

# VIII

# Orlicz and Orlicz–Sobolev Spaces

## Introduction

**8.1** In this final chapter we present some recent results involving replacement of the spaces $L^p(\Omega)$ with more general spaces $L_A(\Omega)$ in which the role usually played by the convex function $t^p$ is assumed by more general convex functions $A(t)$. The spaces $L_A(\Omega)$, called *Orlicz spaces*, are studied in depth in the monograph by Krasnosel'skii and Rutickii [34] and also in the doctoral thesis by Luxemburg [42] to either of which the reader is referred for a more complete development of the material outlined below. The former also contains examples of applications of Orlicz spaces to certain problems in nonlinear analysis.

Following Krasnosel'skii and Rutickii [34] we use the class of "$N$-functions" as defining functions $A$ for Orlicz spaces. This class is not as wide as the class of Young's functions used by Luxemburg [42] (see also O'Neill [55]); for instance, it excludes $L^1(\Omega)$ and $L^\infty(\Omega)$ from the class of Orlicz spaces However, $N$-functions are simpler to deal with and are adequate for our purposes. Only once, in the proof of Theorem 8.35, is it necessary to refer to a more general Young's function.

If the role played by $L^p(\Omega)$ in the definition of the Sobolev space $W^{m,p}(\Omega)$ is assigned instead to an Orlicz space $L_A(\Omega)$, the resulting space is denoted by $W^m L_A(\Omega)$ and called an *Orlicz–Sobolev space*. Many properties of Sobolev spaces have been extended to Orlicz–Sobolev spaces, mainly by Donaldson and Trudinger [22]. We present some of these results in this chapter.

It is also of some interest to note that a gap in the Sobolev imbedding theorem 5.4 can be filled by consideration of Orlicz spaces. Specifically, Case B of that theorem provides no "best" target space for imbeddings of $W^{m,p}(\Omega)$ with $\Omega$ a "regular" domain in $\mathbb{R}^n$ and $mp = n$. We have $W^{m,p}(\Omega) \to L^q(\Omega)$ for $p \le q < \infty$ but $W^{m,p}(\Omega) \not\to L^\infty(\Omega)$. In Theorem 8.25 an optimal imbedding of $W^{m,p}(\Omega)$ into an Orlicz space is constructed. This result is due to Trudinger [66].

## *N*-Functions

**8.2** Let $a$ be a real valued function defined on $[0, \infty)$ and having the following properties:

(a) $a(0) = 0$, $a(t) > 0$ if $t > 0$, $\lim_{t \to \infty} a(t) = \infty$;
(b) $a$ is nondecreasing, that is, $s > t \ge 0$ implies $a(s) \ge a(t)$;
(c) $a$ is right continuous, that is, if $t \ge 0$, then $\lim_{s \to t+} a(s) = a(t)$.

Then the real valued function $A$ defined on $[0, \infty)$ by

$$A(t) = \int_0^t a(\tau)\, d\tau \tag{1}$$

is called an *N-function*.

It is not difficult to verify that any such $N$-function $A$ has the following properties:

(i) $A$ is continuous on $[0, \infty)$;
(ii) $A$ is strictly increasing, that is, $s > t \ge 0$ implies $A(s) > A(t)$;
(iii) $A$ is convex, that is, if $s$, $t \ge 0$ and $0 < \lambda < 1$, then

$$A(\lambda s + (1-\lambda)t) \le \lambda A(s) + (1-\lambda) A(t);$$

(iv) $\lim_{t \to 0+} A(t)/t = 0$, $\lim_{t \to \infty} A(t)/t = \infty$;
(v) if $s > t > 0$, then $A(s)/s > A(t)/t$.

Properties (i), (iii), and (iv) could have been used to define $N$-function since they imply the existence of a representation of $A$ in the form (1) with $a$ having the required properties (a)–(c).

The following are examples of $N$-functions:

$$\begin{aligned} A(t) &= t^p, && 1 < p < \infty, \\ A(t) &= e^t - t - 1, && \\ A(t) &= e^{(t^p)} - 1, && 1 < p < \infty, \\ A(t) &= (1+t)\log(1+t) - t. && \end{aligned}$$

Evidently $A(t)$ is represented by the area under the graph $\sigma = a(\tau)$ from $\tau = 0$ to $\tau = t$ as shown (Fig. 8). Rectilinear segments in the graph of $A$ correspond to intervals of constancy of $a$, and angular points in the graph of $A$ correspond to discontinuities (i.e., vertical jumps) in the graph of $a$.

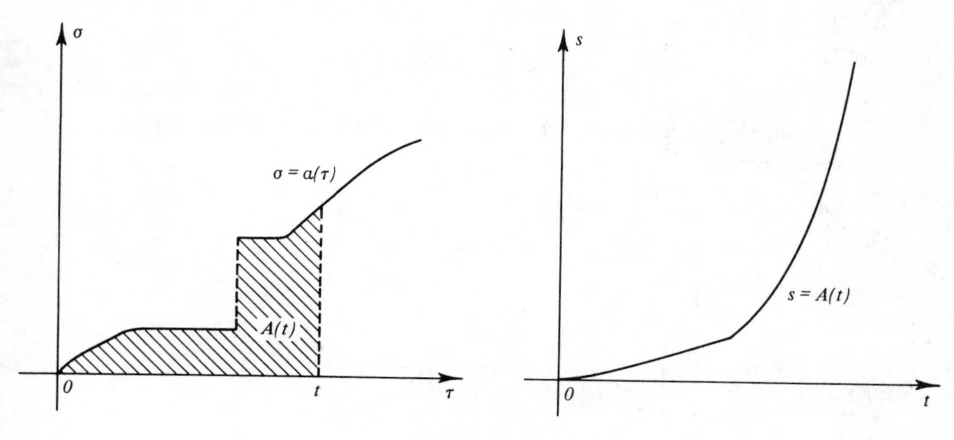


FIG. 8

**8.3** Given $a$ satisfying (a)–(c), we define

$$\tilde{a}(s) = \sup_{a(t) \leq s} t. \tag{2}$$

It is readily checked that the function $a$ so defined also satisfies (a)–(c) and that $a$ can be recovered from $\tilde{a}$ via

$$a(t) = \sup_{\tilde{a}(s) \leq t} s. \tag{3}$$

(If $a$ is strictly increasing, then $\tilde{a} = a^{-1}$.) The $N$-functions $A$ and $\tilde{A}$ given by

$$A(t) = \int_0^t a(\tau)\, d\tau, \qquad \tilde{A}(s) = \int_0^s \tilde{a}(\sigma)\, d\sigma \tag{4}$$

are said to be *complementary*; each is the complement of the other. Examples of such complementary pairs are:

$$A(t) = t^p/p, \qquad \tilde{A}(s) = s^{p'}/p', \qquad 1 < p < \infty, \quad (1/p) + (1/p') = 1;$$

$$A(t) = e^t - t - 1, \qquad \tilde{A}(s) = (1+s)\log(1+s) - s.$$

$\tilde{A}(s)$ is represented by the area to the left of the graph $\sigma = a(\tau)$ [or more correctly $\tau = \tilde{a}(\sigma)$] from $\sigma = 0$ to $\sigma = s$ as shown in Fig. 9. Evidently we have

$$st \leq A(t) + \tilde{A}(s), \tag{5}$$

which is known as *Young's inequality*. Equality holds in (5) if and only if either $t = \tilde{a}(s)$ or $s = a(t)$. Writing (5) in the form

$$\tilde{A}(s) \geq st - A(t)$$

and noting that equality occurs when $t = \tilde{a}(s)$, we have

$$\tilde{A}(s) = \max_{t \geq 0}(st - A(t)).$$

This relationship could have been used as the definition of the $N$-function $\tilde{A}$ complementary to $A$.

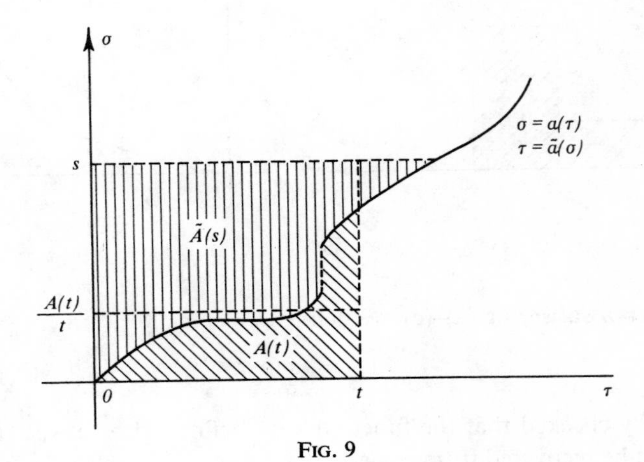


FIG. 9

Since $A$ and $\tilde{A}$ are strictly increasing they have inverses and (5) implies that for every $t \geq 0$

$$A^{-1}(t)\tilde{A}^{-1}(t) \leq A(A^{-1}(t)) + \tilde{A}(\tilde{A}^{-1}(t)) = 2t.$$

On the other hand, $A(t) \leq ta(t)$ so that, considering Fig. 9 again, we have for $t > 0$

$$\tilde{A}(A(t)/t) < (A(t)/t)t = A(t). \tag{6}$$

Replacing $A(t)$ by $t$ in (6), we obtain

$$\tilde{A}(t/A^{-1}(t)) < t.$$

Hence, for any $t > 0$ we have

$$t < A^{-1}(t)\,\tilde{A}^{-1}(t) \leq 2t. \tag{7}$$

**8.4** We shall require certain partial-ordering relationships among *N*-functions. If $A$ and $B$ are two *N*-functions, we say that $B$ *dominates* $A$ *globally* provided there exists a positive constant $k$ such that

$$A(t) \leq B(kt) \tag{8}$$

holds for all $t \geq 0$. Similarly $B$ *dominates* $A$ *near infinity* if there exist positive constants $k$ and $t_0$ such that (8) holds for all $t \geq t_0$. The two *N*-functions $A$ and $B$ are *equivalent globally* (or *near infinity*) if each dominates the other globally (or near infinity). Thus $A$ and $B$ are equivalent near infinity if and only if there exist positive constants $k_1$, $k_2$, and $t_0$ such that if $t \geq t_0$, then $B(k_1 t) \leq A(t) \leq B(k_2 t)$. Such will certainly be the case if

$$0 < \lim_{t\to\infty} \frac{B(t)}{A(t)} < \infty.$$

If $A$ and $B$ have respective complementary *N*-functions $\tilde{A}$ and $\tilde{B}$, then $B$ dominates $A$ globally (or near infinity) if and only if $\tilde{A}$ dominates $\tilde{B}$ globally (or near infinity). Similarly $A$ and $B$ are equivalent if and only if $\tilde{A}$ and $\tilde{B}$ are.

**8.5** If $B$ dominates $A$ near infinity and $A$ and $B$ are not equivalent near infinity, then we say $A$ *increases essentially more slowly than* $B$ near infinity. This is the case if and only if for every $k > 0$

$$\lim_{t\to\infty} \frac{A(kt)}{B(t)} = 0. \tag{9}$$

The reader may verify that (9) is in turn equivalent to the condition

$$\lim_{t\to\infty} \frac{B^{-1}(t)}{A^{-1}(t)} = 0.$$

Let $1 < p < \infty$. We shall hereafter denote by $A_p$ the *N*-function

$$A_p(t) = t^p/p, \qquad 0 \leq t < \infty. \tag{10}$$

If $1 < p < q < \infty$, then $A_p$ increases essentially more slowly than $A_q$ near infinity. However, $A_q$ does not dominate $A_p$ globally.

**8.6** An *N*-function $A$ is said to satisfy a *global* $\Delta_2$*-condition* if there exists a positive constant $k$ such that for every $t \geq 0$,

$$A(2t) \leq kA(t). \tag{11}$$

It is readily seen that this will be the case if and only if for every $r > 1$ there exists a positive constant $k = k(r)$ such that for all $t \geq 0$

$$A(rt) \leq kA(t). \tag{12}$$

Similarly $A$ is said to satisfy a $\Delta_2$-*condition near infinity* if there exists $t_0 > 0$ such that (11) [or equivalently (12) with $r > 1$] holds for all $t \geq t_0$. Evidently $t_0$ may be replaced by any smaller positive number $t_1$, for if $t_1 \leq t \leq t_0$, then

$$A(rt) \leq [A(rt_0)/A(t_1)]\,A(t).$$

If $A$ satisfies a $\Delta_2$-condition globally (or near infinity) and if $B$ is equivalent to $A$ globally (or near infinity), then $B$ also satisfies such a $\Delta_2$-condition. Clearly the $N$-function $A_p(t) = t^p/p$, $1 < p < \infty$, satisfies a global $\Delta_2$-condition. It may be verified that $A$ satisfies a $\Delta_2$-condition globally (or near infinity) if and only if there exists a finite constant $c$ such that

$$(1/c)\,t\alpha(t) \leq A(t) \leq t\alpha(t)$$

holds for all $t \geq 0$ (or for all $t \geq t_0 > 0$) where $A$ is given by (1).

## Orlicz Spaces

**8.7** Let $\Omega$ be a domain in $\mathbb{R}^n$ and let $A$ be an $N$-function. The *Orlicz class* $K_A(\Omega)$ is the set of all (equivalence classes modulo equality a.e. in $\Omega$ of) measurable functions $u$ defined on $\Omega$ and satisfying

$$\int_\Omega A(|u(x)|)\,dx < \infty.$$

Since $A$ is convex $K_A(\Omega)$ is always a convex set of functions but it may not be a vector space; for instance there may exist $u \in K_A(\Omega)$ and $\lambda > 0$ such that $\lambda u \notin K_A(\Omega)$.

We call the pair $(A, \Omega)$ $\Delta$-*regular* if either

(a) $A$ satisfies a global $\Delta_2$-condition, or
(b) $A$ satisfies a $\Delta_2$-condition near infinity and $\Omega$ has finite volume.

**8.8 LEMMA** $K_A(\Omega)$ is a vector space (under pointwise addition and scalar multiplication) if and only if $(A, \Omega)$ is $\Delta$-regular.

PROOF Since $A$ is convex we have:

(i) $\lambda u \in K_A(\Omega)$ provided $u \in K_A(\Omega)$ and $|\lambda| \leq 1$, and
(ii) if $u \in K_A(\Omega)$ implies $\lambda u \in K_A(\Omega)$ for every complex $\lambda$, then $u, v \in K_A(\Omega)$ implies $u + v \in K_A(\Omega)$.

It follows that $K_A(\Omega)$ is a vector space if and only if $u \in K_A(\Omega)$ and $|\lambda| > 1$ implies $\lambda u \in K_A(\Omega)$.

If $A$ satisfies a global $\Delta_2$-condition and $|\lambda| > 1$, then we have by (12) for

$u \in K_A(\Omega)$

$$\int_\Omega A(|\lambda u(x)|)\,dx \le k(|\lambda|)\int_\Omega A(|u(x)|) < \infty.$$

Similarly, if $A$ satisfies a $\Delta_2$-condition near infinity and $\operatorname{vol}\Omega < \infty$, we have for $|\lambda| > 1$, $u \in K_A(\Omega)$, and some $t_0 > 0$

$$\int_\Omega A(|\lambda u(x)|)\,dx = \left(\int_{\{x:|u(x)|\ge t_0\}} + \int_{\{x:|u(x)|<t_0\}}\right)(|\lambda u(x)|)\,dx$$

$$\le k(|\lambda|)\int_\Omega A(|u(x)|)\,dx + A(|\lambda|\,t_0)\operatorname{vol}\Omega < \infty.$$

In either case $K_A(\Omega)$ is seen to be a vector space.

Now suppose that $(A, \Omega)$ is not $\Delta$-regular and, if $\operatorname{vol}\Omega < \infty$, that $t_0 > 0$ is given. There exists a sequence $\{t_j\}$ of positive numbers such that

(i) $A(2t_j) \ge 2^j A(t_j)$, and
(ii) $t_j \ge t_0 > 0$ if $\operatorname{vol}\Omega < \infty$.

Let $\{\Omega_j\}$ be a sequence of mutually disjoint, measurable subsets of $\Omega$ such that

$$\operatorname{vol}\Omega_j = \begin{cases} 1/2^j A(t_j) & \text{if } \operatorname{vol}\Omega = \infty \\ A(t_0)\operatorname{vol}\Omega/2^j A(t_j) & \text{if } \operatorname{vol}\Omega < \infty. \end{cases}$$

Let

$$u(x) = \begin{cases} t_j & \text{if } x \in \Omega_j \\ 0 & \text{if } x \in \Omega \sim \left(\bigcup_{j=1}^{\infty}\Omega_j\right). \end{cases}$$

Then

$$\int_\Omega A(|u(x)|)\,dx = \sum_{j=1}^{\infty} A(t_j)\operatorname{vol}\Omega_j$$

$$= \begin{cases} 1 & \text{if } \operatorname{vol}\Omega = \infty \\ A(t_0)\operatorname{vol}\Omega & \text{if } \operatorname{vol}\Omega < \infty. \end{cases}$$

But

$$\int_\Omega A(|2u(x)|) \ge \sum_{j=1}^{\infty} 2^j A(t_j)\operatorname{vol}\Omega_j = \infty.$$

Thus $K_A(\Omega)$ is not a vector space. ∎

**8.9** The *Orlicz space* $L_A(\Omega)$ is defined to be the linear hull of the Orlicz class $K_A(\Omega)$, that is, the smallest vector space (under pointwise addition and scalar multiplication) containing $K_A(\Omega)$. Evidently $L_A(\Omega)$ consists of all scalar multiples $\lambda u$ of elements $u \in K_A(\Omega)$. Thus $K_A(\Omega) \subset L_A(\Omega)$; these sets being equal if and only if $(A, \Omega)$ is $\Delta$-regular.

The reader may verify that the functional

$$\|u\|_A = \|u\|_{A,\Omega} = \inf\left\{k > 0 : \int_\Omega A\left(\frac{|u(x)|}{k}\right) dx \le 1\right\} \tag{13}$$

is a norm on $L_A(\Omega)$. (This norm is due to Luxemburg [42].) For $\|u\|_A > 0$ the infimum in (13) is attained; in fact, letting $k$ decrease toward $\|u\|_A$ in the inequality

$$\int_\Omega A\left(\frac{|u(x)|}{k}\right) dx \le 1, \tag{14}$$

we obtain by monotone convergence

$$\int_\Omega A\left(\frac{|u(x)|}{\|u\|_A}\right) dx \le 1. \tag{15}$$

Equality may fail to hold in (15) but if equality holds in (14), then $k = \|u\|_A$.

**8.10 THEOREM** $L_A(\Omega)$ is a Banach space with respect to the norm (13).

The completeness proof is quite similar to that for the space $L^p(\Omega)$ given in Theorem 2.10. The details are left to the reader. [We remark that if $1 < p < \infty$ and $A_p$ is given by (10), then

$$L^p(\Omega) = L_{A_p}(\Omega) = K_{A_p}(\Omega).$$

Moreover, $\|u\|_{A_p,\Omega} = p^{-1/p}\|u\|_{p,\Omega}$.]

**8.11** If $A$ and $\tilde{A}$ are complementary $N$-functions, a generalized version of Hölder's inequality

$$\left|\int_\Omega u(x)\,v(x)\,dx\right| \le 2\,\|u\|_{A,\Omega}\,\|v\|_{\tilde{A},\Omega} \tag{16}$$

can be obtained by applying Young's inequality (5) to $|u(x)|/\|u\|_A$ and $|v(x)|/\|v\|_{\tilde{A}}$ and integrating over $\Omega$.

The following elementary imbedding theorem is an analog for Orlicz spaces of Lemma 2.8.

**8.12 THEOREM** The imbedding $L_B(\Omega) \to L_A(\Omega)$ holds if and only if either

(a) $B$ dominates $A$ globally, or
(b) $B$ dominates $A$ near infinity and $\operatorname{vol}\Omega < \infty$.

PROOF If $A(t) \le B(kt)$ for all $t \ge 0$, and if $u \in L_B(\Omega)$, then

$$\int_\Omega A\left(\frac{|u(x)|}{k\,\|u\|_B}\right) dx \le \int_\Omega B\left(\frac{|u(x)|}{\|u\|_B}\right) dx \le 1.$$

Thus $u \in L_A(\Omega)$ and $\|u\|_A \le k\,\|u\|_B$.

If $\operatorname{vol}\Omega < \infty$, let $t_1 = A^{-1}((2\operatorname{vol}\Omega)^{-1})$. If $B$ dominates $A$ near infinity, then there exist positive $t_0$ and $k$ such that $A(t) \le B(kt)$ for $t \ge t_0$. Evidently we have for $t \ge t_1$

$$A(t) \le K_1 B(kt),$$

where $K_1 = \max(1, A(t_0)/B(kt_1))$. If $u \in L_B(\Omega)$ is given, let $\Omega'(u) = \{x \in \Omega : |u(x)/2K_1 k\,\|u\|_B < t_1\}$ and $\Omega''(u) = \Omega \sim \Omega'(u)$. Then

$$\begin{aligned}\int_\Omega A\left(\frac{|u(x)|}{2K_1 k\,\|u\|_B}\right) dx &= \left(\int_{\Omega'(u)} + \int_{\Omega''(u)}\right) A\left(\frac{|u(x)|}{2K_1 k\,\|u\|_B}\right) dx \\ &\le \frac{1}{2\operatorname{vol}\Omega}\int_{\Omega'(u)} dx + K_1 \int_{\Omega''(u)} B\left(\frac{|u(x)|}{2K_1\|u\|_B}\right) dx \\ &\le \frac{1}{2} + \frac{1}{2}\int_\Omega B\left(\frac{|u(x)|}{\|u\|_B}\right) \le 1.\end{aligned}$$

Thus $u \in L_A(\Omega)$ and $\|u\|_A \le 2kK_1\,\|u\|_B$.

Conversely, suppose neither hypothesis (a) nor (b) holds. Then there exist points $t_j > 0$ such that

$$A(t_j) \ge B(jt_j), \qquad j = 1, 2, \ldots.$$

If $\operatorname{vol}\Omega < \infty$, we may assume in addition that

$$t_j \ge (1/j)\,B^{-1}(1/\operatorname{vol}\Omega).$$

Let $\Omega_j$ be a subdomain of $\Omega$ having volume $1/B(jt_j)$, and let

$$u_j(x) = \begin{cases} jt_j & \text{if } x \in \Omega_j \\ 0 & \text{if } x \in \Omega \sim \Omega_j. \end{cases}$$

Then

$$\int_\Omega A(|u_j(x)|/j)\,dx \ge \int_\Omega B(|u_j(x)|)\,dx = 1$$

so that $\|u_j\|_B = 1$ but $\|u_j\|_A \ge j$. Thus $L_B(\Omega)$ is not imbedded in $L_A(\Omega)$. ▮

**8.13** A sequence $\{u_j\}$ of functions in $L_A(\Omega)$ is said to *converge in mean* to $u \in L_A(\Omega)$ if

$$\lim_{j\to\infty} \int_\Omega A(|u_j(x) - u(x)|)\,dx = 0.$$

Convexity of $A$ implies that for $0 < \varepsilon \le 1$ we have

$$\int_\Omega A(|u_j(x)-u(x)|)\,dx \le \varepsilon \int_\Omega A(|u_j(x)-u(x)|/\varepsilon)\,dx$$

from which it is clear that norm convergence in $L_A(\Omega)$ implies mean convergence. The converse holds, that is, mean convergence implies norm convergence if and only if the pair $(A, \Omega)$ is $\Delta$-regular. The proof is similar to that of Lemma 8.8 and is left to the reader.

**8.14** Let $E_A(\Omega)$ denote the closure in $L_A(\Omega)$ of the space of functions $u$ which are bounded on $\Omega$ and have bounded support in $\overline{\Omega}$. If $u \in K_A(\Omega)$, the sequence $\{u_j\}$ given by

$$u_j(x) = \begin{cases} u(x) & \text{if } |u(x)| \le j \text{ and } |x| \le j, \quad x \in \Omega \\ 0 & \text{otherwise} \end{cases} \tag{17}$$

converges a.e. on $\Omega$ to $u$. Since $A(|u(x)-u_j(x)|) \le A(|u(x)|)$, we have by dominated convergence that $u_j$ converges to $u$ in mean in $L_A(\Omega)$. Therefore if $(A, \Omega)$ is $\Delta$-regular, then $E_A(\Omega) = K_A(\Omega) = L_A(\Omega)$. If $(A, \Omega)$ is not $\Delta$-regular, we have

$$E_A(\Omega) \subset K_A(\Omega) \subsetneqq L_A(\Omega) \tag{18}$$

so that $E_A(\Omega)$ is a proper closed subspace of $L_A(\Omega)$ in this case. To verify the first inclusion of (18) let $u \in E_A(\Omega)$ be given. Let $v$ be a bounded function with bounded support such that $\|u-v\|_A < \frac{1}{2}$. Using the convexity of $A$ and (15), we obtain

$$\frac{1}{\|2u-2v\|_A}\int_\Omega A(|2u(x)-2v(x)|)\,dx \le \int_\Omega A\left(\frac{|2u(x)-2v(x)|}{\|2u-2v\|_A}\right)dx \le 1,$$

whence $2u-2v \in K_A(\Omega)$. Since $2v$ clearly belongs to $K_A(\Omega)$ and $K_A(\Omega)$ is convex we have $u = \frac{1}{2}(2u-2v)+\frac{1}{2}(2v)$ belongs to $K_A(\Omega)$.

**8.15 LEMMA** $E_A(\Omega)$ is the maximal linear subspace of $K_A(\Omega)$.

PROOF Let $S$ be a linear subspace of $K_A(\Omega)$ and let $u \in S$. Then $\lambda u \in K_A(\Omega)$ for every scalar $\lambda$. If $\varepsilon > 0$ and $u_j$ is given by (17), then $u_j/\varepsilon$ converges to $u/\varepsilon$ in mean in $L_A(\Omega)$ as noted in Section 8.14. Hence for sufficiently large values of $j$

$$\int_\Omega A\,(|u_j(x)-u(x)|/\varepsilon)\,dx \le 1$$

and therefore $u_j$ converges to $u$ in $L_A(\Omega)$. Thus $S \subset E_A(\Omega)$. ∎

**8.16 THEOREM** Let $\Omega$ have finite volume and suppose that the $N$-function $A$ increases essentially more slowly than $B$ near infinity. Then

$$L_B(\Omega) \to E_A(\Omega).$$

PROOF Since $L_B(\Omega) \to L_A(\Omega)$ is already established we need only show that $L_B(\Omega) \subset E_A(\Omega)$. Since $L_B(\Omega)$ is the linear hull of $K_B(\Omega)$ and $E_A(\Omega)$ is the maximal linear subspace of $K_A(\Omega)$, it is sufficient to show that $\lambda u \in K_A(\Omega)$ whenever $u \in K_B(\Omega)$ and $\lambda$ is a scalar. But there exists a positive number $t_0$ such that $A(|\lambda| t) \le B(t)$ for all $t \ge t_0$. Thus

$$\begin{aligned}\int_\Omega A(|\lambda u(x)|)\,dx &= \left\{\int_{\{x:|u(x)|\le t_0\}} + \int_{\{x:|u(x)|>t_0\}}\right\} A(|\lambda u(x)|)\,dx \\ &\le A(|\lambda| t_0)\operatorname{vol}\Omega + \int_\Omega B(|u(x)|)\,dx < \infty\end{aligned}$$

whence the theorem follows. ∎

## Duality in Orlicz Spaces

**8.17 LEMMA** For fixed $v \in L_{\tilde{A}}(\Omega)$ the linear functional $L_v$ defined by

$$L_v(u) = \int_\Omega u(x)\,v(x)\,dx \tag{19}$$

belongs to $[L_A(\Omega)]'$. Denoting by $\|L_v\|$ its norm in that space, we have

$$\|v\|_{\tilde{A}} \le \|L_v\| \le 2\,\|v\|_{\tilde{A}}. \tag{20}$$

PROOF It follows by Hölder's inequality (16) that if $u \in L_A(\Omega)$, then

$$|L_v(u)| \le 2\,\|u\|_A\,\|v\|_{\tilde{A}}.$$

Thus $L_v$ is bounded on $L_A(\Omega)$ and the second inequality of (20) holds.

To establish the first inequality we may assume that $v \ne 0$ in $L_{\tilde{A}}(\Omega)$ so that $\|L_v\| = K > 0$. Let

$$u(x) = \begin{cases} \tilde{A}\left(\dfrac{|v(x)|}{K}\right)\Big/\dfrac{v(x)}{K} & \text{if } v(x) \ne 0 \\ 0 & \text{if } v(x) = 0. \end{cases}$$

If $\|u\|_A > 1$, then for sufficiently small $\varepsilon > 0$ we have

$$\frac{1}{\|u\|_A - \varepsilon}\int_\Omega A(|u(x)|)\,dx \ge \int_\Omega A\left(\frac{|u(x)|}{\|u\|_A - \varepsilon}\right) > 1.$$

Letting $\varepsilon \to 0+$ we obtain, using inequality (6),

$$\|u\|_A \le \int_\Omega A(|u(x)|)\, dx = \int_\Omega A\left(\tilde{A}\left(\frac{|v(x)|}{K}\right)\bigg/\frac{|v(x)|}{K}\right) dx$$

$$< \int_\Omega \tilde{A}\left(\frac{v(x)}{K}\right) dx = \frac{1}{\|L_v\|}\int_\Omega u(x)\, v(x)\, dx \le \|u\|_A.$$

This contradiction shows that $\|u\|_A \le 1$. Now

$$\|L_v\| = \sup_{\|u\|_A \le 1} |L_v(u)| \ge \|L_v\| \left|\int_\Omega \tilde{A}\left(\frac{|v(x)|}{\|L_v\|}\right) dx\right|$$

so that

$$\int_\Omega \tilde{A}\left(\frac{|v(x)|}{\|L_v\|}\right) dx \le 1. \tag{21}$$

Thus $\|v\|_{\tilde{A}} \le \|L_v\|$. ∎

We remark that the lemma holds also when $L_v$ is restricted to act on $E_A(\Omega)$. To obtain the first inequality of (20) in this case take $\|L_v\|$ to be the norm of $L_v$ in $[E_A(\Omega)]'$ and replace $u$ in the above proof by $\chi_n u$ where $\chi_n$ is the characteristic function of $\Omega_n = \{x \in \Omega : |x| \le n \text{ and } |u(x)| \le n\}$. Evidently $\chi_n u$ belongs to $E_A(\Omega)$, $\|\chi_n u\|_A \le 1$, and (21) becomes

$$\int_\Omega \chi_n(x)\, \tilde{A}\left(\frac{|v(x)|}{\|L_v\|}\right) dx \le 1.$$

Since $\chi_n(x)$ increases to unity a.e. in $\Omega$ we obtain by monotone convergence

$$\int_\Omega \tilde{A}\left(\frac{|v(x)|}{\|L_v\|}\right) dx \le 1$$

so that $\|v\|_{\tilde{A}} \le \|L_v\|$ as before.

**8.18 THEOREM** The dual space $[E_A(\Omega)]'$ of $E_A(\Omega)$ is isomorphic and homeomorphic to $L_{\tilde{A}}(\Omega)$.

PROOF We have already shown that any element $v \in L_{\tilde{A}}(\Omega)$ determines via (19) a bounded linear functional on $L_A(\Omega)$, and so also on $E_A(\Omega)$, having in either case norm differing from $\|v\|_{\tilde{A}}$ by at most a factor of 2. It remains to be shown that every bounded linear functional on $E_A(\Omega)$ is of the form $L_v$ for some such $v$.

Let $L \in [E_A(\Omega)]'$ be given. We define a complex measure $\lambda$ on the measurable subsets of $\Omega$ having finite volume by setting

$$\lambda(S) = L(\chi_S),$$

$\chi_S$ being the characteristic function of $S$. Since

$$\int_\Omega A(|\chi_S(x)| A^{-1}(1/\operatorname{vol} S))\, dx = \int_S (1/\operatorname{vol} S)\, dx = 1 \tag{22}$$

we have

$$|\lambda(S)| \le \|L\| \|\chi_S\|_A = \|L\| [1/A^{-1}(1/\operatorname{vol} S)].$$

Since the right side tends to zero with vol $S$, the measure $\lambda$ is absolutely continuous with respect to Lebesgue measure and by the Radon–Nikodym theorem 1.47, $\lambda$ may be expressed in the form

$$\lambda(S) = \int_S v(x)\, dx,$$

where $v$ is integrable on $\Omega$. Thus

$$L(u) = \int_\Omega u(x) v(x)\, dx$$

holds for measurable, simple functions $u$.

If $u \in E_A(\Omega)$ a sequence of measurable, simple functions $u_j$ converging a.e. to $u$ can be found such that $|u_j(x)| \le |u(x)|$ on $\Omega$. Since $|u_j(x) v(x)|$ converges a.e. to $|u(x) v(x)|$, Fatou's Lemma 1.44 yields

$$\begin{aligned}\left|\int_\Omega u(x) v(x)\, dx\right| &\le \sup_j \int_\Omega |u_j(x) v(x)|\, dx = \sup_j |L(|u_j| \operatorname{sgn} v)| \\ &\le \|L\| \sup_j \|u_j\|_A \le \|L\| \|u\|_A.\end{aligned}$$

It follows that the linear functional

$$L_v(u) = \int_\Omega u(x) v(x)\, dx$$

is bounded on $E_A(\Omega)$ whence $v \in L_{\tilde{A}}(\Omega)$ by the remark following Lemma 8.17. Since $L_v$ and $L$ assume the same values on the measurable simple functions, a set which [see the proof of Theorem 8.20(a)] is dense in $E_A(\Omega)$, they agree on $E_A(\Omega)$ and the theorem is proved. ∎

A simple application of the Hahn–Banach extension theorem shows that if $E_A(\Omega)$ is a proper subspace of $L_A(\Omega)$ [that is, if $(A, \Omega)$ is not $\Delta$-regular], then there exists a bounded linear functional $L$ on $L_A(\Omega)$ not given by (19) for any $v \in L_{\tilde{A}}(\Omega)$. We have as an immediate consequence the following theorem.

**8.19 THEOREM** $L_A(\Omega)$ is reflexive if and only if both $(A, \Omega)$ and $(\tilde{A}, \Omega)$ are $\Delta$-regular.

We omit any discussion of uniform convexity for Orlicz spaces. This subject is treated in Luxemburg's thesis [42].

## Separability and Compactness Theorems

We next generalize the approximation Theorems 2.13, 2.15, and 2.19.

**8.20 THEOREM** (a) $C_0(\Omega)$ is dense in $E_A(\Omega)$.

(b) $E_A(\Omega)$ is separable.

(c) If $J_\varepsilon$ is the mollifier introduced in Section 2.17, then for each $u \in E_A(\Omega)$ we have $\lim_{\varepsilon \to 0+} J_\varepsilon * u = u$ in $E_A(\Omega)$.

(d) $C_0^\infty(\Omega)$ is dense in $E_A(\Omega)$.

PROOF Part (a) is proved by the same method used in Theorem 2.13. In approximating $u \in E_A(\Omega)$ first by simple functions we may assume that $u$ is bounded on $\Omega$ and has bounded support. This is required for the dominated convergence argument used to show that the simple functions converge in norm to $u$ in $E_A(\Omega)$. (The details are left to the reader.)

Part (b) follows from part (a) by the same proof as given for Theorem 2.15.

Consider (c). If $u \in E_A(\Omega)$, let $u$ be extended to $\mathbb{R}^n$ so as to vanish identically outside $\Omega$. Let $v \in L_{\tilde{A}}(\Omega)$. Then

$$\left| \int_\Omega (J_\varepsilon * u(x) - u(x)) v(x)\, dx \right| \le \int_{\mathbb{R}^n} J(y)\, dy \int_\Omega |u(x-\varepsilon y) - u(x)| |v(x)|\, dx$$

$$\le 2 \|v\|_{\tilde{A},\Omega} \int_{|y| \le 1} J(y) \|u_{\varepsilon y} - u\|_{A,\Omega}\, dy$$

by Hölder's inequality (16), where $u_{\varepsilon y}(x) = u(x - \varepsilon y)$. Thus by (20) and Theorem 8.18,

$$\|J_\varepsilon * u - u\|_{A,\Omega} = \sup_{\|v\|_{\tilde{A},\Omega} \le 1} \left| \int_\Omega (J_\varepsilon * u(x) - u(x)) v(x)\, dx \right|$$

$$\le 2 \int_{|y| \le 1} J(y) \|u_{\varepsilon y} - u\|_{A,\Omega}\, dy.$$

Given $\delta > 0$ we can find $\tilde{u} \in C_0(\Omega)$ such that $\|u - \tilde{u}\|_{A,\Omega} < \delta/6$. Evidently $\|u_{\varepsilon y} - \tilde{u}_{\varepsilon y}\|_{A,\Omega} < \delta/6$ and for sufficiently small $\varepsilon$, $\|\tilde{u}_{\varepsilon y} - \tilde{u}\|_{A,\Omega} < \delta/6$ for every $y$ with $|y| \le 1$. Thus $\|J_\varepsilon * u - u\|_{A,\Omega} < \delta$ and (c) is established.

Part (d) is an immediate consequence of (a) and (c). ∎

We remark that $L_A(\Omega)$ is not separable unless $L_A(\Omega) = E_A(\Omega)$, that is, unless $(A, \Omega)$ is $\Delta$-regular. A proof of this fact may be found in the work of Krasnosel'skii and Rutickii [34, Chapter II, Theorem 10.2].

**8.21** A sequence $u_j$ of measurable functions is said to *converge in measure* on $\Omega$ to the function $u$ provided that for each $\varepsilon > 0$ and $\delta > 0$ there exists an integer $M$ such that if $j > M$, then

$$\operatorname{vol}\{x \in \Omega : |u_j(x) - u(x)| > \varepsilon\} \le \delta.$$

Clearly, in this case there also exists an integer $N$ such that if $j, k \ge N$, then

$$\operatorname{vol}\{x \in \Omega : |u_j(x) - u_k(x)| \ge \varepsilon\} \le \delta.$$

**8.22 THEOREM** Let $\Omega$ have finite volume and suppose that the $N$-function $B$ increases essentially more slowly than $A$ near infinity. If the sequence $\{u_j\}$ is bounded in $L_A(\Omega)$ and convergent in measure on $\Omega$, then it is convergent in $L_B(\Omega)$.

PROOF Fix $\varepsilon > 0$ and let $v_{j,k}(x) = [u_j(x) - u_k(x)]/\varepsilon$. Clearly $\{v_{j,k}\}$ is bounded in $L_A(\Omega)$; say $\|v_{j,k}\|_{A,\Omega} \le K$. Now there exists a positive number $t_0$ such that if $t > t_0$, then

$$B(t) \le \tfrac{1}{4} A(t/K).$$

Let $\delta = 1/4B(t_0)$ and set

$$\Omega_{j,k} = \{x \in \Omega : |v_{j,k}(x)| \ge B^{-1}(1/2 \operatorname{vol} \Omega)\}.$$

Since $\{u_j\}$ converges in measure there exists an integer $N$ such that if $j, k \ge N$, then $\operatorname{vol} \Omega_{j,k} \le \delta$. Set

$$\Omega'_{j,k} = \{x \in \Omega_{j,k} : |v_{j,k}(x)| \ge t_0\}, \qquad \Omega''_{j,k} = \Omega_{j,k} \sim \Omega'_{j,k}.$$

For $j, k \ge M$ we have

$$\int_\Omega B(|v_{j,k}(x)|)\, dx = \left(\int_{\Omega \sim \Omega_{j,k}} + \int_{\Omega'_{j,k}} + \int_{\Omega''_{j,k}}\right) B(|v_{j,k}(x)|)\, dx$$

$$\le \frac{\operatorname{vol} \Omega}{2 \operatorname{vol} \Omega} + \frac{1}{4} \int_{\Omega'_{j,k}} A\left(\frac{|v_{j,k}(x)|}{K}\right) dx + \delta B(t_0) \le 1.$$

Hence $\|u_j - u_k\|_{B,\Omega} \le \varepsilon$ and so $\{u_j\}$ converges in $L_B(\Omega)$. ∎

The following theorem will prove useful when we wish to extend the Rellich–Kondrachov Theorem 6.2 to imbeddings of Orlicz–Sobolev spaces.

**8.23 THEOREM** Let $\Omega$ have finite volume and suppose that the $N$-function $B$ increases essentially more slowly than $A$ near infinity. Then any bounded subset $S$ of $L_A(\Omega)$ which is precompact in $L^1(\Omega)$ is also precompact in $L_B(\Omega)$.

PROOF Evidently $L_A(\Omega) \to L^1(\Omega)$ since $\Omega$ has finite volume. If $\{u_j^*\}$ is a sequence in $S$, it has a subsequence $\{u_j\}$ convergent in $L^1(\Omega)$; say $u_j \to u$ in $L^1(\Omega)$. Let $\varepsilon, \delta > 0$. Then there exists an integer $N$ such that if $j \geq N$, then $\|u_j - u\|_{1,\Omega} \leq \varepsilon\delta$. It follows that $\operatorname{vol}\{x \in \Omega : |u_j(x) - u(x)| \geq \varepsilon\} \leq \delta$ so $\{u_j\}$ converges in measure on $\Omega$ and hence also in $L_B(\Omega)$. ▮

## A Limiting Case of the Sobolev Imbedding Theorem

**8.24** If $mp = n$ and $p > 1$, the Sobolev imbedding Theorem 5.4 provides no best (i.e., smallest) target space into which $W^{m,p}(\Omega)$ can be imbedded. In fact we have in this case, for suitably regular $\Omega$,

$$W^{m,p}(\Omega) \to L^q(\Omega), \qquad p \leq q < \infty,$$

but (see Example 5.26)

$$W^{m,p}(\Omega) \not\to L^\infty(\Omega).$$

If the class of target spaces for the imbedding is enlarged to include Orlicz spaces, then a best target space can be found. We consider first bounded domains $\Omega$. The case $m = 1$ of the following theorem was established by Trudinger [66].

**8.25 THEOREM** Let $\Omega$ be a bounded domain in $\mathbb{R}^n$ having the cone property. Let $mp = n$ and $p > 1$. Set

$$A(t) = \exp[t^{n/(n-m)}] - 1 = \exp[t^{p/(p-1)}] - 1. \tag{23}$$

Then there exists the imbedding

$$W^{m,p}(\Omega) \to L_A(\Omega).$$

PROOF Let $x \in \Omega$ and let $C$ be a finite cone contained in $\Omega$ having vertex at $x$. Let $u \in C^m(\overline{C})$. Applying Taylor's formula

$$f(1) = \sum_{j=0}^{m-1} \frac{f^{(j)}(0)}{j!} + \frac{1}{(m-1)!}\int_0^1 (1-t)^{m-1} f^{(m)}(t)\, dt$$

to the function $f(t) = u(y + t(x-y))$, and noting that

$$f^{(j)}(t) = \sum_{|\alpha| = j} \frac{j!}{\alpha!} D^\alpha u(y + t(x-y))(x-y)^\alpha,$$

we obtain

$$|u(x)| \le \sum_{|\alpha| \le m-1} \frac{1}{\alpha!} |D^\alpha u(y)| |x-y|^{|\alpha|}$$

$$+ \sum_{|\alpha| = m} \frac{m}{\alpha!} |x-y|^m \int_0^1 (1-t)^{m-1} |D^\alpha u(y + t(x-y))| \, dt.$$

Let $V$ be the volume and $h$ the height of $C$. Let $(\rho, \theta)$ denote spherical polar coordinates of $y \in C$ referred to $x$ as origin so that $C$ is specified by $0 < \rho < h$, $\theta \in \Sigma$, and the volume element $dy$ can be written in the form $\rho^{n-1}\omega(\theta) \, d\rho \, d\theta$. Then

$$|u(x)| = \frac{1}{V} \int_C |u(x)| \, dy \le \frac{1}{V} \sum_{|\alpha| \le m-1} \frac{h^{|\alpha|}}{\alpha!} \int_C |D^\alpha u(y)| \, dy$$

$$+ \frac{1}{V} \sum_{|\alpha| = m} \frac{m}{\alpha!} \int_\Sigma \omega(\theta) \, d\theta \int_0^h \rho^{n+m-1} \, d\rho \int_0^1 (1-t)^{m-1} |D^\alpha u((1-t)\rho, \theta)| \, dt$$

$$\le K_1 \left\{ \|u\|_{m-1,1,C} + \sum_{|\alpha| = m} \int_\Sigma \omega(\theta) \, d\theta \int_0^h \rho^{n-1} \, d\rho \int_0^\rho \sigma^{m-1} |D^\alpha u(\sigma, \theta)| \, d\sigma \right\}$$

$$= K_1 \left\{ \|u\|_{m-1,1,C} + \sum_{|\alpha| = m} \int_\Sigma \omega(\theta) \, d\theta \int_0^h \sigma^{m-1} |D^\alpha u(\sigma, \theta)| \, d\sigma \int_0^h \rho^{n-1} \, d\rho \right\}$$

$$\le K_2 \left\{ \|u\|_{m-1,1,C} + \sum_{|\alpha| = m} \int_C \frac{|D^\alpha u(z)|}{|z-x|^{n-m}} \, dz \right\}.$$

By density the above inequality holds for all $u \in W^{m,1}(C)$. In particular, for any $u \in W^{m,p}(\Omega)$, and for almost all $x \in \Omega$, we have

$$|u(x)| \le K_2 \left\{ \|u\|_{m-1,1,\Omega} + \sum_{|\alpha| = m} \int_\Omega \frac{|D^\alpha u(y)|}{|x-y|^{n-m}} \, dy \right\},$$

where $K_2$ depends on $m, n$ and the height $h$ and volume $V$ of the cone determining the cone property for $\Omega$.

We wish to estimate $\|u\|_{0,s}$ for arbitrary $s > 1$. Accordingly, if $v \in L^{s'}(\Omega)$ where $s' = s/(s-1)$, then

$$\int_\Omega |u(x) v(x)| \, dx$$

$$\le K_2 \|u\|_{m-1,1} \int_\Omega |v(x)| \, dx + K_2 \sum_{|\alpha| = m} \int_\Omega \int_\Omega \frac{|D^\alpha u(y)| |v(x)|}{|x-y|^{n-m}} \, dy \, dx$$

$$\le K_2 \|u\|_{m-1,1} \|v\|_{0,s'} (\mathrm{vol}\,\Omega)^{1/s}$$

*equation continues*

$$+ K_2 \sum_{|\alpha|=m} \left\{ \int_\Omega \int_\Omega \frac{|v(x)|}{|x-y|^{n-(m/s)}}\, dy\, dx \right\}^{1-(1/p)}$$

$$\times \left\{ \int_\Omega \int_\Omega \frac{|D^\alpha u(y)|^p \, |v(x)|}{|x-y|^{(n-m)/s}}\, dy\, dx \right\}^{1/p}.$$

Now we have from Lemma 5.47 that if $0 \le \nu < n$, then

$$\int_\Omega \frac{1}{|x-y|^\nu}\, dx \le K_3(\nu, n)(\mathrm{vol}\,\Omega)^{1-(\nu/n)}.$$

In fact a review of the proof of that lemma shows that $K_3(\nu, n) = K_4/(n-\nu)$ with $K_4$ depending only on $n$. Hence

$$\begin{aligned}\int_\Omega \int_\Omega \frac{|v(x)|}{|x-y|^{n-(m/s)}}\, dy\, dx &\le K_4 \frac{s}{m} (\mathrm{vol}\,\Omega)^{m/sn} \int_\Omega |v(x)|\, dx \\ &\le K_5 s (\mathrm{vol}\,\Omega)^{(1/sp)+(1/s)} \|v\|_{0,s'}.\end{aligned}$$

Also

$$\begin{aligned}\int_\Omega \int_\Omega \frac{|D^\alpha u(y)|^p \, |v(x)|}{|x-y|^{(n-m)/s}}\, dy\, dx &\le \int_\Omega |D^\alpha u(y)|^p\, dy \cdot \|v\|_{0,s'} \cdot \left\{ \int_\Omega \frac{1}{|x-y|^{n-m}}\, dx \right\}^{1/s} \\ &\le \|D^\alpha u\|_{0,p}^p \|v\|_{0,s'} (K_5 (\mathrm{vol}\,\Omega)^{1/p})^{1/s}.\end{aligned}$$

Hence

$$\begin{aligned}\int_\Omega |u(x)\, v(x)|\, dx &\le K_2 \|u\|_{m-1,1} \|v\|_{0,s'} (\mathrm{vol}\,\Omega)^{1/s} \\ &\quad + K_6 \sum_{|\alpha|=m} s^{(p-1)/p} \|D^\alpha u\|_{0,p} \|v\|_{0,s'} (\mathrm{vol}\,\Omega)^{1/s}.\end{aligned}$$

Since $s^{(p-1)/p} > 1$ and since $W^{m-1,1}(\Omega) \to W^{m,p}(\Omega)$ it follows that

$$\|u\|_{0,s} = \sup_{v \in L^{s'}(\Omega)} \frac{\int_\Omega |u(x)\, v(x)|\, dx}{\|v\|_{0,s'}} \le K_7 s^{(p-1)/p} (\mathrm{vol}\,\Omega)^{1/s} \|u\|_{m,p}.$$

The constant $K_7$ depends only on $m, n$ and the cone determining the cone property for $\Omega$. Setting $s = nk/(n-m) = pk/(p-1)$, we obtain

$$\begin{aligned}\int_\Omega |u(x)|^{pk/(p-1)}\, dx &\le \mathrm{vol}\,\Omega \cdot \left\{ \frac{pk}{p-1} \right\}^k \{K_7 \|u\|_{m,p}\}^{pk/(p-1)} \\ &= \mathrm{vol}\,\Omega \cdot \left\{ \frac{k}{e^{p/(p-1)}} \right\}^k \left\{ eK_7 \left( \frac{p}{p-1} \right)^{(p-1)/p} \|u\|_{m,p} \right\}^{pk/(p-1)}\end{aligned}$$

Since $e^{p/(p-1)} > e$, the series $\sum_{k=1}^{\infty}(1/k!)(k/e^{p/(p-1)})^k$ converges to a finite sum, say $K_8$. Let $K_9 = \max(1, K_8 \operatorname{vol}\Omega)$ and put

$$K_{10} = eK_9K_7[p/(p-1)]^{(p-1)/p}\|u\|_{m,p} = K_{11}\|u\|_{m,p}.$$

Then

$$\int_\Omega \left(\frac{|u(x)|}{K_{10}}\right)^{pk/(p-1)} dx \le \frac{\operatorname{vol}\Omega}{K_9^{pk/(p-1)}}\left(\frac{k}{e^{p/(p-1)}}\right)^k < \frac{\operatorname{vol}\Omega}{K_9}\left(\frac{k}{e^{p/(p-1)}}\right)^k$$

since $K_9 \ge 1$ and $pk/(p-1) > 1$. Expanding $A(t)$ in power series, we now obtain

$$\begin{aligned}\int_\Omega A\left(\frac{|u(x)|}{K_{10}}\right) dx &= \sum_{k=1}^{\infty}\frac{1}{k!}\int_\Omega\left(\frac{|u(x)|}{K_{10}}\right)^{pk/(p-1)} dx \\ &< \frac{\operatorname{vol}\Omega}{K_9}\sum_{k=1}^{\infty}\frac{1}{k!}\left(\frac{k}{e^{p/(p-1)}}\right)^k \le 1.\end{aligned}$$

Hence $u \in L_A(\Omega)$ and

$$\|u\|_A \le K_{10} = K_{11}\|u\|_{m,p},$$

where $K_{11}$ depends on $n, m, \operatorname{vol}\Omega$, and the cone determining the cone property for $\Omega$. ▮

The imbedding established in the above theorem is "best possible" in the sense that if there exists any imbedding of the form

$$W_0^{m,p}(\Omega) \to L_B(\Omega),$$

then $A$ dominates $B$ near infinity. A proof of this fact for the case $m = 1$, $n = p > 1$ can be found in the notes of Hempel and co-workers [30]. The general case is left to the reader as an exercise.

Theorem 8.25 can be generalized to fractional-order spaces. For results in this direction the reader is referred to Grisvard [28] and Peetre [56].

**8.26** If $\Omega$ is unbounded and so (having the cone property) has infinite volume, then the $N$-function $A$ given by (23) may not decrease rapidly enough at zero to allow membership in $L_A(\Omega)$ of every $u \in W^{m,p}(\Omega)$ (where $mp = n$). Let $k_0$ be the smallest integer such that $k_0 \ge p-1$ and define a modified $N$-function $A_0$ by

$$A_0(t) = \exp(t^{p/(p-1)}) - \sum_{j=0}^{k_0-1}(1/j!)t^{jp/(p-1)}.$$

Evidently $A_0$ is equivalent to $A$ near infinity so for any domain $\Omega$ having finite volume, $L_A(\Omega)$ and $L_{A_0}(\Omega)$ coincide and have equivalent norms. However, $A_0$ enjoys the further property that for $0 < r \leq 1$,

$$A_0(rt) \leq r^{k_0 p/(p-1)} A_0(t) \leq r^p A_0(t). \tag{24}$$

We show that if $mp = n$ and $\Omega$ has the cone property (but may be unbounded), then

$$W^{m,p}(\Omega) \to L_{A_0}(\Omega).$$

As in the proof of Lemma 5.14 we may write $\Omega$ as a union of countably many subdomains $\Omega_j$ each having the cone property determined by some fixed cone independent of $j$, satisfying for some constants $K_1$ and $K_2$

$$0 < K_1 \leq \operatorname{vol} \Omega_j \leq K_2,$$

and finally such that for some positive integer $R$ any $R+1$ of the subdomains $\Omega_j$ have empty intersection. It follows by Theorem 8.25 that if $u \in W^{m,p}(\Omega)$, then

$$\|u\|_{A_0,\Omega_j} \leq K_3 \|u\|_{m,p,\Omega_j},$$

where $K_3$ does not depend on $j$. Using (24) with $r = R^{1/p} \|u\|_{m,p,\Omega_j}^{-1} \|u\|_{m,p,\Omega}$ and the finite intersection property of the domains $\Omega_j$, we have

$$\begin{aligned} \int_\Omega A_0\left(\frac{|u(x)|}{R^{1/p} K_3 \|u\|_{m,p,\Omega}}\right) dx &\leq \sum_{j=1}^{\infty} \int_{\Omega_j} A_0\left(\frac{|u(x)|}{R^{1/p} K_3 \|u\|_{m,p,\Omega}}\right) dx \\ &\leq \sum_{j=1}^{\infty} \frac{\|u\|_{m,p,\Omega_j}^p}{R \|u\|_{m,p,\Omega}^p} \leq 1. \end{aligned}$$

Hence $\|u\|_{A_0,\Omega} \leq R^{1/p} K_3 \|u\|_{m,p,\Omega}$ as required.

We remark that if $k_0 > p-1$, the above result can be improved slightly by using in place of $A_0$ the $N$-function $\max(t^p, A_0(t))$.

## Orlicz–Sobolev Spaces

**8.27** For a given domain $\Omega$ in $\mathbb{R}^n$ and a given defining $N$-function $A$ the *Orlicz–Sobolev space* $W^m L_A(\Omega)$ consists of those (equivalence classes of) functions $u$ in $L_A(\Omega)$ whose distributional derivatives $D^\alpha u$ also belong to $L_A(\Omega)$ for all $\alpha$ with $|\alpha| \leq m$. The space $W^m E_A(\Omega)$ is defined in analogous fashion. It may be checked by the same method used in the proof of Theorem 3.2 that $W^m L_A(\Omega)$ is a Banach space with respect to the norm

$$\|u\|_{m,A} = \|u\|_{m,A,\Omega} = \max_{0 \leq |\alpha| \leq m} \|D^\alpha u\|_{A,\Omega}, \tag{25}$$

and that $W^m E_A(\Omega)$ is a closed subspace of $W^m L_A(\Omega)$ and hence also a Banach space under (25). It should be kept in mind that $W^m E_A(\Omega)$ coincides with $W^m L_A(\Omega)$ if and only if $(A, \Omega)$ is $\Delta$-regular. If $1 < p < \infty$ and $A_p(t) = t^p$, then $W^m L_{A_p}(\Omega) = W^m E_{A_p}(\Omega) = W^{m,p}(\Omega)$, the latter space having norm equivalent to that of the former.

As in the case of ordinary Sobolev spaces, $W_0^m L_A(\Omega)$ is taken to be the closure of $C_0^\infty(\Omega)$ in $W^m L_A(\Omega)$. [An analogous definition for $W_0^m E_A(\Omega)$ clearly leads to the same space in all cases.]

Many properties of Orlicz–Sobolev spaces are obtained by very straightforward generalization of the proofs of the same properties for ordinary Sobolev spaces. We summarize some of these in the following theorem and refer the reader to the corresponding results in Chapter III for method of proof. The details can also be found in the article by Donaldson and Trudinger [22].

**8.28 THEOREM** (a) $W^m E_A(\Omega)$ is separable (Theorem 3.5).

(b) If $(A, \Omega)$ and $(\tilde{A}, \Omega)$ are $\Delta$-regular, then $W^m E_A(\Omega) = W^m L_A(\Omega)$ is reflexive (Theorem 3.5).

(c) Each element $L$ of the dual space $[W^m E_A(\Omega)]'$ is given by

$$L(u) = \sum_{0 \le |\alpha| \le m} \int_\Omega D^\alpha u(x)\, v_\alpha(x)\, dx$$

for some functions $v_\alpha \in L_{\tilde{A}}(\Omega)$, $0 \le |\alpha| \le m$ (Theorem 3.8).

(d) $C^\infty(\Omega) \cap W^m E_A(\Omega)$ is dense in $W^m E_A(\Omega)$ (Theorem 3.16).

(e) If $\Omega$ has the segment property, then $C^\infty(\overline{\Omega})$ is dense in $W^m E_A(\Omega)$ (Theorem 3.18).

(f) $C_0^\infty(\mathbb{R}^n)$ is dense in $W^m E_A(\mathbb{R}^n)$. Thus $W_0^m L_A(\mathbb{R}^n) = W^m E_A(\mathbb{R}^n)$ (Theorem 3.18).

## Imbedding Theorems for Orlicz–Sobolev Spaces

**8.29** Imbedding results analogous to those obtained for the spaces $W^{m,p}(\Omega)$ in Chapters V and VI can also be formulated for the Orlicz–Sobolev spaces $W^m L_A(\Omega)$ and $W^m E_A(\Omega)$. The first results in this direction were obtained by Dankert [20] and by Donaldson. A fairly general imbedding theorem along the lines of Theorems 5.4 and 6.2 was presented by Donaldson and Trudinger [22] and we develop it below.

As was the case with ordinary Sobolev spaces, most of the imbedding results are obtained for domains having the cone property. Exceptions are those yielding (generalized) Hölder continuity estimates; these require the strong local Lipschitz property. Some results below are obtained only for bounded domains. The method used in extending the analogous results for ordinary Sobolev spaces to unbounded domains (see Lemma 5.14) does not appear to extend in a straightforawrd manner when general Orlicz spaces are involved. In this sense the imbedding picture is still incomplete.

**8.30** We concern ourselves for the time being with imbeddings of $W^1L_A(\Omega)$; the imbeddings of $W^mL_A(\Omega)$ are summarized in Theorem 8.40. In the following $\Omega$ is always understood to be a domain in $\mathbb{R}^n$.

Let $A$ be a given $N$-function. We shall always suppose that

$$\int_0^1 \frac{A^{-1}(t)}{t^{(n+1)/n}}\, dt < \infty, \tag{26}$$

replacing, if necessary, $A$ by another $N$-function equivalent to $A$ near infinity. [If $\Omega$ has finite volume, (26) places no restrictions on $A$ from the point of view of imbedding theory since $N$-functions equivalent near infinity determine identical Orlicz spaces.]

Suppose also that

$$\int_1^\infty \frac{A^{-1}(t)}{t^{(n+1)/n}}\, dt = \infty. \tag{27}$$

For instance, if $A = A_p$ given by (10), then (27) holds precisely when $p \leq n$. With (27) satisfied we define the *Sobolev conjugate* $A_*$ of $A$ by setting

$$A_*^{-1}(t) = \int_0^t \frac{A^{-1}(\tau)}{\tau^{(n+1)/n}}\, d\tau, \qquad t \geq 0. \tag{28}$$

It may readily be checked that $A_*$ is an $N$-function. If $1 < p < n$, we have, setting $q = np/(n-p)$,

$$A_{p*}(t) = q^{1-q}p^{-q/p}A_q(t).$$

It is also readily seen for the case $p = n$ that $A_{n*}(t)$ is equivalent near infinity to the $N$-function $e^t - t - 1$.

Before stating the first imbedding theorem we prepare the following technical lemma that will be needed in the proof.

**8.31 LEMMA** Let $u \in W_{\text{loc}}^{1,1}(\Omega)$ and let $f$ satisfy a Lipschitz condition on $\mathbb{R}$. Then $g \in W_{\text{loc}}^{1,1}(\Omega)$ where $g(x) = f(|u(x)|)$, and

$$D_j g(x) = f'(|u(x)|)\operatorname{sgn} u(x) \cdot D_j u(x).$$

PROOF Since $|u| \in W_{\text{loc}}^{1,1}(\Omega)$ and $D_j|u(x)| = \operatorname{sgn} u(x) \cdot D_j u(x)$ it is sufficient to establish the lemma for positive, real-valued functions $u$ so that $g(x) = f(u(x))$. Let $\phi \in \mathcal{D}(\Omega)$. Letting $\{e_j\}_{j=1}^n$ be the usual basis in $\mathbb{R}^n$, we obtain

$$-\int_{\Omega} f(u(x))\, D_j \phi(x)\, dx = -\lim_{h\to 0}\int_{\Omega} f(u(x))\frac{\phi(x)-\phi(x-he_j)}{h}\, dx$$
$$= \lim_{h\to 0}\int_{\Omega}\frac{f(u(x+he_j))-f(u(x))}{h}\phi(x)\, dx$$
$$= \lim_{h\to 0}\int_{\Omega} Q(x,h)\frac{u(x+he_j)-u(x)}{h}\phi(x)\, dx,$$

where, since $f$ is Lipschitz, for each $h$ the function $Q(\cdot, h)$ is defined a.e. on $\Omega$ by

$$Q(x,h) = \begin{cases} \dfrac{f(u(x+he_j))-f(u(x))}{u(x+he_j)-u(x)} & \text{if } \; u(x+he_j) \neq u(x) \\ f'(u(x)) & \text{otherwise.} \end{cases}$$

Moreover, $\|Q(\cdot, h)\|_{\infty,\Omega} \le K$ for some constant $K$ independent of $h$. A well-known theorem in functional analysis assures us that for some sequence of values of $h$ tending to zero, $Q(\cdot, h)$ converges to $f'(u(\cdot))$ in the weak-star topology of $L^\infty(\Omega)$. On the other hand, since $u \in W^{1,1}(\operatorname{supp}\phi)$ we have

$$\lim_{h\to 0}\frac{u(x+he_j)-u(x)}{h}\phi(x) = D_j u(x)\cdot\phi(x)$$

in $L^1(\operatorname{supp}\phi)$. It follows that

$$-\int_{\Omega} f(u(x))\, D_j\phi(x) = \int_{\Omega} f'(u(x))\, D_j(x)\,\phi(x)\, dx,$$

which evidently implies the lemma. ▌

**8.32 THEOREM** Let $\Omega$ be bounded and have the cone property in $\mathbb{R}^n$. If (26) and (27) hold, then

$$W^1L_A(\Omega) \to L_{A_*}(\Omega).$$

Moreover, if $B$ is any $N$-function increasing essentially more slowly than $A_*$ near infinity, then the imbedding

$$W^1L_A(\Omega) \to L_B(\Omega)$$

is compact.

PROOF The function $s = A_*(t)$ as defined by (28) satisfies the differential equation

$$A^{-1}(s)\frac{ds}{dt} = s^{(n+1)/n}, \tag{29}$$

and hence, by virtue of the left inequality of (7),

$$\frac{ds}{dt} \leqslant s^{1/n}\tilde{A}^{-1}(s).$$

Therefore $\sigma(t) = [A_*(t)]^{(n-1)/n}$ satisfies the differential inequality

$$\frac{d\sigma}{dt} \leqslant \frac{n-1}{n}\tilde{A}^{-1}((\sigma(t))^{n/(n-1)}). \tag{30}$$

Let $u \in W^1L_A(\Omega)$ and suppose for the moment that $u$ is bounded on $\Omega$ and is not zero in $L_A(\Omega)$. Then $\int_\Omega A_*(|u(x)|/\lambda)\,dx$ decreases continuously from infinity to zero as $\lambda$ increases from zero to infinity, and accordingly assumes the value unity for some positive value of $\lambda$. Thus

$$\int_\Omega A_*\left(\frac{|u(x)|}{K}\right)dx = 1, \qquad K = \|u\|_{A_*}. \tag{31}$$

Let $f(x) = \sigma(|u(x)|/K)$. Evidently $u \in W^{1,1}(\Omega)$ and $\sigma$ is Lipschitz on the range of $|u|/K$ so that, by Lemma 8.31, $f$ belongs to $W^{1,1}(\Omega)$. By Theorem 5.4 we have $W^{1,1}(\Omega) \to L^{n/(n-1)}(\Omega)$ and so

$$\begin{aligned}\|f\|_{0,n/(n-1)} &\leqslant K_1\left\{\sum_{j=1}^{n}\|D_jf\|_{0,1} + \|f\|_{0,1}\right\}\\ &= K_1\left\{\sum_{j=1}^{\infty}\frac{1}{K}\int_\Omega \sigma'\left(\frac{|u(x)|}{K}\right)|D_ju(x)|\,dx + \int_\Omega \sigma\left(\frac{|u(x)|}{K}\right)dx\right\}.\end{aligned} \tag{32}$$

By (31) and Hölder's inequality (16), we obtain

$$\begin{aligned}1 &= \left\{\int_\Omega A_*\left(\frac{|u(x)|}{K}\right)dx\right\}^{(n-1)/n} = \|f\|_{0,n/(n-1)}\\ &\leqslant \frac{2K_1}{K}\sum_{j=1}^{n}\left\|\sigma'\left(\frac{|u|}{K}\right)\right\|_{\tilde{A}}\|D_ju\|_A + K_1\int_\Omega \sigma\left(\frac{|u(x)|}{K}\right)dx.\end{aligned} \tag{33}$$

Making use of (30), we have

$$\begin{aligned}\left\|\sigma'\left(\frac{|u|}{K}\right)\right\|_{\tilde{A}} &\leqslant \frac{n-1}{n}\left\|\tilde{A}^{-1}\left(\left(\sigma\left(\frac{|u|}{K}\right)\right)^{n/(n-1)}\right)\right\|_{\tilde{A}}\\ &= \frac{n-1}{n}\inf\left\{\lambda > 0 : \int_\Omega \tilde{A}\left(\frac{\tilde{A}^{-1}(A_*(|u(x)|/K))}{\lambda}\right)dx \leqslant 1\right\}.\end{aligned}$$

Suppose $\lambda > 1$. Then

$$\int_\Omega \tilde{A}\left(\frac{\tilde{A}^{-1}(A_*(|u(x)|/K))}{\lambda}\right)dx \leqslant \frac{1}{\lambda}\int_\Omega A_*\left(\frac{|u(x)|}{K}\right)dx = \frac{1}{\lambda} < 1.$$

Thus

$$\left\| \sigma'\left(\frac{|u|}{K}\right) \right\|_{\tilde{A}} \leqslant \frac{n-1}{n}. \tag{34}$$

Let $g(t) = A_*(t)/t$ and $h(t) = \sigma(t)/t$. It is readily checked that $h$ is bounded on finite intervals and $\lim_{t\to\infty} g(t)/h(t) = \infty$. Thus there exists a constant $t_0$ such that if $t \geqslant t_0$, then $h(t) \leqslant g(t)/2K_1$. Putting $K_2 = K_1 \sup_{0\leqslant t\leqslant t_0} h(t)$, we have, for all $t \geqslant 0$,

$$\sigma(t) \leqslant (1/2K_1)\, A_*(t) + (K_2/K_1)\, t.$$

Hence

$$\begin{aligned} K_1 \int_\Omega \sigma\left(\frac{|u(x)|}{K}\right) dx &\leqslant \frac{1}{2} \int_\Omega A_*\left(\frac{|u(x)|}{K}\right) dx + \frac{K_2}{K} \int_\Omega \|u(x)\|\, dx \\ &\leqslant \frac{1}{2} + \frac{K_3}{K} \|u\|_A, \end{aligned} \tag{35}$$

where $K_3 = 2K_2 \|1\|_{\tilde{A}} < \infty$ since $\Omega$ has finite volume.

Combining (33)–(35), we obtain

$$1 \leqslant (2K_1/K)(n-1)\, \|u\|_{1,A} + \tfrac{1}{2} + (K_3/K)\, \|u\|_A,$$

so that

$$\|u\|_{A_*} = K \leqslant K_4 \|u\|_{1,A}. \tag{36}$$

We note that $K_4$ can depend on $n$, $A$, $\operatorname{vol}\Omega$, and the cone determining the cone property for $\Omega$.

To extend (36) to arbitrary $u \in W^1L_A(\Omega)$ let

$$u_k(x) = \begin{cases} u(x) & \text{if } |u(x)| \leqslant k \\ k \operatorname{sgn} u(x) & \text{if } |u(x)| > k. \end{cases} \tag{37}$$

Evidently $u_k$ is bounded and belongs to $W^1L_A(\Omega)$ by Lemma 8.31. Moreover, $\|u_k\|_{A_*}$ increases with $k$ but is bounded by $K_4 \|u\|_{1,A}$. Thus $\lim_{k\to\infty} \|u_k\|_{A_*} = K$ exists and $K \leqslant K_4 \|u\|_{1,A}$. By Fatou's lemma

$$\int_\Omega A_*\left(\frac{|u(x)|}{K}\right) dx \leqslant \lim_{k\to\infty} \int_\Omega A_*\left(\frac{|u_k(x)|}{K}\right) dx \leqslant 1$$

whence $u \in L_{A_*}(\Omega)$ and (36) holds.

Since $\Omega$ has finite volume we have

$$W^1L_A(\Omega) \to W^{1,1}(\Omega) \to L^1(\Omega),$$

the latter imbedding being compact by Theorem 6.2. A bounded subset $S$ of $W^1L_A(\Omega)$ is bounded in $L_{A_*}(\Omega)$ and precompact in $L^1(\Omega)$, and hence precompact in $L_B(\Omega)$ by Theorem 8.23 whenever $B$ increases essentially more slowly than $A$ near infinity. ▮

Theorem 8.32 extends to arbitrary (even unbounded) domains $\Omega$ provided $W$ is replaced by $W_0$.

**8.33 THEOREM** Let $\Omega$ be any domain in $\mathbb{R}^n$. If the $N$-function $A$ satisfies (26) and (27), then

$$W_0{}^1L_A(\Omega) \to L_{A_*}(\Omega).$$

Moreover, if $\Omega_0$ is a bounded subdomain of $\Omega$, then the imbeddings

$$W_0{}^1L_A(\Omega) \to L_B(\Omega_0)$$

exist and are compact for any $N$-function $B$ increasing essentially more slowly than $A$ near infinity.

PROOF If $u \in W_0{}^1L_A(\Omega)$, then the function $f$ in the above proof can be approximated in $W^{1,1}(\Omega)$ by elements of $C_0{}^\infty(\Omega)$. By Sobolev's inequality (Section 5.11), (32) holds with the term $\|f\|_{0,1}$ absent from the right side. Therefore (35) is not needed and the proof does not require that $\Omega$ has finite volume. The cone property is not required either since Sobolev's inequality holds for all $u \in C_0{}^\infty(\mathbb{R}^n)$. The compactness arguments are similar to those above. ▮

**8.34 REMARK** Theorem 8.32 is not optimal in the sense that for some $A$, $L_{A_*}$ is not necessarily the smallest Orlicz space into which $W^1L_A(\Omega)$ can be imbedded. For instance if $A(t) = A_n(t) = t^n/n$, then, as noted earlier, $A_*(t)$ is equivalent near infinity to $e^t - t - 1$. However this $N$-function increases essentially more slowly near infinity than does $\exp(t^{n/(n-1)}) - 1$ so that Theorem 8.25 gives a sharper result than does Theorem 8.32. Donaldson and Trudinger [22] assert that Theorem 8.32 can be improved by the methods of Theorem 8.25 provided $A$ dominates near infinity every $A_p$ with $p < n$, but that Theorem 8.32 gives optimal results if for some $p < n$, $A_p$ dominates $A$ near infinity.

**8.35 THEOREM** Let $\Omega$ have the cone property in $\mathbb{R}^n$. Let $A$ be an $N$-function for which

$$\int_1^\infty \frac{A^{-1}(\tau)}{\tau^{(n+1)/n}}\, d\tau < \infty \tag{38}$$

Then

$$W^1L_A(\Omega) \to C_B(\Omega) = C(\Omega) \cap L^\infty(\Omega).$$

PROOF Let $C$ be a finite cone contained in $\Omega$. We shall show that there exists a constant $K_1$ depending on $n$, $A$, and the dimensions of $C$ such that

$$\|u\|_{\infty,C} \leqslant K_1 \|u\|_{1,A,C}. \tag{39}$$

In so doing, we may assume without loss of generality that $A$ satisfies (26), for if not, and if $B$ is an $N$-function satisfying (26) and equivalent to $A$ near infinity, then $W^1L_A(C) \to W^1L_B(C)$ with imbedding constant depending on $A$, $B$, and vol $C$ by Theorem 8.12. Since $B$ satisfies (38) we would have

$$\|u\|_{\infty,C} \leqslant K_2 \|u\|_{1,B,C} \leqslant K_3 \|u\|_{1,A,C}.$$

Now $\Omega$ can be expressed as a union of congruent copies of some such finite cone $C$ so that (39) clearly implies

$$\|u\|_{\infty,\Omega} \leqslant K_1 \|u\|_{1,A,\Omega}. \tag{40}$$

Since $A$ is assumed to satisfy (26) and (38) we have

$$\int_0^\infty \frac{A^{-1}(t)}{t^{(n+1)/n}}\,dt = K_4 < \infty.$$

Let

$$\Lambda^{-1}(t) = \int_0^t \frac{A^{-1}(\tau)}{\tau^{(n+1)/n}}\,d\tau.$$

Then $\Lambda^{-1}$ maps $[0,\infty)$ in a one-to-one way onto $[0, K_4)$ and has convex inverse $\Lambda$. We extend the domain of definition of $\Lambda$ to $[0,\infty)$ by setting $\Lambda(t) = \infty$ for $t \geqslant K_4$. The function $\Lambda$, which is a Young's function (see Luxemburg [42] or O'Neill [55]), is not an $N$-function as defined in Section 8.2 but nevertheless the Luxemburg norm

$$\|u\|_{\Lambda,C} = \inf\left\{k > 0 : \int_\Omega \Lambda(|u(x)|/k)\,dx \leqslant 1\right\}$$

is easily seen to be a norm on $L^\infty(C)$ equivalent to the usual norm; in fact,

$$(1/K_4)\|u\|_{\infty,C} \leqslant \|u\|_{\Lambda,C} \leqslant [1/\Lambda^{-1}(1/\mathrm{vol}\,C)]\,\|u\|_{\infty,C}. \tag{41}$$

Moreover, $s = \Lambda(t)$ satisfies the differential equation (29) so that the proof of Theorem 8.32 can be carried over in this case to yield, for $u \in W^1L_A(C)$,

$$\|u\|_{\Lambda,C} \leqslant K_5 \|u\|_{1,A,C}. \tag{42}$$

Inequality (39) now follows from (41) and (42).

By Theorem 8.28(d) an element $u \in W^1E_A(\Omega)$ can be approximated in norm by functions continuous on $\Omega$. It follows from (40) that $u$ must coincide a.e. in $\Omega$ with a continuous function. (See the first paragraph of the proof of Lemma 5.15.)

Suppose that an $N$-function $B$ can be constructed so that $B(t) = A(t)$ near zero, $B$ increases essentially more slowly than $A$ near infinity, and

$$\int_1^\infty \frac{B^{-1}(t)}{t^{(n+1)/n}}\,dt \leqslant 2\int_1^\infty \frac{A^{-1}(t)}{t^{(n+1)/n}}\,dt < \infty.$$

Then by Theorem 8.16, $u \in W^1 L_A(C)$ implies $u \in W^1 E_B(C)$ so that $W^1 L_A(\Omega) \subset C(\Omega)$ as required.

It remains, therefore, to construct such an $N$-function $B$. Let $1 < t_1 < t_2 < \cdots$ be such that

$$\int_{t_k}^\infty \frac{A^{-1}(t)}{t^{(n+1)/n}}\,dt = \frac{1}{2^{2k}}\int_1^\infty \frac{A^{-1}(t)}{t^{(n+1)/n}}\,dt.$$

We define a sequence $\{s_k\}$ with $s_k \geqslant t_k$, and the function $B^{-1}(t)$, inductively as follows.

Let $s_1 = t_1$ and $B^{-1}(t) = A^{-1}(t)$ for $0 \leqslant t \leqslant s_1$. Having chosen $s_1, s_2, \ldots, s_{k-1}$ and defined $B^{-1}(t)$ for $0 \leqslant t \leqslant s_{k-1}$, we continue $B^{-1}(t)$ to the right of $s_{k-1}$ along a straight line with slope $(A^{-1})'(s_{k-1}-)$ (which always exists since $A^{-1}$ is concave) until a point $t_k'$ is reached where $B^{-1}(t_k') = 2^{k-1}A^{-1}(t_k')$. Such $t_k'$ exists because $\lim_{t\to\infty} A^{-1}(t)/t = 0$. If $t_k' \geqslant t_k$, let $s_k = t_k'$. Otherwise let $s_k = t_k$ and extend $B^{-1}$ from $t_k'$ to $s_k$ by setting $B^{-1}(t) = 2^{k-1}A^{-1}(t)$. Evidently $B^{-1}$ is concave and $B$ is an $N$-function. Moreover, $B(t) = A(t)$ near zero and since

$$\lim_{t\to\infty} \frac{B^{-1}(t)}{A^{-1}(t)} = \infty,$$

$B$ increases essentially more slowly than $A$ near infinity. Finally,

$$\begin{aligned}
\int_1^\infty \frac{B^{-1}(t)}{t^{(n+1)/n}}\,dt &\leqslant \int_1^{s_1} \frac{A^{-1}(t)}{t^{(n+1)/n}}\,dt + \sum_{k=2}^\infty \int_{s_{k-1}}^{s_k} \frac{2^{k-1}A^{-1}(t)}{t^{(n+1)/n}}\,dt \\
&\leqslant \int_1^\infty \frac{A^{-1}(t)}{t^{(n+1)/n}}\,dt + \sum_{k=2}^\infty 2^{k-1}\int_{t_{k-1}}^\infty \frac{A^{-1}(t)}{t^{(n+1)/n}}\,dt \\
&= 2\int_1^\infty \frac{A^{-1}(t)}{t^{(n+1)/n}}\,dt.
\end{aligned}$$

as required. ∎

**8.36 THEOREM** Let $\Omega$ be a domain in $\mathbb{R}^n$ having the strong local Lipschitz property. If the $N$-function $A$ satisfies

$$\int_1^\infty \frac{A^{-1}(t)}{t^{(n+1)/n}}\,dt < \infty, \tag{43}$$

then there exists a constant $K$ such that for any $u \in W^1L_A(\Omega)$ (which may be assumed continuous by the previous theorem) and all $x, y \in \Omega$ we have

$$|u(x)-u(y)| \leqslant K\|u\|_{1,A,\Omega} \int_{|x-y|^{-n}}^{\infty} \frac{A^{-1}(t)}{t^{(n+1)/n}}\, dt. \tag{44}$$

PROOF We establish (44) for the case when $\Omega$ is a cube of unit edge; the extension to more general strongly Lipschitz domains can be carried out just as in the proof of Lemma 5.17. As in that lemma we let $\Omega_\sigma$ denote a parallel subcube of $\Omega$ and obtain for $x \in \overline{\Omega}_\sigma$

$$\left| u(x) - \frac{1}{\sigma^n} \int_{\Omega_\sigma} u(z)\, dz \right| \leqslant \frac{\sqrt{n}}{\sigma^{n-1}} \int_0^1 t^{-n}\, dt \int_{\Omega_{t\sigma}} |\operatorname{grad} u(z)|\, dz.$$

By (22), $\|1\|_{\tilde{A},\Omega_{t\sigma}} = 1/\tilde{A}^{-1}(t^{-n}\sigma^{-n})$. It follows by Hölder's inequality and (7) that

$$\begin{aligned}
\int_{\Omega_{t\sigma}} |\operatorname{grad} u(z)|\, dz &\leqslant 2\,\|\operatorname{grad} u\|_{A,\Omega_{t\sigma}} \|1\|_{\tilde{A},\Omega_{t\sigma}} \\
&\leqslant 2\,\|u\|_{1,A,\Omega}/\tilde{A}^{-1}(t^{-n}\sigma^{-n}) \\
&\leqslant 2\sigma^n t^n\, \|u\|_{1,A,\Omega} A^{-1}(t^{-n}\sigma^{-n}).
\end{aligned}$$

Hence

$$\begin{aligned}
\left| u(x) - \frac{1}{\sigma^n} \int_{\Omega_\sigma} u(z)\, dz \right| &\leqslant 2\sqrt{n}\,\sigma\, \|u\|_{1,A,\Omega} \int_0^1 A^{-1}\!\left(\frac{1}{t^n\sigma^n}\right) dt \\
&= \frac{2}{\sqrt{n}} \|u\|_{1,A,\Omega} \int_{\sigma^{-n}}^{\infty} \frac{A^{-1}(\tau)}{\tau^{(n+1)/n}}\, d\tau.
\end{aligned} \tag{45}$$

If $x, y \in \Omega$ and $\sigma = |x-y| < 1$, there exists such a subcube $\Omega_\sigma$ with $x, y \in \overline{\Omega}_\sigma \subset \Omega$. Using (45) for $x$ and $y$, we obtain

$$|u(x)-u(y)| \leqslant \frac{4}{\sqrt{n}} \|u\|_{1,p,\Omega} \int_{|x-y|^{-n}}^{\infty} \frac{A^{-1}(t)}{t^{(n+1)/n}}\, dt.$$

For $|x-y| \geqslant 1$, (44) follows from (40) and (43). ∎

**8.37** Let $M$ denote the class of positive, continuous, increasing functions of $t > 0$ which tend to zero as $t$ decreases to zero. If $\mu \in M$, the space $C_\mu(\overline{\Omega})$, consisting of functions $u \in C(\overline{\Omega})$ for which the norm

$$\|u; C_\mu(\overline{\Omega})\| = \|u; C(\overline{\Omega})\| + \sup_{\substack{x,y\in\Omega \\ x\neq y}} \frac{|u(x)-u(y)|}{\mu(|x-y|)}$$

is finite, is a Banach space under that norm. The theorem above asserts that if (43) holds, then

$$W^1L_A(\Omega) \to C_\mu(\overline{\Omega}) \qquad \text{with} \quad \mu(t) = \int_{t^{-n}}^{\infty} \frac{A^{-1}(\tau)}{\tau^{(n+1)/n}}\, d\tau. \tag{46}$$

If $\mu, \nu \in M$ are such that $\mu/\nu \in M$, then for bounded $\Omega$ we have, as in Theorem 1.31, that the imbedding

$$C_\mu(\overline{\Omega}) \to C_\nu(\overline{\Omega})$$

is compact. Hence so also is

$$W^1L_A(\Omega) \to C_\nu(\overline{\Omega}),$$

if $\mu$ is given by (46).

The following is a trace imbedding theorem which generalizes (the case $m = 1$ of) Lemma 5.19.

**8.38 THEOREM** Let $\Omega$ be a bounded domain in $\mathbb{R}^n$ having the cone property, and let $\Omega^k$ denote the intersection of $\Omega$ with a $k$-dimensional plane in $\mathbb{R}^n$. Let $A$ be an $N$-function for which (26) and (27) hold, and let $A_*$ be given by (28). Let $1 \leqslant p < n$ where $p$ is such that the function $B$ given by $B(t) = A(t^{1/p})$ is an $N$-function. If either $n-p < k \leqslant n$, or $p = 1$, and $n-1 \leqslant k \leqslant n$, then

$$W^1L_A(\Omega) \to L_{A_*^{k/n}}(\Omega^k),$$

where $A_*^{k/n}(t) = [A_*(t)]^{k/n}$.

Moreover, if $p > 1$ and $C$ is an $N$-function increasing essentially more slowly than $A_*^{k/n}$ near infinity, then the imbedding

$$W^1L_A(\Omega) \to L_C(\Omega^k) \tag{47}$$

is compact.

PROOF The problem of verifying that $A_*^{k/n}$ is an $N$-function is left to the reader. Let $u \in W^1L_A(\Omega)$ be a bounded function. Then

$$\int_{\Omega^k} A_*^{k/n}\left(\frac{|u(y)|}{K}\right) dy = 1, \qquad K = \|u\|_{A_*^{k/n}, \Omega^k}. \tag{48}$$

We wish to show that

$$K \leqslant K_1 \|u\|_{1,A,\Omega} \tag{49}$$

with $K_1$ independent of $u$. Since (49) is known to hold for the special case

$k = n$ (Theorem 8.32) we may assume without loss of generality that

$$K \geqslant \|u\|_{A_*,\Omega} = \|u\|_{A_*^{n/n},\Omega^n}. \tag{50}$$

Let $\omega(t) = [A_*(t)]^{1/q}$ where $q = np/(n-p)$. By Lemma 5.19 (the case $m = 1$) we have

$$\begin{aligned}\left\|\omega\left(\frac{|u|}{K}\right)\right\|^p_{kp/(n-p),\Omega^k} &\leqslant K_2\left\{\sum_{j=1}^n \left\|D_j\omega\left(\frac{|u|}{K}\right)\right\|^p_{p,\Omega} + \left\|\omega\left(\frac{|u|}{K}\right)\right\|^p_{p,\Omega}\right\}\\ &= K_2\left\{\frac{1}{K^p}\sum_{j=1}^n \int_\Omega \left|\omega'\left(\frac{|u(x)|}{K}\right)\right|^p |D_j u(x)|^p\,dx\right.\\ &\qquad \left.+\int_\Omega \left|\omega\left(\frac{|u(x)|}{K}\right)\right|^p dx\right\}.\end{aligned}$$

Using (48) and noting that $\||v|^p\|_{B,\Omega} \leqslant \|v\|^p_{A,\Omega}$, we obtain

$$\begin{aligned}1 &= \left(\int_{\Omega^k}\left(A_*\left(\frac{|u(y)|}{K}\right)\right)^{k/n} dy\right)^{(n-p)/k} = \left\|\omega\left(\frac{|u|}{K}\right)\right\|^p_{kp/(n-p),\Omega^k}\\ &\leqslant \frac{2K_2}{K^p}\sum_{j=1}^n \left\|\left(\omega'\left(\frac{|u|}{K}\right)\right)^p\right\|_{\tilde{B},\Omega} \||D_j u|^p\|_{B,\Omega} + K_2\left\|\omega\left(\frac{|u|}{K}\right)\right\|^p_{p,\Omega}\\ &\leqslant \frac{2nK_2}{K^p}\left\|\left(\omega'\left(\frac{|u|}{K}\right)\right)^p\right\|_{\tilde{B},\Omega} \|u\|^p_{1,A,\Omega} + K_2\left\|\omega\left(\frac{|u|}{K}\right)\right\|^p_{p,\Omega}.\end{aligned} \tag{51}$$

Now $B^{-1}(t) = [A^{-1}(t)]^p$ and so, using (29) and (7), we have

$$\begin{aligned}[\omega'(t)]^p &= (1/q^p)[A_*(t)]^{p(1-q)/q}[A_*'(t)]^p\\ &= (1/q^p)A_*(t)[1/B^{-1}(A_*(t))] \leqslant (1/q^p)\tilde{B}^{-1}(A_*(t)).\end{aligned}$$

It follows by (50) that

$$\int_\Omega \tilde{B}\left(\left(\frac{\omega'(|u(x)|/K)}{1/q}\right)^p\right) dx \leqslant \int_\Omega A_*\left(\frac{|u(x)|}{K}\right) dx \leqslant 1.$$

So

$$\|(\omega'(|u|/K))^p\|_{\tilde{B},\Omega} \leqslant 1/q^p. \tag{52}$$

Now set $g(t) = A_*(t)/t^p$ and $h(t) = (\omega(t)/t)^p$. It is readily checked that $\lim_{t\to\infty} g(t)/h(t) = \infty$. In order to see that $h(t)$ is bounded near zero let $s = A_*(t)$ and consider

$$(h(t))^{1/p} = \frac{(A_*(t))^{1/q}}{t} = \frac{s^{(1/p)-(1/n)}}{\displaystyle\int_0^s \frac{A^{-1}(\tau)}{\tau^{(n+1)/n}}\,d\tau} \leqslant \frac{s^{1/p}}{\displaystyle\int_0^s \frac{[B^{-1}(\tau)]^{1/p}}{\tau}\,d\tau}.$$

Since $B$ is an $N$-function $\lim_{\tau\to\infty+} B^{-1}(\tau)/\tau = \infty$. Hence for sufficiently small values of $t$ we have

$$(h(t))^{1/p} \leqslant \frac{s^{1/p}}{\int_0^s \tau^{-1+1/p}\,d\tau} = \frac{1}{p}.$$

Therefore there exists a constant $K_3$ such that for $t \geqslant 0$

$$(\omega(t))^p \leqslant (1/2K_2)\,A_*(t) + K_3\,t^p.$$

Using (50), we now obtain

$$\begin{aligned}\left\|\omega\left(\frac{|u|}{K}\right)\right\|_{p,\Omega}^p &\leqslant \frac{1}{2K_2}\int_\Omega A_*\left(\frac{|u(x)|}{K}\right)dx + \frac{K_3}{K^p}\int_\Omega |u(x)|^p\,dx \\ &\leqslant \frac{1}{2K_2} + \frac{2K_3}{K^p}\,\|\,|u|^p\|_{B,\Omega}\,\|1\|_{\tilde{B},\Omega} \\ &\leqslant \frac{1}{2K_2} + \frac{K_4}{K^p}\,\|u\|_{A,\Omega}^p. \end{aligned} \tag{53}$$

From (51)–(53) there follows the inequality

$$1 \leqslant \frac{2nK_2}{K^p}\cdot\frac{1}{q^p}\,\|u\|_{1,A,\Omega}^p + \frac{1}{2} + \frac{K_4 K_2}{K^p}\,\|u\|_{A,\Omega}^p$$

and hence (49). The extension of (49) to arbitrary $u \in W^1L_A(\Omega)$ now follows as in the proof of Theorem 8.32.

Since $B(t) = A(t^{1/p})$ is an $N$-function and $\Omega$ is bounded we have $W^1L_A(\Omega) \to W^{1,p}(\Omega) \to L^1(\Omega^k)$, the latter imbedding being compact by Theorem 6.2 (provided $p > 1$). The compactness of (47) now follows by Theorem 8.23. ▮

**8.39** We conclude this chapter with the general Orlicz–Sobolev imbedding theorem of Donaldson and Trudinger [22]. For a given $N$-function $A$ we define a sequence of $N$-functions $B_0, B_1, B_2, \ldots$ as follows:

$$B_0(t) = A(t)$$

$$(B_k)^{-1}(t) = \int_0^t \frac{(B_{k-1})^{-1}(\tau)}{\tau^{(n+1)/n}}\,d\tau, \qquad k = 1, 2, \ldots.$$

At each stage we assume that

$$\int_0^1 \frac{(B_k)^{-1}(\tau)}{\tau^{(n+1)/n}}\,d\tau < \infty, \tag{54}$$

replacing $B_k$, if necessary, by another $N$-function equivalent to it near infinity and satisfying (54). Let $J = J(A)$ be the smallest nonnegative integer such that

$$\int_1^\infty \frac{(B_J)^{-1}(\tau)}{\tau^{(n+1)/n}}\, d\tau < \infty.$$

Evidently $J(A) \leqslant n$.

If $\mu$ belongs to the class $M$ defined in Section 8.37, we define the space $C_\mu{}^m(\overline{\Omega})$ to consist of those functions $u \in C^m(\overline{\Omega})$ for which $D^\alpha u \in C_\mu(\overline{\Omega})$. The space $C_\mu{}^m(\overline{\Omega})$ is a Banach space with respect to the norm

$$\|u; C_\mu{}^m(\overline{\Omega})\| = \max_{|\alpha| \leq m} \|D^\alpha u; C_\mu(\overline{\Omega})\|.$$

**8.40 THEOREM** Let $\Omega$ be a bounded domain in $\mathbb{R}^n$ having the cone property. Let $A$ be an $N$-function.

(a) If $m \leqslant J(A)$, then $W^m L_A(\Omega) \to L_{B_m}(\Omega)$ and the imbedding $W^m L_A(\Omega) \to L_C(\Omega)$ is compact for any $N$-function $C$ increasing essentially more slowly than $B_m$ near infinity.

(b) If $m > J(A)$, then $W^m L_A(\Omega) \to C_B(\Omega) = C(\Omega) \cap L^\infty(\Omega)$.

(c) If also $\Omega$ has the strong local Lipschitz property and if $m > J = J(A)$, then $W^m L_A(\Omega) \to C_\mu^{m-J-1}(\overline{\Omega})$ where

$$\mu(t) = \int_{t^{-n}}^\infty \frac{(B_J)^{-1}(\tau)}{\tau^{(n+1)/n}}\, d\tau.$$

Moreover, the imbeddings $W^m L_A(\Omega) \to C^{m-J-1}(\overline{\Omega})$ and $W^m L_A(\Omega) \to C_\nu^{m-J-1}(\overline{\Omega})$ are compact provided $\nu \in M$ and $\mu/\nu \in M$.

**8.41 REMARK** The above theorem follows in a straightforward way from Theorems 8.32, 8.35, and 8.36. Moreover, if we replace $L_A$ by $E_A$ in part (a), we get $W^m E_A(\Omega) \to E_{B_m}(\Omega)$ since the sequence $\{u_k\}$ defined by (37) in the proof of Theorem 8.32 converges to $u$ if $u \in W^1 E_A(\Omega)$. Theorem 8.40 holds without restriction on $\Omega$ if $W^m L_A(\Omega)$ is everywhere replaced by $W_0{}^m L_A(\Omega)$.

# References


1. R. A. Adams, Some integral inequalities with applications to the imbedding of Sobolev spaces defined over irregular domains, *Trans. Amer. Math. Soc.* **178** (1973), 401–429.
2. R. A. Adams, Capacity and compact imbeddings, *J. Math. Mech.*, **19** (1970), 923–929.
3. R. A. Adams and J. Fournier, Compact imbedding theorems for functions without compact support, *Canad. Math. Bull.*, **14** (1971), 305–309.
4. R. A. Adams and J. Fournier, Some imbedding theorems for Sobolev spaces, *Canad. J. Math.* **23** (1971), 517–530.
5. R. D. Adams, N. Aronszajn, and K. T. Smith, Theory of Bessel potentials, part II, *Ann. Inst. Fourier* (*Grenoble*), **17** (No. 2) (1967), 1–135.
6. S. Agmon, "Lectures on Elliptic Boundary Value Problems," Van Nostrand-Reinhold, Princeton, New Jersey, 1965.
7. N. Aronszajn and K. T. Smith, Theory of Bessel potentials, Part I, *Ann. Inst. Fourier* (*Grenoble*) **11** (1961), 385–475.
8. N. Aronszajn, F. Mulla and P. Szeptycki, On spaces of potentials connected with $L^p$-classes, *Ann. Inst. Fourier* (*Grenoble*) **13** (No. 2) (1963), 211–306.
9. O. V. Besov, On a certain family of functional spaces. Imbedding and continuation theorems, *Dokl. Akad. Nauk SSSR*, **126** (1959), 1163–1165.
10. O. V. Besov, On some conditions of membership in $L^p$ for derivatives of periodic functions *Naučn. Dokl. Vyss. Skoly. Fiz.–Mat. Nauki* (1959), 13–17.
11. F. E. Browder, On the eigenfunctions and eigenvalues of general linear elliptic operators, *Proc. Nat. Acad. Sci. USA* **39** (1953), 433–439.
12. F. E. Browder, On the spectral theory of elliptic differential operators I, *Math. Ann.* **142** (1961), 22–130.
13. P. L. Butzer and H. Berens, "Semigroups of Operators and Approximation," Springer-Verlag, Berlin, 1967.

14. A. P. Calderón, Lebesgue spaces of differentiable functions and distributions, *Proc. Sym. Pure Math.*, **4** (1961), 33–49. Amer. Math. Soc., Providence, Rhode Island.
15. A. P. Calderón, Intermediate spaces and interpolation, the complex method, *Studia Math.* **24** (1964), 113–190.
16. A. P. Calderón and A. Zygmund, On the existence of certain singular integrals, *Acta Math.* **88** (1952), 85–139.
17. C. W. Clark, The Hilbert–Schmidt property for embedding maps between Sobolev spaces, *Canad. J. Math.* **18** (1966), 1079–1084.
18. C. W. Clark, "Introduction to Sobolev spaces," (Seminar notes) Univ. of British Columbia, Vancouver, 1968.
19. J. A. Clarkson, Uniformly convex spaces, *Trans. Amer. Math. Soc.* **40** (1936), 396–414.
20. G. Dankert, "Sobolev imbedding theorems in Orlicz spaces," (Thesis). Univ. of Cologne, 1966.
21. J. Deny and J. L. Lions, Les espaces du type de Beppo Levi, *Ann. Inst. Fourier (Grenoble)* **5** (1955), 305–370.
22. T. K. Donaldson and N. S. Trudinger, Orlicz–Sobolev spaces and imbedding theorems, *J. Functional Analysis*, **8** (1971), 52–75.
23. G. Ehrling, On a type of eigenvalue problem for certain elliptic differential operators, *Math. Scand.* **2** (1954), 267–285.
24. E. Gagliardo, Properietà di alcune classi di funzioni in piu variabili, *Ricerche Mat.* **7** (1958), 102–137.
25. E. Gagliardo, Ulteriori properietà di alcune classi di funzioni in più variabili, *Ricerche Mat.* **8** (1959), 24–51.
26. I. G. Globenko, Embedding theorems for a region with null salient points, *Dokl. Akad. Nauk SSSR* **132** (1960), 251–253 English trans.: *Soviet Math. Dokl.* **1** (1960), 517–519
27. I. G. Globenko, Some questions in the theory of imbedding for domains with singularities on the boundary, *Mat. Sb.*, **57** (99) (1962), 201–224.
28. P. Grisvard, Commutativité de deux foncteurs d'interpolation et applications, *J. Math. Pures Appl.* **45** (1966) 143–290.
29. G. H. Hardy, J. E. Littlewood, and G. Polya, "Inequalities," Cambridge Univ. Press, London and New York, 1943.
30. J. A. Hempel, G. R. Morris, and N. S. Trudinger, On the sharpness of a limiting case of the Sobolev imbedding theorem, *Bull. Austral. Math. Soc.* **3** (1970), 369–373.
31. M. Hestenes, Extension of the range of a differentiable function, *Duke J. Math.* **8** (1941) 183–192.
32. E. Hewitt and K. Stromberg, "Real and Abstract Analysis," Springer-Verlag, New York, 1969.
33. V. I. Kondrachov, Certain properties of functions in the spaces $L^p$, *Dokl. Akad. Nauk SSSR* **48** (1945), 535–538.
34. M. A. Kranosel'skii and Ya. B. Rutickii, "Convex Functions and Orlicz Spaces," Noordhoff, Groningen, The Netherlands, 1961.
35. L. Lichenstein, Eine elementare Bemerkung zur reelen Analysis, *Math. Z.* **30** (1929), 794–795.
36. J. L. Lions, Une construction d'espaces d'interpolation, *C.R. Acad. Sci. Paris* **83** (1960), 1853–1855.
37. J. L. Lions, Sur les espaces d'interpolation; dualité, *Math. Scand.* **9** (1961), 147–177.
38. J. L. Lions, Théorèmes de trace et d'interpolation (IV), *Math. Ann.* **151** (1963), 42–56.
39. J. L. Lions, "Problèmes aux Limites dans les Équations aux Derivées Partielles," (Seminar notes), Univ. of Montreal Press, 1965.

40. J. L. Lions and E. Magenes, Problemi ai limiti non omogenei (III), *Ann. Scuola Norm. Sup. Pisa* **15** (1961), 41–103.
41. P. I. Lizorkin, Boundary properties of functions from "weight" classes, *Dokl. Akad. Nauk SSSR* **132** (1960), 514–517 English trans.: *Soviet Math. Dokl* **1** (1960), 589–593.
42. W. Luxemburg, "Banach function spaces," (Thesis), Technische Hogeschool te Delft, The Netherlands, 1955.
43. K. Maurin, Abbildungen vom Hilbert–Schmidtschen Typus und ihre Anwendungen, *Math. Scand.* **9** (1961), 359–371.
44. V. G. Maz'ja, Classes of domains and imbedding theorems for function spaces, *Dokl. Akad. Nauk SSSR* **133** (1960), 527–530 English transl.: *Soviet Math. Dokl.* **1** (1960), 882–885.
45. V. G. Maz'ja, $p$-conductance and theorems of imbedding certain function spaces into the space $C$, *Dokl. Akad. Nauk SSSR* **140** (1961), 299–302 [English transl.: *Soviet Math. Dokl.* **2** (1961), 1200–1203].
46. N. Meyers and J. Serrin, $H = W$, *Proc. Nat. Acad. Sci. USA* **51** (1964), 1055–1056.
47. C. B. Morrey, Functions of several variables and absolute continuity, II, *Duke J. Math.* **6** (1940), 187–215.
48. M. E. Munroe, "Measure and Integration," 2nd ed. Addison-Wesley, Reading, Massachusetts 1971.
49. S. M. Nikol'skii, Inequalities for entire functions of exponential type and their application to the theory of differentiable functions of several variables, *Trudy Mat. Inst. Steklov* **38** (1951), 244–278 [English transl.: *Amer. Math. Soc. Transl.* (2), **80** (1969), 1–38].
50. S. M. Nikol'skii, Properties of certain classes of functions of several variables on differentiable manifolds, *Mat. Sb.* **33** (75) (1953), 261–326 [English transl.: *Amer. Math. Soc. Transl.* (2), **80** (1969), 39–118].
51. S. M. Nikol'skii, Extension of functions of several variables preserving differential properties, *Mat. Sb.* **40** (82) (1956), 243–268 [English transl.: *Amer. Math. Soc. Transl.* (2), **83** (1969), 159–188].
52. S. M. Nikol'skii, On imbedding, continuation and approximation theorems for differentiable functions of several variables, *Russian Math. Surveys* **16** (1961) 55–104.
53. L. Nirenberg, Remarks on strongly elliptic partial differential equations, *Comm. Pure Appl. Math.* **8** (1955), 649–675.
54. L. Nirenberg, An extended interpolation inequality, *Ann. Scuola Norm. Sup. Pisa* **20** (1966) 733–737.
55. R. O'Neill, Fractional intergration in Orlicz spaces, *Trans. Amer. Math. Soc.* **115** (1965), 300–328.
56. J. Peetre, Espaces d'interpolation et théorème de Soboleff, *Ann Inst. Fourier* (*Grenoble* **16** (1966), 279–317.
57. F. Rellich, Ein Satz über mittlere Konvergenz, *Göttingen Nachr.* (1930), 30–35.
58. W. Rudin, "Real and Complex Analysis," McGraw-Hill, New York, 1966.
59. W. Rudin, "Functional Analysis," McGraw-Hill, New York, 1973.
60. L. Schwartz, "Théorie des Distributions," Hermann, Paris, 1966.
61. R. T. Seeley, Extension of $C^\infty$-functions defined in a half-space, *Proc. Amer. Math. Soc.* **15** (1964), 625–626.
62. S. L. Sobolev, On a theorem of functional analysis, *Mat. Sb.* **46** (1938), 471–496.
63. S. L. Sobolev, "Applications of Functional Analysis in Mathematical Physics," Leningrad 1950 [English transl.: *Amer. Math. Soc., Transl., Math Mono.* **7** (1963).]
64. S. L. Sobolev and S. M. Nikol'skii, Imbedding theorems, *Izdat. Akad. Nauk SSSR, Leningrad* (1963), 227–242 [English transl.: *Amer. Math. Soc. Transl.* (2), **87** (1970), 147–173.

64a. E. M. Stein, "Singular Integrals and Differentiability Properties of Functions," (Princeton Math. Series, Vol. 30), Chapt. V, Princeton Univ. Press, Princeton, New Jersey, 1970.
65. E. M. Stein and G. Weiss, "Introduction to Fourier Analysis on Euclidean Spaces" (Princeton Math. Series, Vol. 32), Princeton Univ. Press, Princeton, New Jersey, 1972.
66. N. S. Trundinger, On imbeddings into Orlicz spaces and some applications, *J. Math. Mech.* **17** (1967), 473–483.
67. S. V. Uspenskii, An imbedding theorem for S. L. Sobolev's classes $W_p^r$ of fractional order, *Dokl. Akad. Nauk SSSR*, **130** (1960) 992–993 [English transl.: *Soviet Math. Dokl.* **1** (1960), 132–133].
68. S. V. Uspenskii, Properties of the classes $W_p^r$ with fractional derivatives on differentiable manifolds, *Dokl. Akad. Nauk SSSR* **132** (1960), 60–62 [English trans.: *Soviet Math. Dokl.* **1** (1960), 495–497].
69. K. Yosida, "Functional Analysis," Springer-Verlag, Berlin, 1965.
70. S. Zaidman, "Équations Différentielles Abstraites," (Seminar notes), Univ. of Montréal Press, 1966.

# Index

K



L



M



N



O



P

A 5
B 6
C 7
D 8
E 9
F 0
G 1
H 2
I 3
J 4